AF614745

Methods in Molecular Biology™

Series Editor
John M. Walker
School of Life Sciences
University of Hertfordshire
Hatfield, Hertfordshire, AL10 9AB, UK

For other titles published in this series, go to
www.springer.com/series/7651

Live Cell Imaging

Methods and Protocols

Edited by

Dmitri B. Papkovsky

University College Cork, Cork, Ireland

Editor
Dmitri B. Papkovsky
Department of Biochemistry
University College Cork
Cavanagh Pharmacy Bldg.
College Road
Cork
Ireland
d.papkovsky@ucc.ie

ISSN 1064-3745 e-ISSN 1940-6029
ISBN 978-1-60761-403-6 e-ISBN 978-1-60761-404-3
DOI 10.1007/978-1-60761-404-3

Library of Congress Control Number: 2009939132

Printed on acid-free paper

springer.com

Preface

Live cell imaging has now become a routine tool in biomedical and life science research. It is hard to imagine an active academic research department, pharmaceutical or biotechnology company without access to this technology and without using it on a regular basis. Over the last decade, major progress in this area has been achieved, making this core biochemical, cell and molecular biology techniques even more versatile, affordable, and mature. On the other hand, we continue witnessing numerous new, breakthrough developments which advance this technology even further, extending its capabilities and measurement standards. A variety of advanced-imaging methodologies, probe chemistries, experimental procedures, dedicated instruments, integrated systems, and a large number of new applications have come to the fore very recently. One can mention, for example, ultra-high resolution methods breaking the canonical diffraction limits, multi-photon excitation imaging and sample manipulation (e.g., (un)caging, permeabilization), new chemically and genetically engineered probes for key markers and parameters of cellular function, multi-color imaging, specialized detection formats, custom-built systems employing new optoelectronics and engineering solutions, user-friendly multi-mode microscopes, software, and data analysis algorithms. All this provide unprecedented opportunities for the real-time investigation of live objects, including individual cells, sub-cellular organelles, and even individual molecules, with high level of detail and information content. Being until recently a privilege of large institutions and centralized facilities, live cell imaging systems are now spreading into small labs, while sophisticated high content imaging stations are being deployed to screening labs.

At the same time, the wide and ever increasing range of imaging techniques and applications necessitates regular updates for existing users as well as an up-to-date introduction and some general guidance for newcomers to this area. This volume of the *Methods in Molecular Biology* series provides a comprehensive compendium of experimental approaches to live cell imaging in the form of several overview chapters followed by representative examples and case studies covering different aspects of the methodology. The 21 chapters of this volume are prepared by leaders in these fields, and the outstanding contribution of the authors is gratefully acknowledged. The book provides a range of state-of-the-art protocols extensively validated in complex biological studies. It highlights new experimental and instrumental opportunities and helps researchers to select appropriate imaging methods for their specific biological questions and measurement tasks. Each method also highlights the potential challenges and experimental artefacts which are likely to appear and which unfortunately are still not very uncommon. We believe that this volume will contribute to the further development and dissemination of this fundamentally important technology which spans across many disciplines including molecular and cell biology, chemistry, physics, optics, engineering, cell physiology, and medicine.

Dmitri B. Papkovsky

Contents

Contributors

MARCEL AMELOOT • *Cell Physiology Group, Biomedical Research Institute, Hasselt University, Diepenbeek, Belgium*

CORINA BALUT • *Cell Biology and Physiology Department, School of Medicine, University of Pittsburgh, Pittsburgh, PA, USA*

SZILVIA BARON • *Laboratory of Ca^{2+}-transport ATPases, Department of Molecular Cell Biology, Katholieke Universiteit Leuven, Belgium*

JEREMY BONOR • *Department of Biological Sciences, University of Delaware, Newark, DE, USA*

STEPHEN A. BOPPART • *Biophotonics Imaging Laboratory, Beckman Institute for Advanced Science and Technology, University of Illinois at Urbana-Champaign, Urbana, IL, USA*

JOSEPH BRZOSTOWSKI • *Laboratory of Immunogenetics, National Institute of Allergy and Infectious Diseases, National Institutes of Health, Rockville, MD, USA*

GILLES CHARVIN • *Laboratoire Joliot-Curie & Laboratoire de Physique, Ecole Normale Supérieure, Lyon, France; The Rockefeller University, New York, NY, USA*

YAN CHEN • *Department of Surgery, The University of Hong Kong, Hong Kong, P. R. China*

FREDERICK CROSS • *The Rockefeller University, New York, NY, USA*

RUSLAN I. DMITRIEV • *Biochemistry Department, University College Cork, Cork, Ireland*

ALEXANDER EGNER • *Department of NanoBiophotonics, Max Planck Institute for Biophysical Chemistry, Goettingen, Germany*

JOACHIM FANDREY • *Institut für Physiologie, Universität Duisburg-Essen, Essen, Germany*

ANDREAS FERCHER • *Biochemistry Department, University College Cork, Cork, Ireland*

RONALD S. FLANNAGAN • *Program in Cell Biology, The Hospital for Sick Children, Toronto, ON, Canada*

JULIA GERASIMENKO • *Department of Physiology, Biomedical School, University of Liverpool, Liverpool, UK*

OLEG GERASIMENKO • *Department of Physiology, Biomedical School, University of Liverpool, Liverpool, UK*

BENEDIKT W. GRAF • *Biophotonics Imaging Laboratory, Beckman Institute for Advanced Science and Technology, University of Illinois at Urbana-Champaign, Urbana, IL, USA*

SERGIO GRINSTEIN • *Program in Cell Biology, The Hospital for Sick Children, Toronto, ON, Canada; Department of Biochemistry and Institute of Medical Sciences, University of Toronto, Toronto, ON, Canada*

SCOTT A. HILDERBRAND • *Center for Molecular Imaging Research, Massachusetts General Hospital/Harvard Medical School, Charlestown, MA, USA*

STEFAN JAKOBS • *Mitochondrial Structure and Dynamics/Department of NanoBiophotonics, Max Planck Institute for Biophysical Chemistry, Goettingen, Germany*

WEN-HONG LI • *Departments of Cell Biology and Biochemistry, University of Texas Southwestern Medical Center, Dallas, TX, USA*

TERRY MCCANN • *TJM Consultancy, Kent, UK*
ARTHUR MILLIUS • *Cardiovascular Research Institute and Department of Biochemistry, University of California, San Francisco, CA, USA*
DANIEL NEUMANN • *Mitochondrial Structure and Dynamics/Department of NanoBiophotonics, Max Planck Institute for Biophysical Chemistry, Goettingen, Germany*
ANJA NOHE • *Department of Biological Sciences, University of Delaware, Newark, DE, USA*
MARTIN OHEIM • *INSERM, U603, Paris, France; CNRS, UMR8154, Paris, France; Laboratory of Neurophysiology and New Microscopies, University Paris Descartes, Paris, France*
CATHERINE OIKONOMOU • *The Rockefeller University, New York, NY, USA*
TOMAS C. O'RIORDAN • *Luxcel Biosciences Ltd., BioTransfer Unit, UCC, Cork, Ireland*
DMITRI B. PAPKOVSKY • *Biochemistry Department, University College Cork, Cork, Ireland*
MADDY PARSONS • *Randall Division of Cell and Molecular Biophysics, King's College London, London, UK*
TAO PENG • *Department of Chemistry, The University of Hong Kong, Hong Kong, P. R. China*
JULIE PERROY • *Functional Genomic Institute, Department of Neurobiology, Unité mixte de recherche 5203 Centre National de la Recherche Scientifique, Unité 661 Institut National de la Santé et de la Recherche Médicale, Université Montpellier I & II, Montpellier, France*
ELIZABETH PHAM • *Institute of Biomaterials and Biomedical Engineering, University of Toronto, Toronto, ON, Canada*
SUSAN K. PIERCE • *Laboratory of Immunogenetics, National Institute of Allergy and Infectious Diseases, National Institutes of Health, Rockville, MD, USA*
ROMAN SCHMIDT • *Department of NanoBiophotonics, Max Planck Institute for Biophysical Chemistry, Goettingen, Germany*
JIAN-GANG SHEN • *School of Chinese Medicine, The University of Hong Kong, Hong Kong, P. R. China*
ILSE SMETS • *Department PHL-Bio, PHL University College, Diepenbeek, Belgium*
HAE WON SOHN • *Laboratory of Immunogenetics, National Institute of Allergy and Infectious Diseases, National Institutes of Health, Rockville, MD, USA*
PAUL STEELS • *Cell Physiology Group, Biomedical Research Institute, Hasselt University, Diepenbeek, Belgium*
KENJI SUGIMOTO • *Live Cell Imaging Institute, Osaka Prefecture University, Sakai, Osaka, Japan; Laboratory of Applied Molecular Biology, Division of Bioscience and Informatics, Graduate School of Life and Environmental Sciences, Osaka Prefecture University, Sakai, Osaka, Japan*
ZHEN-NING SUN • *Department of Chemistry, The University of Hong Kong, Hong Kong, P. R. China*
PAUL KWONG-HANG TAM • *Department of Surgery, The University of Hong Kong, Hong Kong, P. R. China*
PAVEL TOLAR • *Laboratory of Immunogenetics, National Institute of Allergy and Infectious Diseases, National Institutes of Health, Rockville, MD, USA*
SHIGENOBU TONE • *Department of Biochemistry, Kawasaki Medical School, Okayama, Japan*

KEVIN TRUONG • *Institute of Biomaterials and Biomedical Engineering and Edward S. Rogers Sr. Department of Electrical and Computer Engineering, University of Toronto, Toronto, ON, Canada*

MARTIN VANDEVEN • *Cell Physiology Group, Biomedical Research Institute, Hasselt University Diepenbeek, Belgium*

HUA-LI WANG • *Department of Chemistry, The University of Hong Kong, Hong Kong, P. R. China*

MANUS W. WARD • *Department of Physiology and Medical Physics, Royal College of Surgeons in Ireland, Dublin, Ireland*

ORION D. WEINER • *Cardiovascular Research Institute and Department of Biochemistry, University of California, San Francisco, CA, USA*

DANIEL C. WORTH • *Randall Division of Cell and Molecular Biophysics, King's College London, London, UK*

CHRISTOPH WOTZLAW • *Institut für Physiologie, Universität Duisburg-Essen, Essen, Germany*

CHRISTIAN A. WURM • *Mitochondrial Structure and Dynamics/Department of NanoBiophotonics, Max Planck Institute for Biophysical Chemistry, Goettingen, Germany*

DAN YANG • *Department of Chemistry, The University of Hong Kong, Hong Kong, P. R. China*

ALEXANDER V. ZHDANOV • *Biochemistry Department, University College Cork, Cork, Ireland*

Part I

General Principles and Overview

Chapter 1

Instrumentation for Live-Cell Imaging and Main Formats

Martin Oheim

Abstract

Unlike immunofluorescence confocal microscopy of fixed samples or microscopic surface analysis in material sciences that both involve largely indestructible samples, life-cell imaging focuses on live cells. Imaging live specimen is by definition minimally invasive imaging, and photon efficiency is the primordial concern, even before issues of spatial, temporal or, spectral resolution, of acquisition speed and image contrast come in. Beyond alerting the reader that good live-cell images are often not the crisp showcase images that you know from the front page, this chapter is concerned with providing a fresh look on one of the routine instruments in modern biological research. Irrespective of whether you are a young researcher setting up your own lab or a senior investigator choosing equipment for a new project, at some stage you will most likely face decision making on what (fluorescence) imaging set-up to buy. In as much as this choice is about a long-lived and often relatively costly piece of equipment and, more importantly, impacts on your future experimental program, this choice can be a tricky one. It involves considering a multitude of parameters, some of which are discussed here.

Key words: Fluorescence, live-cell imaging, microscopy, instrumentation.

1. Introduction

Fluorescence microscopy has evolved from an add-on contrast mode of the laboratory light microscope to a puzzling multitude of formats probing different aspects of molecular fluorophores. Classically requiring nothing else but a bright white light source for wide-field illumination (often termed a "burner"), a set of fluorophore-specific filters housed and purchased as a pre-assembled "cube," and an imaging detector ("camera"), the rapid technological evolution of scientific instruments has involved virtually all elements of the fluorescence microscope.

D.B. Papkovsky (ed.), *Live Cell Imaging*, Methods in Molecular Biology 591,
DOI 10.1007/978-1-60761-404-3_1, © Humana Press, a part of Springer Science+Business Media, LLC 2010

Fundamental choices for the user concern the illumination source, where arc-lamps are increasingly being replaced by lasers or high-power light-emitting diodes (LEDs) as discrete-wave-band illumination devices, pulsed or continuous-wave excitation, point scanning vs. whole-field excitation, filter-based or dispersion-based fluorescence band selection, integrating vs. photon-counting detectors, multi-channel or spectral detection, intensity of lifetime detection, to name only a few. To these options concerning instrumentation add those coming from the rapid progress in the synthesis and generation of molecular fluorophores, photolabile caged compounds, fluorescent and photo-switchable proteins, and photoactivated ion channels that often call for specific add-ons and imaging modalities. Moreover, the detection of intrinsic signals (autofluorescence, scattered light, higher harmonic generation) offers interesting alternatives to conventional fluorescence imaging depending on the introduction of exogenous probes.

For a novice, it might appear difficult to navigate through this diversity of instrumentation, formats, and probes and to have an educated choice among the variety of equipment or software available. This chapter, without attempting to be complete, is meant to provide the groundwork for choosing and evaluating instrumentation for live-cell imaging. Emphasis is on principles and constraints imposed by the different techniques rather than on a detailed discussion of specific equipment.

1.1. Further Reading and Web Resources

It is beyond the scope of this introduction to provide a detailed discussion of the ever increasing number of different formats of fluorescence microscopy. Fluorescence microscopy in its many variants is a standard theme in undergraduate and graduate courses, and a number of excellent reviews and textbooks are devoted to this subject; see below for a selection. To these, we have added the online resources provided by the different scientific societies as well as companies. We also alert the reader to the many excellent hands-on training courses that are held each summer and which – at least for the more prestigious ones – combine excellent theoretical training with the possibility to get your hands on the most recent pieces of equipment and thus provide valuable information before decision making about which piece of equipment to get for your own lab. Other important sources of first-hand information are the numerous cost-free optics and photonics journals.

1.2. Selected Fluorescence Textbooks

Although by no means complete, these recent (re-)editions of classic books provide an in-depth coverage of many aspects of fluorescence microscopy techniques, with a specific emphasis on biological and live-cell imaging.

1. Lakowicz, J. R., Principles of Fluorescence Spectroscopy, Springer, Heidelberg, New York, 3rd edition, 2006
2. Pawley, J. B. (ed.), Handbook of Confocal Microscopy, Springer, Heidelberg, New York, 3rd edition, 2006
3. Goldman, R. D. and Spector, D. L. (eds.) Live Cell Imaging – A Laboratory Manual. CSHL Press, Cold Spring Habor, 2005
4. Imaging in Neuroscience and Development – A Laboratory Manual. CSHL Press, Cold-Spring Habor, 2005

1.3. Web-Based Resources

Almost all microscope suppliers now offer free online tutorials that cover many aspects of microscopy: resolution, contrast generation, microscopic optics, and basics of fluorescence microscopy. They also point the reader toward related courses (often organized in partnership with the companies) and review articles.

1. Olympus Microscopy resource center: http://www.olympusmicro.com/primer/java/index.html
2. Zeiss Microscopy, http://www.zeiss.com/, and then link to "Technical Information"
3. Nikon Microscopy, http://www.microscopyu.com/tutorials/, and, specifically on confocal microscopy, http://www.microscopyu.com/articles/confocal/
4. Leica Microsystems, http://www.leica-microsystems.com/ website, then link to "Leica Scientific Forum"
5. Molecular Expressions Images from the Microscope, National High Magnetic Field Laboratory (NHMFL), Tallahassee, http://micro.magnet.fsu.edu/

1.4. Courses

There is an ever-increasing number of courses that permit both theoretical training and hands-on experience. Here are some of the better known ones:

1. Marine Biological Laboratory (MBL), Woods Hole, http://www.mbl.edu/education/
2. Cold Spring Habor Laboratory, Cold Spring Habor, http://meetings.cshl.edu/courses.html
3. NIH Bio-trac courses, Bethesda, http://www.biotrac.com/pages/courses.html
4. Live-Cell Microscopy Course, UBC, Vancouver, http://www.3dcourse.ubc.ca
5. Quantitative Fluorescence Microscopy Course, Mount Desert Island Biological Laboratory (MDIBL), http://www.cbi.pitt.edu/qfm/index.html

6. Marine Biological Association, Plymouth, The Microelectrode Techniques for Cell Physiology, http://www.mba.ac.uk/events.php
7. European Molecular Biology Organization, Practical Courses, EMBO Course on Light Microscopy in Living Cells, Heidelberg, http://www.mba.ac.uk/events.php

A more comprehensive list is found by linking to http://www.olympusfluoview.com/resources/courses.html

1.5. Free Photonics Journals

Another valuable source of information for beginning as well as confirmed microscopists is freely available monthly journals. A particularity of the photonics market, there are quite a few of them. Although often in close (sometimes all-too-close) proximity with advertiser and manufacturer opinion, these publications are a showcase of recent developments in optics, microscopy, and biophotonics. They provide up-to-date information on new equipment, notable technical achievements, and provide an excellent overview of trade fairs and meetings, both to come (which is good if you are to chose components for your microscope), or in the form of brief synopses, reviewing recent trends. While not replacing the academic literature, they certainly broaden your horizon and keep you connected to the often rapidly evolving technology at no extra cost.

1. Photonics Spectra
2. Europhotonics
3. Biophotonics International, all three from Photonics Media, Laurin Publishing, http://www.photonics.com
4. BioTechniques, from Informa Healthcare, http://www.biotechniques.com
5. Optics & Laser Europe, IOP Publishing, http://www.optics.org
6. Photonik international (*German*) from AT-Fachverlag, http://www.photonik.de
7. Photoniques, de la Société Française d'Optique, http://www.photoniques.com

1.6. Keep Your Cells Alive

When imaging live samples, the first question arising is how to use the limited photon budget without compromising sample viability. Careful controls should be made to develop a quantitative notion of how the used intensities affect the biological process under study. The result is often surprising because photodamage starts gradually before obvious signs occur. For example, two studies investigating how ample was two-photon photodamage when imaging intracellular free calcium ($[Ca^{2+}]_i$) (1, 2) found that the slope and kinetics of neuronal calcium signals were attenuated much earlier than electrophysiological signs of photodamage occurred. Thus, before starting your imaging, begin by thinking how you can

- optimize the photon yield (i.e., the number of collected signal photons vs. photons injected into the sample),
- avoid excitation wavelengths at which the sample absorption (and hence heating and photodamage) is strong,
- minimize the applied dye concentration,
- avoid producing crisp showcase images but aim for live cells,
- repeat the same experiment using different excitation intensities and using different emission bands,
- choose the imaging format that excels in your specific application.

This last point is crucial because it involves a choice of equipment and may lead you to the conclusion that the ideal experiment is impossible with existing material and can only be realized through an external collaboration. **Figure 1.1** provides a scheme (3) that helps rationalizing this decision making. Although oversimplistic, this scheme is useful because it points at the limitations of particular techniques and brings up useful parameters that should go into your consideration, e.g., background rejection, optical sectioning, imaging speed, or penetration depth (to only name a few).

1.7. Do You Really Need a Microscope?

Free yourself from the acquired wisdom that the choice is among the diverse upright and inverted scopes offered by the "big four," Zeiss, Leica, Nikon, and Olympus, and that the decision is merely a question of preference, compatibility with existing equipment, or the best commercial offer. Instead, ask yourself the following questions:

- Do I really need a microscope?
- Do I need eyepieces, bulky microscope bodies, inaccessible and unchangeable intermediate optics, and a limited flexibility governed by the elements to select from the suppliers' catalogue?
- You probably already have a good routine microscope for cell culture, patch-clamping, or checking immunofluorescence labeling before going to your facility's confocal or routine imaging. But do you need to combine different imaging formats as expensive (and often sub-optimal) add-ons on the same instrument?
- Or can you do better, with dedicated, small, and inexpensive set-ups?

A number of companies now offer modular solutions that allow you to configure the microscope the best way according to your needs. These solutions provide an interesting alternative

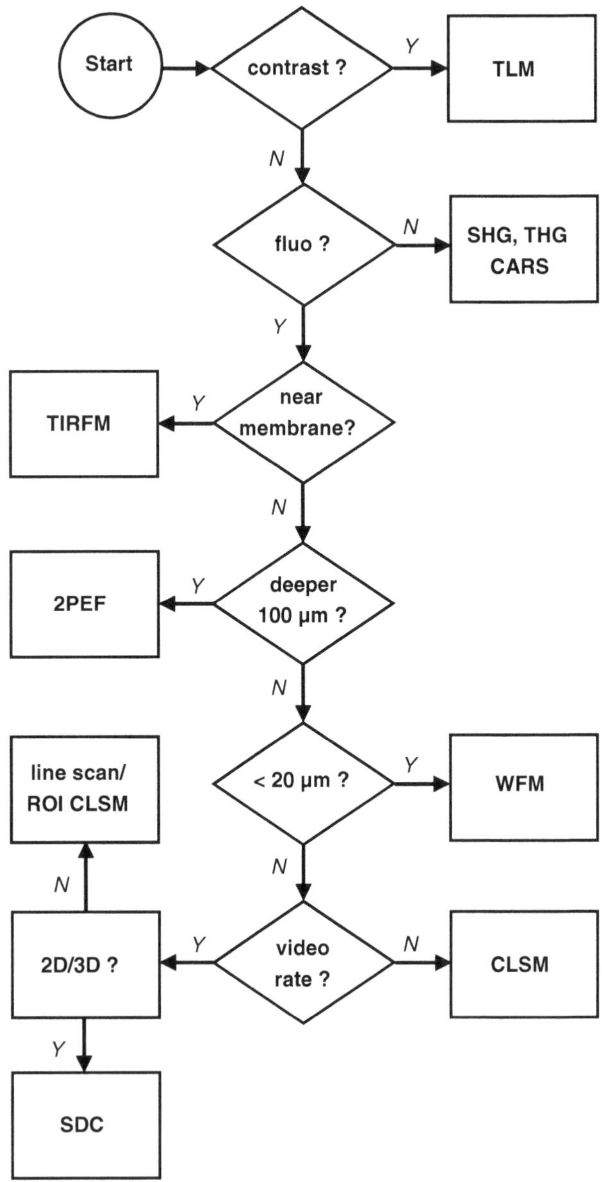

Fig. 1.1. *Abbreviations:* Y/N – yes/no; TLM – transmitted-light microscopies; SHG, THG – second (third) harmonic generation; CARS – coherent anti-Stokes Raman scattering; TIRFM – total internal reflection fluorescence microscopy; WFM – wide-field microscopy; 2PEF – two-photon excitation fluorescence; CLSM –confocal laser scanning microscopy; ROI – region of interest; SDC – spinning-disk confocal.

to the often somewhat finicky custom microscopes assembled from optical bench systems and also provide a means to generate hybrids between commercial microscope components and custom optomechanics. All of these approaches have in common that they offer full control of what you put in your microscope and allow replacing specific components that become limiting in a given application. We later discuss two examples of such custom microscopes built in our own laboratory.

Below, some of such modular microscopes the author is currently aware of are listed.

1. Olympus BXFM series (part of the BX51/BX61 upright microscope series). Most components of this upright microscope are sold individually, permitting to build your own structures.
2. Zeiss Axio Scope 1.Vario, a more recent but similar, component-based upright and highly modular microscope originally designed for the material sciences, http://www.zeiss.de/C12567BE0045ACF1/Contents-Frame/8FE44E3197A08FEBC125742E005BD1E1.
3. Already somewhat more reductionist is the SliceScope from Scientifica. It is a commercial minimally stripped down microscope body (without eyepieces) for combined electrophysiology and DIC/fluorescence imaging, http://www.scientifica.uk.com.
4. A similar system, equipped with Dodt contrast, is available from Siskiyou, http://www.siskiyou.com/imaging_system_mrk200-infrared-fluorescence.shtml.
5. TILL Photonics, offers a highly modular automated microscope that even goes below a bench or a robotic sample handler. Here the approach is rather to rethink the microscope body and to evolve in the direction of screening-by-imaging, http://www.till-photonics.com/Products/imic.php. With their now selling YANUS-4 laser scan head driven by all-digital smartmove boards, they equally offer a building block for constructing your own confocal or two-photon laser scanning microscope.
6. Somewhat more integrated, Prairie Technologies' Ultima is a confocal attachment with (optionally) two sets of galvanotmetric mirrors or acousto-optic deflectors for combined imaging and uncaging, http://www.prairie-technologies.com/ultima.htm.
7. Becker & Hickl has a confocal scan head (DCS-120) and detector modules for both intensity and fluorescence lifetime measurements, http://www.becker-hickl.de/.
8. Even further toward DIY, the opto-mecanical component supplier Linos (formerly Spinder und Hoyer) has its classical microbench, a 40 by 40 mm cage and rail system with components (eyepieces, revolver, lenses, C-mounts, apertures, ...) ready to build a microscope from scratch, http://www.linos.com/pages/home/shop-mechanik/banksysteme/mikrobank/.

Also exists in a 20 × 20-mm nanobench version. Unfortunately, the future of the even larger macrobench (150 × 150 mm) is with a question mark.

9. AHF (http://ahf.de) has the missing part for the microbench: a 45° holder for a 1-mm 25 × 63-mm (Zeiss or Olympus) standard dichroic mirror.
10. Interestingly, Thorlabs offers a fair number of cage system components that are largely compatible with the Microbench, thus considerably enlarging the choice of optical components, http://www.thorlabs.com/navigation.cfm?Guide_ID=2002.

2. Instrumentation

2.1. Understanding the Building Blocks of the Laboratory Microscope

It is quite instructive to forget what you know about laboratory microscopes for a moment. Modern fluorescence microscopes are highly modular and can be thought of as a box with lots of arms sticking in and out (**Fig. 1.2**). Excitation arms can be different channels of epi-illumination, a laser injected through a side port for total internal fluorescence or photoswitching, a spinning-disc confocal attachment, or a pulsed UV-lamp for flash-photolysis.

For a given excitation channel i and fluorophore j, excitation is fully described in terms of the source spectral emission S(λ), transmission of the excitation filter EX(λ), and the reflectivity diachroic beamsplitter (1–BS(λ)), which, for a give excitation channel, are multiplied along the excitation optical path:

$$\mathrm{ex}_i(\lambda) = \mathrm{S}(\lambda) \cdot \mathrm{EX}_i(\lambda) \cdot (1 - \mathrm{BS}(\lambda)) \qquad [1]$$

combined with the sample molar extinction ε_j and absorbance spectrum $E_j(\lambda)$ and integrated over λ to give an excitation spectral function:

$$\xi_{ij} = \int d\lambda\, \mathrm{ex}_i(\lambda) \cdot E_j(\lambda) \cdot \varepsilon_j. \qquad [2]$$

On the emission site, one proceeds analogously for each detection arm k by multiplying the fluorophore quantum yield and emission spectrum, spectral transmission of the dichroic and emission filters, and the detector spectral sensitivity:

$$\mathrm{em}_k(\lambda) = \mathrm{BS}(\lambda) \cdot \mathrm{EM}_k(\lambda) \cdot \mathrm{D}(\lambda). \qquad [3]$$

Upon integration,

$$\xi'_{jk} = \int d\lambda \phi \cdot F_j(\lambda) \cdot \mathrm{em_k}(\lambda) \qquad [4]$$

Finally, the product of eqs. [**2**] and [**1.4**],

$$\xi_{ijk} \equiv \xi_{ij}\xi'_{jk} \quad [5]$$

measures the signal of one mole/l of fluorophore *j* viewed through channel *k* upon excitation in channel *i*, and similarly for all permutations *ijk*.

Optimizing the photon yield now consists in maximizing the detection spectral function, while balancing the excitation spectral function with that of other fluorophores present in the sample.

This balancing can be achieved not only by choosing appropriate filters but also by considering the source and detector spectra and the microscope intermediate optics and objective lens. Their transmission can conveniently be combined with that of the beamsplitter, thereby accounting for the double passage of excitation and emission light through the microscope.

The following list provides a quick overview of the principal choices available and briefly discusses their respective advantages and disadvantages with respect to the scheme shown in **Fig. 1.2**.

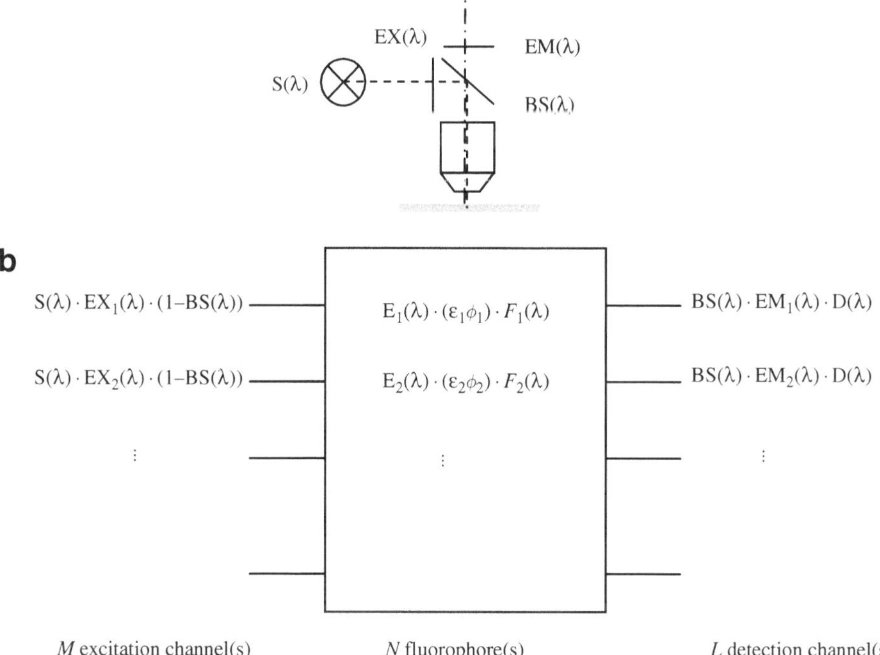

Fig. 1.2. (**a**) Schematic respresentation of an upright epifluorescence microscope. S – source; EX – excitation filter; BS – diachroic beamsplitter; EM – emission filter; D – detector. See main text for details. (**b**) Box plot of the different excitation and emission channels along with their spectral throughput, obtained by multiplying instrument and fluorophore spectral properties along the excitation and detection arms. See main text for details.

Light sources: the choice is no longer only between Hg- and Xe-burners but involves a large variety of sources including arc-lamp based monochromators, lasers (gas, diode-pumped solid state, DPSS), laser diodes, white light and color LEDs, and photonic crystal fibers for supercontinuum generation. These vary in

- wavelength, tuning range, bandwidth and power,
- luminous density,
- noise and long-term stability,
- cost,
- coherence, polarization,
- pulsed vs. continous-wave (CV) operation.

Excitation filters: typically, band-pass filters are used to narrow down the source spectrum and select specific wavelength bands, clean up laser lines, attenuate illumination intensity, or select a polarization direction. Main choices are between hard- and soft-coated interference filters as well as holographic notch filters. Important selection criteria include the following:

- their in-band transmission and off-band rejection (optical density),
- bandwidth or line-width,
- tolerance to high illuminating intensities (burnout, hole-burning).

  A similar reasoning must be made for the *dichroic mirror* that, for the more common long-pass, separates excitation and emission light by reflecting the excitation light onto the sample and transmitting the collected fluorescence. Chief parameters to consider in addition to center wavelength and steepness are as follows:

- Spectral holes: many dichroics perform nicely close to the transition wavelength, but display large variations and spectral holes at remote wavelength. "Extended range" ("XR") dichroics can be a solution, as are the new hard-coated filters that typically outperform the older soft coatings.
- Transmitted excitation light. Although often not a problem, low-light applications can suffer from the residual transmitted light that can be orders of magnitude more intense than the collected fluorescence. Stacking dichroics or matched long-pass filters can be an answer.
- Finally, particularly for multi-color scanning microscopies, the angle-dependence of the cut-on of the dichroic can result in surprising chromatic changes depending on whether paraxial regions or the periphery of the field-of-view is imaged. Typically, the cut-on wavelength changes as

$$\lambda(\theta) = \lambda_0\sqrt{1 - (sin\theta/n_{\text{eff}})^2}, \qquad [6]$$

For angles θ increasing beyond 45°, n_{eff} is determined by the dichroic of the order of 1.7–1.86 for most of the coatings, but varies for *p*- or *s*-polarized light so that their average must typically be considered for the collected fluorescence.

The *microscope intermediate optical path* (tube lens, scan lens, epi-illuminator, projection lenses in the case of compound magnifiers) and the *objective* are often not considered in detail, until unanticipated losses occur. For example, both UV and near-IR transmission become a problem, e.g., in fura-2 Ca^{2+} imaging, flash photolysis, or photoactivation, as well as two- or three-photon microscopy using far-IR excitation.

- UV transmission. With the increasing demand for highly corrected objectives (and tube lenses), high refractive index glasses and optical cement are often used chromatic corrections that limit UV transmission. Also, with microelectronics getting smaller and smaller, the formerly used UV-light inspection techniques often fail because of resolution limitations. Thus, dedicated UV-transmitting optics is getting rarer and rarer.
- Another often overlooked aspect is limiting intermediate apertures (filter cubes, lenses, irises) that reduce the collected light fraction of scattered fluorescence photons in two-photon microscopy. While scattered photons are usually rejected in fluorescence imaging (because they are not assigned to the pixel of their origin and hence blur the image) they constitute useful signal in 2PEF microscopy that seeks to optimize the light collection. Thus, keeping the product of *d* $\sin(\theta_{\text{eff}})$ constant is of crucial importance for not using light. Here, *d* is the size of the imaged spot and θ_{eff} is the half-angle of the effective numerical aperture.

Much of the same reasoning is true for choosing and optimizing detectors. Know your microscope is the rule toward getting optimal results.

2.2. The Performance Triangle

Irrespective of all these parameters, you can only distribute your collected photons once. Thus, whenever you take an image of a live cell, you take from the budget of photons that your sample emits before irreversibly undergoing photodamage and photodestruction. If you invest them into higher spatial resolution, higher temporal resolution or larger image contrast (signal-to-noise ratio) is your choice. And it is often a difficult one.

Resolution vs. magnification. On most available microscopes, resolution is diffraction-limited. Thus, the smallest distance that two objects can be close by and still be detected as two is defined by the numerical aperture of the objective, according to Abbe's law. Importantly, the detector spatial resolution (i.e., the pixel size in an imaging detector, or the scan angle for a laser-scanning microscope) must be adapted, according to Nyquist's sampling theorem: two picture elements (pixels) per resel (resolution element), $0.61\lambda/(\mathrm{NA})$. Smaller pixels do not enhance resolution but increase photobleaching by void oversampling.

Superresolution, i.e., imaging beyond the diffraction limit has attracted wide interest over the last years and extensive reviews have been published. By narrowing down the fluorescence excitation volume (through stimulated emission depletion, STED, or structured illumination and image reconstruction methods) or by sequentially imaging individual fluorophores and reconstructing the image from the sum projection (STORM, PALM, and its variants) high spatial frequency information can be obtained. But again, superresolution translates into sampling at higher spatial frequencies, so that at a constant photo budget, either the dye concentration in the sample must be increased or the field-of-view must be reduced. Also, some super-resolution techniques require sample pre-bleaching, photoactivation, or the STED beam in addition to the conventional excitation light. It is safe to say that for many of these exciting developments, a critical evaluation of the photodamage resulting in the live samples still has to be done.

Image contrast comes from the number of meaningful signal photons over the unwanted background in a given fluorescence detection channel. Thus, spectral considerations directly come into play. If contrast is generated by splitting up the signal in many different spectral channels, then each of these channels will contribute to the noise and the signal-to-noise ratio will inevitably drop. Therefore, "multi-"spectral imaging is generally preferable over "hyper-"spectral imaging and a small number of detection channels followed by spectral unmixing outperforms full-blown spectral images (4, 5). Similarly, contrast in a given spectral bin can of course be increased by cranking the laser power up, but this again increases photobleaching and tears from the available photon budget. Therefore, it is useful to keep the "performance triangle" of fluorescence microscopy (*see* **Fig. 1.3**) in mind, any improvement in one of the image parameters comes at the expense of the others.

2.3. Additional Considerations

- *Long-term observation* of live samples critically relies on constant observation conditions, both in terms of the instrument (i.e., maintaining the focal stability and minimizing thermal drift) and of the biological sample (control of physiological temperature, ambient CO_2, and humidity levels). Due to condensation, placing the entire microscope in an incubator is not preferred, but many suppliers offer small-on stage or plexiglass-box incubators that can be fitted to many routine microscopes.

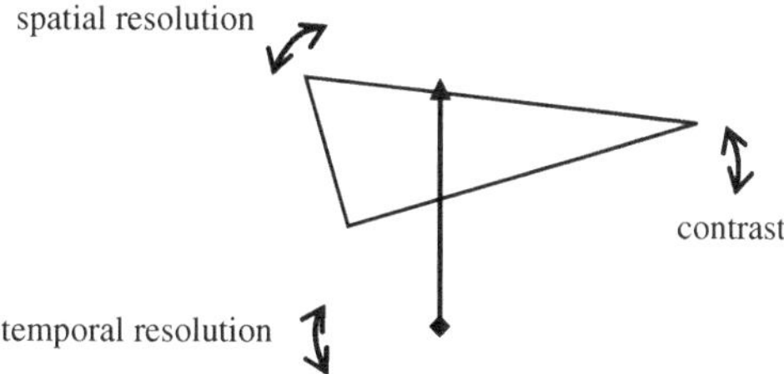

Fig. 1.3. The performance triangle. Distributing a constant photon budget into any of these three imaging parameters inevitably reduces the other two.

- Deciding between a *motorized vs. manual set-up* is not only a question of price and convenience. Many microscopists like the manual and thus more direct focus and stage control, e.g., when placing recording electrodes or local perfusion systems under visual control. On the other hand, many motorized systems now come with a fairly well-developed software, allowing the user to generate look-up tables of stage positions, objective *z*-positions or even automated follow software that keeps patch pipettes already placed above the sample plane in the field-of-view whilst searching for the cell to record from.
- *High-throughput microscopy* is increasingly becoming an option through the ongoing integration of machine vision, robotics, microfluidics, and automated analysis software. Several commercial systems are available, albeit at high cost.
- *Shared set-ups vs. single-user set-ups.* Perform a realistic evaluation of beamtime: which fraction of your experiment time is effectively being used for imaging? Which part is devoted to preparing and installing the sample? Which steps could equally be performed elsewhere to free precious beamtime? Most laboratory microscopes are under-used, but multi-user set-ups require clear shared responsibilities, agreed-on standard operation protocols, and an efficient communication among users. Otherwise, the gain in instrument use will easily be eaten up by problem solving and conflicts of unhappy experimenters.
- *Multi-functionality vs. dedicated set-up.* Beyond budgetary and space constraints, this often is a question of the type of experiments you have in mind. All too complicated microscopes are expensive, error-prone, and rarely all contrast modes are being used in the same experiment. Thus, "small is beautiful" is a guideline that more often than not gives good returns, particularly when set-ups are shared or even open to external users. Trained personnel and regular maintenance will make all the difference.

- *Optimal optical performance* in a given imaging format often involves custom equipment or add-ons to commercial microscopes. Typically being fairly labor-intensive, alignment sensitive, and often run with custom software that resembles beta-versions, these "expert systems" are less user-friendly but outperform standard equipment. Many recent imaging formats, including light-sheet based illumination, HILO (6, 7), but also versatile STED or 2PEF microscopes, still require building your own setup. At the same time, instrument development and building is typically longer than hoped for, so that the researcher has to evaluate the need for quick results against instrument performance.
- *Laser safety* is obviously a concern, particularly with home-built apparatus. While interlocks, beam stops, and protective shutters are mandatory in commercial microscopes, custom set-ups are often more reminiscent of open optical bench systems and do not comply with legal guidelines. Hence, developers and experimenters should stay in close contact and new users should be briefed about risks.

3. Concluding Remarks

Neither does this introduction replace a careful reading of the original papers describing different imaging formats nor does it replace making your own experience with the equipment you bought. But it alerts the reader to consider some parameters that are not so obvious when looking at microscope brochures and reading the often very condensed "materials and methods" sections. If there is a simple conclusion, then it is this: know your microscope. It pays for your research and your next budget.

References

1. Hopt, A. and E. Neher, *Highly nonlinear photodammage in two-photon fluorescence microscopy.* Biophys. J., 2001. **80**(4): 2029–36.
2. Koester, H.J., et al., *Ca^{2+} fluorescence imaging with pico- and femtosecond two-photon excitation: signal and photodamage.* Biophys. J., 1999. 77(4): 2226–36.
3. Frigault, M.M., et al., *Live-cell imaging: tips and tools.* Biophys. J., 2009. **96**(3): Supplement 1, 30a.
4. Nadrigny, F., et al., *Detecting fluorescent protein expression and co-localization on single secretory vesicles with linear spectral unmixing.* Eur. J. Biophys., 2006. **35**(6): 533–47.
5. Neher, R.A. and E. Neher, *Optimizing imaging parameters for the separation of multiple labels in a fluorescence image.* J. Microsc., 2004. **213**(1): 46–62.
6. Tokunaga, M., N. Imamoto, and K. Sakata-Sogawa, *Highly inclined thin illumination enables clear single-molecule imaging in cells.* Nat. Methods, 2008. **5**: 159–61.
7. van't Hoff, M., V. de Sars, and M. Oheim, *A programmable light engine for quantitative single-molecule TIRF and HILO imaging.* Opt. Express, 2008. **16**(22): 18495–504.

Chapter 2

Labels and Probes for Live Cell Imaging: Overview and Selection Guide

Scott A. Hilderbrand

Abstract

Fluorescence imaging is an important tool for molecular biology research. There is a wide array of fluorescent labels and activatable probes available for investigation of biochemical processes at a molecular level in living cells. Given the large number of potential imaging agents and numerous variables that can impact the utility of these fluorescent materials for imaging, selection of the appropriate probes can be a difficult task. In this report an overview of fluorescent imaging agents and details on their optical and physical properties that can impact their function are presented.

Key words: Fluorescence, fluorescent labels, fluorogenic probes, sensors, microscopy, imaging.

1. Introduction

Fluorescence imaging is a vital tool for the investigation of biological processes in the fields of cell, molecular, and systems biology. Its development has had a profound impact on our ability to decipher how these systems function at the cellular and molecular level. The development of fluorescence microscopy as an investigative tool has its origins in the 1850s with the first descriptions of "refrangible radiations" from biological materials by George Stokes (1). These radiations were later named fluorescence. However, the use of fluorescence as a diagnostic tool in microscopy would remain undeveloped until the construction of the first UV light microscopes by August Köhler in 1904 (2). Not long after the work of Köhler, the first purpose-built fluorescence microscopes were prepared, but it was not until the 1960s

D.B. Papkovsky (ed.), *Live Cell Imaging*, Methods in Molecular Biology 591,
DOI 10.1007/978-1-60761-404-3_2, © Humana Press, a part of Springer Science+Business Media, LLC 2010

that these instruments became commonplace. Some of the first fluorescence microscopy experiments focused on observation of the intrinsic fluorescence of the biological samples under investigation (3). These investigations were invaluable for expanding our understanding of physiology, but they provided little insight on the function of biochemical and other physiological processes at a molecular level. For a more in depth investigation of these processes within the cell, a switch from intrinsic to extrinsic fluorophores is necessary. Today, numerous fluorescent materials are available for use in fluorescence microscopy.

Fluorescent compounds suitable for live cell imaging can be divided into two broad categories: labels and responsive probes. Fluorescent labels are imaging agents whose fluorescence signal remains constant. Good labels are typified by stable optical properties that do not vary significantly as a function of their local environment. These fluorescent species are often coupled with targeting groups or have genetically controlled expression. Responsive probes do not rely on preferential uptake or targeting. These sensors rely on changes in fluorescence intensity, wavelength, or lifetime for their function, and can be small molecule, polymer, or nanoparticle based. In this report we will provide an overview of current fluorescent labels and probes for use in live cell imaging of molecular processes.

2. Fluorescent Labels

The first advances toward the development of modern fluorescent labels are credited to the immunologist, Albert Coons in the 1940s. In his early research, he developed fluorescein isothiocyanate (FITC) (4), which remains one of the most ubiquitous fluorescent labels today, for coupling to antibodies targeted against pneumococcal bacteria. Today, targeted labels are among the most commonly employed fluorescent imaging agents. In addition to antibodies, targeting groups can be proteins, peptides, DNA aptamers, small molecule ligands, or stains for specific macromolecular structures. The emissive reporters in these labels can be fluorophores, fluorescent or bioluminescent proteins, or nanoparticles such as quantum dots. The efficacy of imaging with these compounds is dependent on their specific uptake, sequestration, or expression at a subcellular level.

2.1. Small Molecule Fluorophores

Prior to the development of FITC labels, a limited number of fluorophores with synthetic handles suitable for bioconjugation were available. Many of these early labels were based on dyes with fluorescence excitation in the UV (5). The fluorescence emission signals from these dyes can be difficult to separate from tissue autofluorescence. In contrast, today there is a vast and often confusing array of fluorophore labels available to scientists. These fluorophores span the optical spectrum from the UV to the visible and extend into the near infrared (6–12). Many of the currently available amine reactive fluorophores are summarized in **Fig. 2.1** and the structures of some representative labels are shown in **Fig. 2.2**. Several factors must be considered in choosing the appropriate fluorophores for constructing effective imaging agents. These include method of attachment to the targeting group, excitation and emission wavelengths, brightness, hydrophilicity, and cost.

There are many current chemistries available for the coupling of fluorescent labels to biomolecules and targeting groups (**Fig. 2.3**). The most frequently employed synthetic handle for bioconjugation is the succinimidyl ester, which forms stable amide bonds after reaction with primary and secondary amines. The isothiocyanate group may also be used for coupling to amines, generating a thiourea linkage. In cases where reaction of a succinimidyl ester or isothiocyanate derivatized fluorophore with an amine is not feasible, additional coupling groups are available. Iodoacetamide, maleimide, and dithiol-modified fluorophores are useful for covalent conjugation to thiols. Hydrazine and hydrazide modified dyes can be used for coupling to aldehydes and ketones, forming relatively stable hydrazone linkages. More recently, the development of bioorthogonal coupling schemes has attracted significant interest for preparation of fluorescent probes.

Bioorthogonal couplings rely on use of reaction partners that display little or no reactivity with common biological materials. Two examples of these reactions are the Staudinger ligation (13) and the "click" reaction (14). The Staudinger ligation involves coupling of a methyl ester electrophilic trap with an azide to generate an amide linkage and one equivalent of N_2. This reaction is mediated by oxidation of an adjacent phosphine. The click reaction is a copper(I) catalyzed [3+2] cycloaddition between an azide and an alkyne that results in the formation of a stable triazole product (14). This reaction has excellent potential for use in design of targeted fluorescent probes. However, there are only a few azide or alkyne modified dyes currently available for this reaction, most of which emit in the visible region. The potential utility of the click reaction in biology suggests that in the coming years the selection of azide and alkyne modified dyes is likely to expand greatly. For example, recent efforts have yielded new, efficient synthetic routes to far-red/near infrared emitting cyanine dyes modified with either azide or alkyne groups, one example of which, CyAM-5 alkyne, is shown in **Fig. 2.2** (15). Although highly selective, cytotoxic copper(I) is necessary for the traditional click coupling, and therefore direct use of this reaction in living biological systems has not been possible. The

nm 300 400 500 600 700 800 900

UV

Fluorophore	Abs (nm)	Em (nm)	ε (cm^{-1}M^{-1})	ϕ_f	Ref.	Notes:
Dansyl chloride (DsCl)	337	492	5,300		6	*a*
Pyrenesulfonyl chloride	350	380	28,000		6	*b*
Coumarin-6-sulfonyl chloride	360	405			7	*c*
7-(Diethylamino)coumarin-3-carboxylic acid NHS	445	482			7	*d*
AF 350	346	445	19,000		6	*e*
AF 430	435	535	15,000		6	*e*
ATTO 390	390	475	24,000	0.90	8	*f*
ATTO 425	437	483	45,000	0.90	8	*f*
Dy 415	418	467	34,000		9	*g*

Visible

Fluorophore	Abs (nm)	Em (nm)	ε (cm^{-1}M^{-1})	ϕ_f	Ref.	Notes:
NBD-X	466	535	22,000		6	*h*
FITC (5-isomer)	494	519	77,000		6	*i*
5-Carboxy-fluorescein	494	520	78,000		6	*i*
5(6)-Rhodamineisothiocyanate (TRITC)	544	572	84,000		6	*h*
Carboxy-tetramethylrhodamine (TAMRA)	554	584			7	*j*
Carboxy-X-rhodamine	575	605			7	*j*
AF 488	494	517	73,000	0.92	6	*e*
AF 546	554	570	112,000	0.79	6	*e*
AF 568	578	602	88,000	0.69	6	*e*
AF 594	590	617	92,000		6	*e*
AF 633	621	639	159,000		6	*h*
ATTO 488	498	520	90,000	0.80	8	*f*
ATTO 532	533	560	115,000	0.90	8	*f*
ATTO 550	554	574	120,000	0.80	8	*f*
ATTO 565	565	590	120,000	0.90	8	*f*
ATTO 590	594	624	120,000	0.80	8	*f*
ATTO 633	629	657	130,000	0.64	8	*f*
BODIPY FL	502	510	82,000		6	*h*
BODIPY TMR-X	544	570	60,000		6	*h*
Cy 3	548	562	150,000	0.04	10	*f*
Cy 3.5	581	596	130,000	0.14	10	*f*
Dy 495	493	521	70,000		9	*g*
Dy 547	557	574	150,000		9	*g*
Dy 590	580	599	120,000		9	*g*
Dy 610	610	630	80,000		9	*g*
Dy 631	637	658	200,000		9	*g*

Near-Infrared

Fluorophore	Abs (nm)	Em (nm)	ε (cm^{-1}M^{-1})	ϕ_f	Ref.	Notes:
AF 647	651	672	270,000	0.33	6	*k*
AF 660	668	698	132,000	0.37	6	*k*
AF 680	684	707	183,000	0.36	6	*k*
AF 700	702	723	205,000	0.25	6	*k*
AF 750	753	782	290,000	0.12	6	*k*
ATTO 655	663	684	125,000	0.30	8	*f*
ATTO 680	680	700	125,000	0.30	8	*f*
ATTO 700	700	719	120,000	0.25	8	*f*
BODIPY 650/665-X	646	660	102,000		6	*h*
Cy 5	646	662	250,000	0.27	10	*f*
Cy 5.5	673	692	190,000	0.23	10	*f*
Cy 7	747	774	200,000	0.28	10	*f*
Dy 647	653	672	250,000		9	*g*
Dy 680	690	709	140,000		9	*g*
Dy 700	707	730	140,000		9	*g*
Dy 750	747	776	270,000		9	*g*
Dy 780	782	800	170,000		9	*g*
IRDye 800CW	774	789	240,000		11	*l*
IRDye 680	676	698	140,000		11	*l*
VivoTag 680	670	688	100,000		12	*f*
VivoTag-S 750	750	775	240,000		12	*f*

Notes: *a.* After reaction with alkylamine in CHCl3; *b.* after reaction with alkylamine in MeOH; *c.* after reaction with alkylamine in CH3CN; *d.* phosphate buffer, pH 7.0; *e.* aqueous, pH 7.0; f. aqueous; *g.* EtOH; h. MeOH; *i.* water pH 9; *j.* phosphate buffer, pH 8.0; *k.* quantum yield in PBS, pH 7.2, all other data in MeOH; *l.* PBS, pH 7.4.

Fig. 2.1. Commercial amine reactive fluorophore labels.

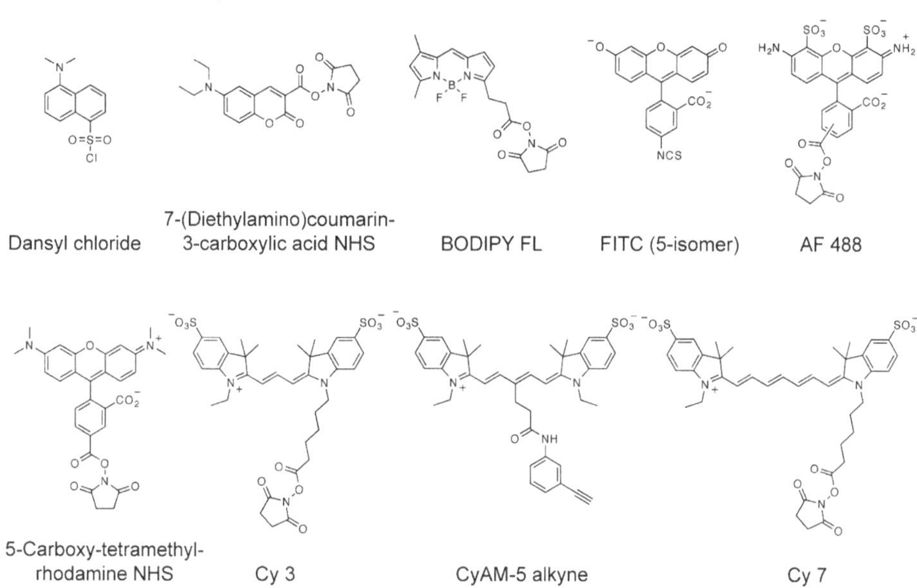

Fig. 2.2. Representative structures of fluorescent labels with emission in the blue (7-(diethylamino)coumarin-3-carboxylic acid NHS), green (BODIPY FL, FITC, and AF 488), orange (5-carboxy-tetramethylrhodamine NHS and Cy 3), far red (CyAM-5 alkyne), and near infrared (Cy 7).

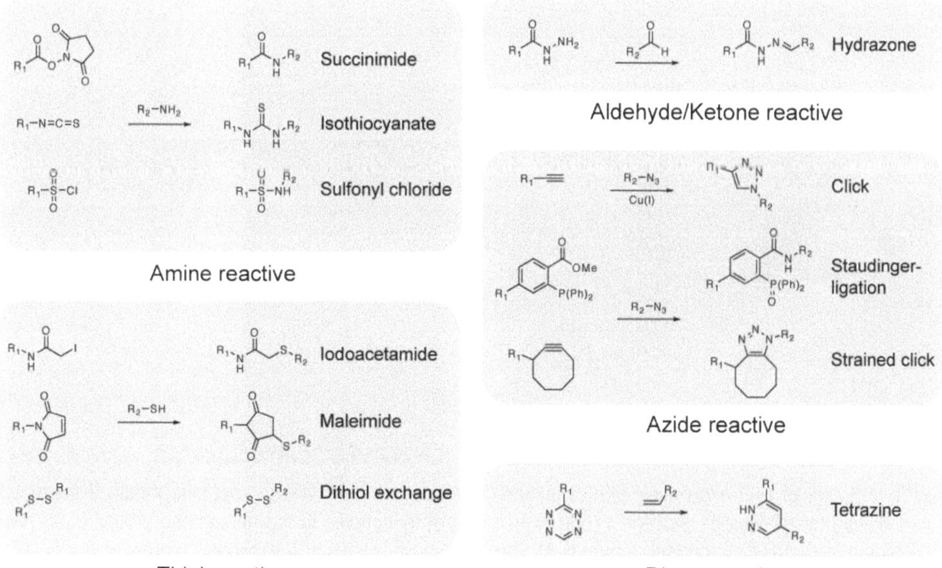

Fig. 2.3. Common coupling chemistries for attachment of fluorescent labels to targeting groups and biomolecules.

issue of copper cytotoxicity in the click reaction has been overcome by Bertozzi and others via preparation of new ring-strained cyclooctyne derivatives that do not require a catalyst (16–18). The coupling of cyclooctyne containing fluorophores with azide-modified sugars has been demonstrated for imaging of surface glycosylation in live cells (16) and zebra fish embryos (19). Another bioorthogonal conjugation strategy compatible with live cells was reported in 2008 (20, 21). This coupling scheme involves use of an inverse-electron demand Diels–Alder cycloaddition between a modified tetrazine and a norbornene dienophile (20). The tetrazine-based coupling shows excellent selectivity in biological media and was used to label SKBR3 breast cancer cells that were pre-treated with norbornene modified trastuzumab (**Fig. 2.4**). The availability of labels for use with classical and bioorthogonal coupling reactions provides a wide selection of methods for attachment of fluorescent reporters to biological targets. The choice of coupling chemistry will be dependent on the specific reactive chemical groups available on the targeting molecule such as amines, thiols, or ketones. When the coupling or labeling reaction must be performed in a biological environment in the presence of live cells, the copper-free click reaction, Staudinger ligation, or the tetrazine cycloaddition reactions are appropriate conjugation methods.

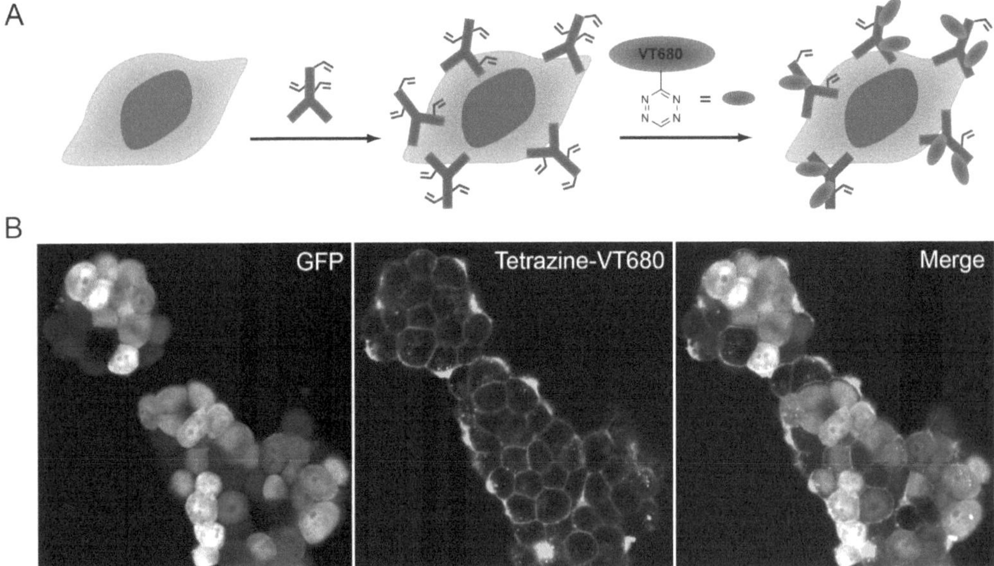

Fig. 2.4. Pre-targeting of GFP expressing SKBR3 human breast cancer cells with norbornene modified trastuzumab antibodies followed by addition of tetrazine-VT680, which covalently couples to the norbornene groups in an inverse electron demand Diels-Alder cycloaddition (**panel A**). **Panel B** shows confocal microscopy images of the cells after pre-targeting and VT-680 treatment in the GFP channel (*left*), VT680 channel (*center*), and the merged image (*right*).

Coupling chemistries have been used to prepare a wide array of imaging agents utilizing antibodies, aptamers, peptides, and small molecules. For example, anti-human epidermal growth factor 2 (HER2) antibodies conjugated with IRDye 800 were used to show antibody binding to HER2 expressing SKBR3 breast cancer cells and for in vivo fluorescence imaging in a mouse model (22). DNA aptamers have also been used to target tumor cells (23). In much the same way, peptides have been coupled to a variety of fluorophores for preparation of several targeted imaging agents. This strategy has been widely used for targeting the $\alpha_{\nu}\beta_{3}$ integrin cell adhesion molecule with the RGD peptide motif for investigation of cancer cells and tissues (24–27). Targeting approaches need not be limited to short peptides. Larger peptides and proteins may also be used for directed delivery of optical reporters. Probes for selective imaging of epidermal growth factor receptor (EGFR) have been prepared via conjugation of Cy5.5 fluorophores to the 6-kDa epidermal growth factor protein. This probe was demonstrated to specifically home in on MDA-MB-468 cancer cells, which have high EGFR expression levels, but not to MDA-MB-435 cells which do not express EGFR (28). Small molecule based targeting strategies have also been employed through use of well-known bioactive small molecules such as folate (29) or through combinatorial approaches (30).

Peptide-based targeting has been expanded to incorporate bacteriophage nanoparticles as multivalent peptide carriers. This allows for facile integration of peptide screening for the cellular target of interest (via use of bacteriophage display libraries) with optical imaging and microscopy techniques. The M13 bacteriophage, commonly used in bacteriophage screening, has randomized peptide libraries displayed on its pIII coat proteins. The bacteriophage particles also contain 2700 copies of the pVIII coat protein, which have their amino termini exposed to the solvent. These amine groups are available for bioconjugation to fluorophores. Therefore, once a phage clone specific for the receptor of interest is identified, it can be modified via standard succinimide or isothiocyanate coupling procedures to prepare a fluorescent targeted imaging probe. This strategy was first demonstrated in 2004 (31) and further expanded for other imaging applications (32–34).

In addition to the bioconjugation strategy and selection of the targeting group, the optical properties of the fluorophore are another important factor in the design of targeted probes for live cell imaging. Although there are many fluorophores with excitation and emission in the UV, these fluorescent labels are not appropriate for certain imaging applications due to concerns regarding exposure of the cells to UV light, which may disrupt normal cell function. UV excitation may also result in higher background fluorescence signal from the sample, arising from the excitation of intrinsic biological fluorophores. Problems may occur with other common dyes, such fluorescein. Fluorescein is a pH-sensitive dye with a fluorogenic pK_a of 6.4; therefore, fluorescent labels containing fluorescein may display distinctly different fluorescence emission intensities depending upon the pH of their local environment. Consequently, for imaging applications where the probe signal will be quantified, fluorescein may not be suitable. Other fluorophores with similar excitation and emission such as BODIPY FL or AF488 (6), which is based on a pH-insensitive rhodamine scaffold, are more appropriate for use in experiments requiring probe quantification. In many applications, labels with fluorescence emission in the NIR are preferred. NIR-emitting fluorophores are not susceptible to interference from biological autofluorescence and are directly translatable to many in vivo imaging applications due to the increased optical transparency of biological tissue between ~650 and 1000 nm (35). Fluorophores can show distinct changes in their fluorescence intensity and/or fluorescence emission wavelength based on the polarity of their local environment. The fluorescence quantum yields and emission wavelengths of dansyl fluorophores are well known to vary with the polarity of the surrounding media. Solvent polarity-based changes in fluorescence emission wavelengths and quantum yields can often be minimized by increasing the polarity of the fluorophore. Fluorophores that show little or no polarity-dependent changes on their optical properties tend to contain one or more solubilizing groups such as sulfonate or carboxylate moieties. For example, the optical properties of the near infrared emitting fluorophore, Cy5.5, which has four sulfonate groups, are relatively insensitive to changes in the local microenvironment.

Many fluorophores can be modified to act as optical switches that are activated by exposure to UV light. Several applications have made use of these photoactivatable or "caged" fluorophores. Caged fluorophores have been employed in dynamic imaging applications where specific temporal and spatial activation of a small population of fluorophore labels is required. These masked fluorophores, such as caged fluorescein, are prepared by reaction of the fluorophore with *o*-nitrobenzylbromide to form the non-fluorescent photoactivatable compound (36). The fluorescence can be activated by irradiation at 365 nm to cleave the *o*-nitrobenzyl group, releasing the free fluorophore (**Fig. 2.5**). In one early demonstration of this approach, microtubule flux in the mitotic spindle was monitored following photoactivation of caged fluorescein-labeled tubulin (36). Similarly, a caged resorufin was used to observe intracellular actin filament dynamics (37). More recently, a series of cell permeable caged coumarin derivatives (38, 39) has been designed for the study of intercellular gap junctions. After intracellular delivery of these caged fluorophores, a small population of the caged coumarins was activated and used as a fluorescent reporter to monitor the migration of the dye molecules through the gap junctions (39).

The brightness of the fluorophore is a key consideration. When targeting cellular components that are expressed in low levels, the fluorescence signal from the optical reporter needs to

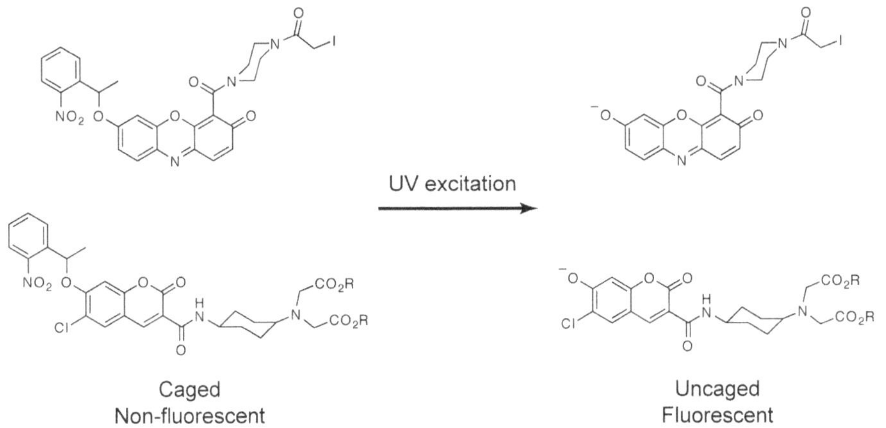

UV excitation

Caged
Non-fluorescent

Uncaged
Fluorescent

Fig. 2.5. Uncaging of non-fluorescent *o*-nitrobenzyl modified resorufin (*top*) and coumarin (*bottom*) derivatives after exposure to UV light.

be bright. The brightness is defined as the product of the fluorescence quantum yield and extinction coefficient of the fluorophore. For imaging low concentrations of cellular targets, weak fluorophores such as those based on NBD or pyrene may not be suitable. Some of the brightest fluorophores emitting in the visible are based on rhodamine or BODIPY scaffolds. Both of these fluorophore classes are typified by quantum yields approaching unity and extinction coefficients of 80,000 $M^{-1}cm^{-1}$ or more. In the far-red/NIR there are many bright fluorophores (6). Common NIR-emitting cyanine dyes are typified by large extinction coefficients often exceeding 200,000 $M^{-1}cm^{-1}$ and quantum yields of 20% or greater (10). However, the fluorescence quantum yields of fluorophores with emission > 800 nm begin to drop off considerably (34). This is the result of the relatively small energy difference between the ground and the excited states of these dyes, which allows for enhanced non-radiative decay of the fluorophore from the excited state.

The polarity of the fluorophore is an important factor in imaging agent design and may have a significant impact on the function of the fluorescent label. Many popular fluorescent labels are highly water-soluble polar species. Examples include AF488, fluorescein, sulforhodamine 101, and most cyanine-based far-red/NIR fluorophores. Imaging agents using polar fluorophores may not be able to cross the cell membrane by passive diffusion processes. Unless a targeted energy dependent transport mechanism is utilized, they are better suited for use as components of fluorescent reporters for imaging cell membrane or extracellular matrix components. Other fluorescent labels, such as DNA stains, rely on the permeability properties of the cell membrane for their function. These charged fluorescent molecules are often unable to penetrate healthy cells with intact membranes. If the membrane is compromised, as occurs with apoptotic or necrotic cells, these dyes are able to enter the cell. One common method for preparing cell-permeable labels relies on the activity of intracellular esterases. The acetate or acetoxymethyl ester derivatives of many xanthene dye derivatives, such as fluorescein, are non-fluorescent, and non-polar, so that they may enter the cell via passive diffusion processes. Once inside the cell, the fluorescence signal of these fluorophores may be unmasked by intracellular esterase activity, which cleaves the acetyl groups from the fluorescein backbone, regenerating fluorescein. The free fluorescein is negatively charged under physiological conditions and therefore becomes trapped inside the cell (**Fig. 2.6**).

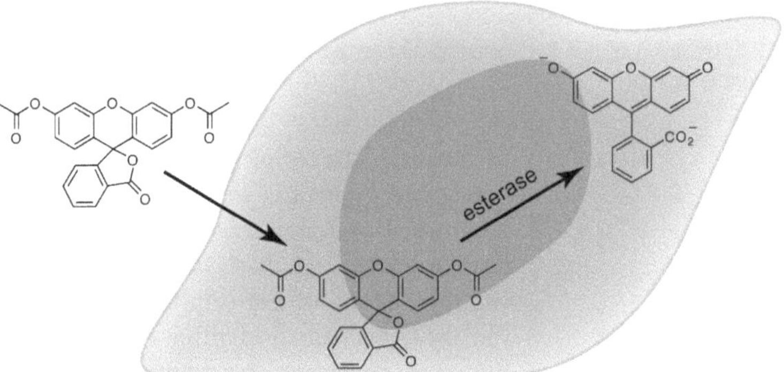

Fig. 2.6. Internalization of non-polar fluorescein diacetate followed by cleavage of the acetate groups by intracellular esterases, releasing polar fluorescein, which is trapped inside the cell.

Many of the more elaborate commercially available fluorophores are expensive, often costing over \$200/mg. Certain imaging applications may require large quantities of probe, especially those involving in vivo microscopy. There are several more affordable fluorophore options with fluorescence emission in the visible range, such as fluorescein and rhodamine isothiocyanate derivatives (7). In contrast, there are few inexpensive commercially available NIR-emitting fluorophores, although efficient and inexpensive routes to prepare conjugatable fluorophores emitting in the NIR from commercially available precursors have been developed. The most common synthetic method is via nucleophilic attack on chloride containing carbocyanine precursors to install carboxylic acid functionality (34, 40, 41). These reactions can often be performed in a simple one-pot procedure with $>$ 90% efficiency and do not require any purification step (34, 42).

2.2. Quantum Dot Labels

Luminescent semiconducting nanocrystals (QDs) are commonly used as labels for imaging at the cellular and subcellular levels (43 , 44). As with small molecule fluorophores, QDs have been used for imaging a variety of cellular and subcellular targets. For example, targeting HER2 receptors on breast cancer cells and cytoplasmic actin and microtubule fibers has been demonstrated (44). Quantum dots are available with amine or carboxylic acid surface groups for bioconjugation reactions and come in a wide range of emission colors from the visible to NIR (6). Unlike organic dyes, quantum dots may be excited over a broad wavelength range with the highest extinction coefficients, often greater than 1,000,000 $M^{-1}cm^{-1}$, observed in the UV. These materials have several advantages over traditional fluorophores. The broad excitation range of QDs allows for simultaneous excitation of multiple quantum dots with different emission wavelengths using a single wavelength light source. Furthermore, QDs are not susceptible to rapid photobleaching under intense excitation, and therefore may be more suitable for confocal and other microscopy techniques, which require prolonged high intensity light exposure. Despite their significant advantages, QDs are not ideal for all imaging applications. Issues concerning QD blinking may complicate single molecule imaging experiments. Many quantum dots materials contain toxic cadmium (45), which was recently shown to leach out of the nanocrystal cores into the surrounding environment under certain biologically relevant conditions (46). In addition, commercially available QDs typically have a hydrodynamic diameter of 20–30 nm, significantly larger than small molecule organic fluorophores. The large size of the quantum dots may be a liability for imaging applications where the size of the fluorescent reporter could interfere with the function of the biological process under investigation. Actin fibers labeled with QDs have a proportionally decreased percent motility when compared to the corresponding AF488 organic fluorophore labeled filaments (47). Additionally, larger QDs may not be suited for monitoring fast diffusing neurotransmitters (48). As a result of their potential limitations for monitoring certain cellular processes, significant effort has been put forth to design improved quantum dots for live cell imaging applications. A large fraction of the typical QD diameter comes from polymer surface coating of the particles; therefore, efforts to decrease the thickness of this coating while maintaining ideal solubility characteristics could open QDs to new potential imaging applications. Following this strategy, QDs employing a short polyethylene glycol modified dihydrolipoic acid head group with a hydrodynamic diameter of 11 nm have been reported (49). Furthermore, the new smaller QDs have been engineered to contain only one site for biological labeling (49). Glutamate receptors labeled with the new, smaller QDs displayed a demonstrably improved ability to diffuse into neuronal synapses in comparison the corresponding commercial QD labeled receptors.

2.3. Genetically Encoded Labels

The researchers Roger Tsien, Martin Chalfie, and Samu Shimomura were recently awarded the 2008 Nobel prize in chemistry for their pioneering research on the identification, cloning, and modification of fluorescent proteins (50). Like QDs, genetically encoded fluorescent or chemiluminescent proteins are becoming commonplace. Recent reviews provide a comprehensive overview (51, 52). As with quantum dots and organic fluorophores, attention has been paid to developing fluorescent proteins in a rainbow of emission colors. Dozens of variants of these fluorescent proteins have been detailed in the literature (51), several of which are in use today with emission in the blue, green, yellow, orange, red, and far red from EGFP, EYFP, mOrange, mCherry, and mPlum, respectively. Unlike QDs and small molecule fluorophores, these species are useful in imaging applications where they can be used to monitor gene expression (53). Fluorescent proteins are also well suited for investigation of chemotaxis. Fluorescent protein expressing cells were used to investigate the role of the hematopoietic protein-1 (HEM-1) complex in cell motility (54, 55). The use of fluorescent proteins has been advantageous for the investigation of cell mitosis after challenge of human MDA cells with the anti-mitotic chemotherapeutics docetaxel (56) and paclitaxel (57).

Bioluminescent enzymes, like fluorescent proteins, are genetically encoded labels, although they require an additional substrate to generate a luminescent signal. Bioluminescent proteins have been isolated from a variety of organisms such as *Photinus pyralis* (firefly) (58), *Renilla reniformis* (sea pansy) (59), and *Pyrophorus plagiophthalamus* (click beetle) (60,61) with emission at ~480, ~560, and ~600 nm, respectively. The firefly and click beetle luciferases use luciferin, whereas the sea pansy luciferase requires colenterazine as a substrate. The lux operon may be used to investigate bacterial systems. This operon encodes both the luciferase and other proteins necessary for synthesis of the luciferin substrate (62, 63).

3. Responsive Probes

The use of targeted fluorescent labels and genetically encoded fluorophores has been invaluable in expanding our understanding of how the molecular machinery of the cell functions. However, these probes do not provide a detailed direct view of the function of many signaling molecules and messengers involved in cellular function. To investigate the interactions of these molecules, activatable or switchable smart probes are necessary.

The phenomenon of fluorescence is a particularly versatile process, with many different parameters that can be utilized for development of activatable probes. These properties include fluorescence intensity shifts, wavelength shifts, chemiluminescence activation, and fluorescence lifetime changes. Of these photophysical properties, most often biochemical probes are based on strategies to develop turn-on or wavelength-shift probes. Optimized fluorogenic probes share many selection criteria with targeted fluorescent labels. Factors to consider include biocompatibility and water solubility of the probes, suitability for extracellular or intracellular delivery, brightness of the fluorophore, and fluorescence excitation and emission wavelengths. In addition to these variables, other circumstances may influence the choice of probe. For example, selectivity of the imaging agent for the enzyme or analyte of interest is an important consideration. It is typically quite difficult to design a fluorogenic probe that displays complete selectivity to the target of interest. Virtually every known probe displays at least some basal activation by competing analytes or enzymes. The magnitude and mode of the fluorescence response is another factor. Typically activatable probes displaying an increase in emission or shifts in the absorption or emission spectra are preferred. Many turn-off fluorescence based sensors have been designed, but these agents are more difficult to use for cell imaging due to complications arising from detecting fluorescence decreases by microscopy. Turn-on probes can often be designed to show extremely strong fluorescence activation, often over 100-fold, but may not be suitable for experiments where quantitative measurements are required. With an off–on fluorescence response, it is difficult to account for baseline fluorescence signal arising from the non-activated probe and variations in the local concentration of the imaging agent. When quantitative measurements are required, sensors with a ratiometric fluorescence response are preferred. These sensors are suitable for quantitative measurements of analyte concentration because they allow for determination of the fluorescence activation in a manner independent of the local probe concentration. In the following sections, an overview of current turn-on and wavelength-shift fluorescence-based probes for bioimaging will be presented.

3.1. Enzyme Activation

Many enzyme activatable probes are based on the well-known phenomenon of self-quenching by organic fluorophores when held in close proximity to each other. An early example of this strategy in a fluorogenic probe suitable for use with live cells is a NIR-emitting fluorescent sensor for cathepsin D activity (**Fig. 2.7**). Cathepsin D is an aspartic protease that is known to be overexpressed in breast cancer cells. The probe consists of a polylysine polymer backbone modified with polyethylene glycol (PEG) polymers on the lysine side chains to improve solubility of the probe.

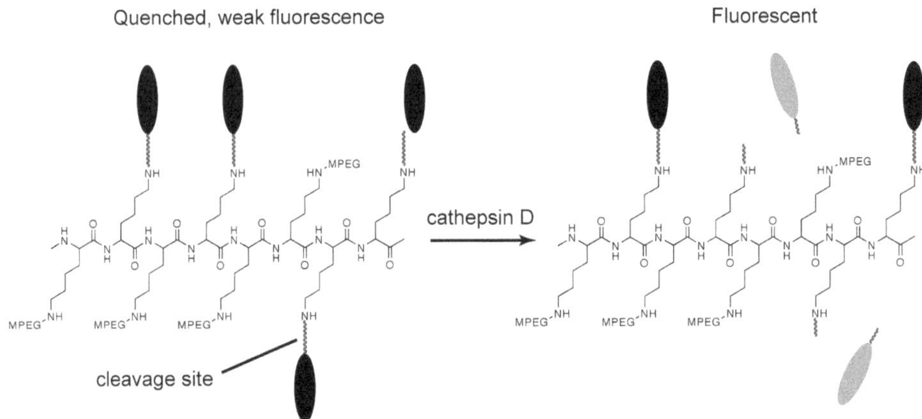

Fig. 2.7. Enzymatic activation of a polylysine based NIR activatable probe for cathepsin D.

In addition, several of the lysine side chains are further modified with a cathepsin D specific cleavage sequence containing NIR-emitting Cy5.5 fluorophores (64). Up to 24 Cy5.5 fluorophores are incorporated per polylysine polymer. The high density of fluorophores allows for efficient self-quenching of the dyes. In this case, more than 99% of the fluorescence emission of the Cy5.5 fluorophores is quenched in the probe (64). Upon cleavage of the probe with cathepsin D, up to 60-fold increase in fluorescence signal is possible. The effectiveness of this probe was demonstrated in vivo by imaging of mice bearing cathepsin D positive tumors where a signal-to-noise ratio of up to 22.8 was reported between tumor and non-target tissue (65). This flexible design strategy for enzyme activatable probes is useful for both endo- and exopeptidase enzymes. In addition to cathepsin D, fluorogenic probes for other enzymes such as cathepsin K, caspase-1, and MMP-2 utilizing this activation strategy have been reported (66–68).

An alternative strategy employed for the design of enzyme activatable probes does not rely on dye–dye quenching interactions. Instead, its fluorescence switching is based on chemical modification of the fluorophore reporter to alter its optical emission properties. One method for achieving this is by alterating the electronic structure of the fluorophore via formation of covalent bonds on portions of the fluorophore that are directly involved in fluorescence emission. For example, many classes of fluorophores such as 7-amino coumarins, rhodamines, fluoresceins, and Nile blue derivatives have amine or phenoxy groups that are part of their conjugated chromophore system and are available for chemical modification. Modification of these groups often has a dramatic effect on the fluorescence emission of the fluorophore. These changes are typified by strong hipsochromic shifts of the absorption maximum of the dye and a concomitant blue shift of the emission maximum. In many cases, the quantum yield of the modified fluorophore is also significantly decreased. Enzyme activatable probes using this strategy can often show greater than 100-fold activation. This approach is useful for detection of enzymatic activity from exopeptidases. Many of these fluorogenic enzyme substrates have been prepared by conjugation of an amine group on 7-amino coumarin or rhodamine 110 to the carboxy terminus of a peptide sequence specific for the enzyme of interest. Formation of the amide bond on the coumarin or rhodamine abolishes the characteristic fluorescence emission at approximately 430 or 520 nm for the coumarin and rhodamine, respectively. Enzyme action on the substrate releases the coumarin and restores its fluorescent signal. A range of fluorogenic probes for peptidases activated by cathepsins (69), caspases (70, 71), elastases (72), and trypsin (72) have been developed using this approach. This strategy has also been adapted to fluorogenic probes for sugars and phosphatases (73).

3.2. Metal Ion Sensing

The design of effective probes for metal ions faces many challenges. The primary concerns are selectivity, metal ion affinity, and fluorescence response. There are numerous activatable and ratiometric fluorescence-based sensors for detection of bio-relevant ions such as Ca^{2+}, Mg^{2+}, Na^{2+}, K^{2+}, Zn^{2+}, Cu^{2+}, Fe^{3+}, and H^{+}. However, with the exception of pH responsive sensors (H^{+} ions), nearly every metal ion probe has side reactivity with analytes other than the targeted metal ion. Therefore the presence or absence of potential interfering ions influences the probe choice. The affinity of the analyte to the probe is another factor. The K_d values for analyte dissociation from the fluorescence-based sensor should be matched to the expected concentration of the ion under investigation to yield optimal response.

Fluorescent probes for calcium ions form one of the most diverse classes of metal ion sensors. The wide array of probes is in part due to the importance of calcium as a signaling molecule in biology. Cellular Ca^{2+} plays many functional and regulatory roles from muscle fiber contraction to signal transduction. Dozens of Ca^{2+} responsive sensors have been detailed in the literature, and a complete review of these probes is beyond the scope of this chapter. For more detailed information on Ca^{2+} probes, there are several excellent literature reviews (74, 75). Calcium ion probes can be divided into two broad categories: intensity based and ratiometric. Probes in both classes have a wide range of reported K_d values. For example, the fluorogenic Oregon Green 488 BAPTA-1, -6F, and -5 N probes, which are all based on the BAPTA (1,2-bis(*o*-aminophenoxy)ethane-*N*,*N*,*N*′,*N*′-tetraacetic acid) chelating group have tunable Ca^{2+} K_d values. By varying the substituents on the BAPTA chelator, the K_d for Ca^{2+} can be altered from 170 nM to 20 μM (**Fig. 2.8**) (6). Many low affinity

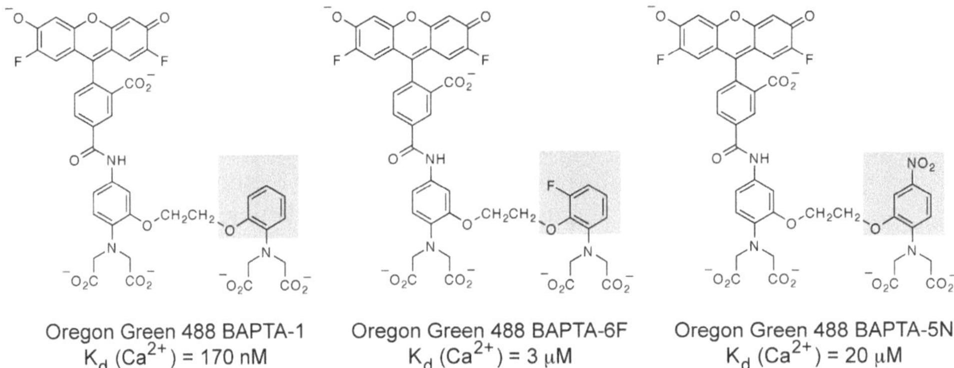

Fig. 2.8. Fluorogenic oregon green-based Ca^{2+} selective probes. The K_d values of these sensors can be tuned by altering the electron-withdrawing properties of the substituents on the BAPTA chelating moieties.

calcium binders also display selectivity for Mg^{2+}. The Mg^{2+} probe Mag-Fluo-4 has K_d values of 4.7 mM and 22 μM for Mg^{2+} and Ca^{2+}, respectively (6). The Mg^{2+} binding of this probe is well matched to the typical sub- to low-millimolar cellular magnesium levels (76). Ratiometric probes for Ca^{2+} can either show a change in absorption wavelength or emission wavelength upon coordination of the divalent ion. Examples include Fura-2 and Indo-1 for absorption and emission wavelength shift sensors, respectively (77). Ratiometric response with both classes of probes is possible, although sensors showing shifts in fluorescence emission are preferable since only one excitation source is necessary. This is particularly important in applications where laser excitation is used or where analysis will be performed by flow cytometry.

In addition to activatable probes for calcium, which were first reported in the early 1980s (78), recent years have seen the development of selective probes for many other metal ions. In the past decade significant attention has been directed toward imaging intracellular zinc to investigate its roles in biological homeostasis and signal transduction. Here we give an overview of several widely applied sensors for zinc. For a more complete survey of current Zn^{2+} selective probes, please see one of several recent reviews (79–81). Much of the pioneering work on design of efficient fluorescence-based zinc sensors originated in the Lippard and Nagano laboratories. The Lippard lab has reported a series of fluorescein-based probes for zinc using the dipicolylamine chelating group (82–85). The earliest of these probes, Zinpyr-1 displays a threefold increase in emission upon binding Zn^{2+} and has a 0.7 nM binding affinity (82). Sensors with decreased Zn^{2+} affinity have subsequently been prepared via replacement of one of the pyridyl arms of the dipicolylamine chelating motif with thiophene or thioether coordinating groups, decreasing the binding affinity

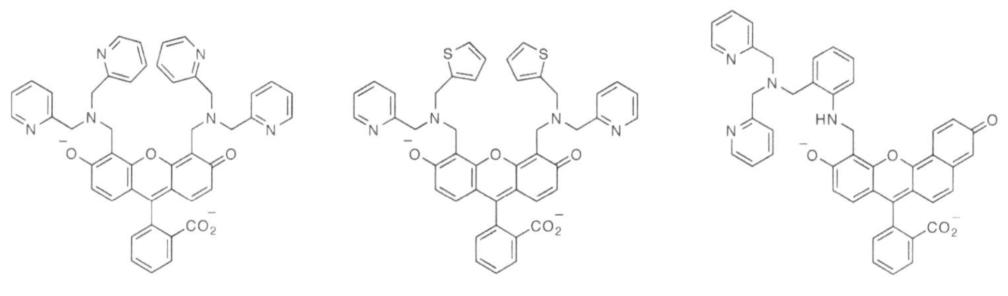

ZP2 K_d (Zn^{2+}) = 0.5 nM   ZS5 K_d (Zn^{2+}) = 1.5 μM   ZNP1 K_d (Zn^{2+}) = 0.55 nM

Fig. 2.9. Representative Zn^{2+} selective sensors. ZP2 and ZS5 are intensity based turn-on probes with sub nanomolar and micromolar Zn^{2+} affinities, respectively. ZNP1 is a ratiometric sensor with dual emission at 545 and 624 nm.

from the low nM to μM (**Fig. 2.9**) (86). One of these low-affinity probes, ZS5 was used to visualize glutamate-mediated Zn(II) uptake in dendrites and Zn(II) release resulting from nitrosative stress (86). As with Lippard, Nagano has focused on design of fluorescent Zn^{2+} sensors based on the fluorescein scaffold. Probes of the ZnAF family have low fluorescence background and strong activation of up to 69-fold upon Zn^{2+} coordination (87, 88). These sensors have been used to visualize Zn^{2+} release in the rat hippocampus (88) and to monitor presynaptic Zn^{2+} pools (89). Systematic modification of the dipicolylamine chelating moiety on these probes has enabled preparation of ZnAF probes with K_d values ranging from 2.7 nM to 600 μM (90). The ratiometric Zn^{2+} probes FuraZin and IndoZin, which are based on the related Fura and Indo Ca^{2+} sensors, (91) were used to monitor intracellular zinc uptake (92). Both FuraZin and IndoZin are excited at short wavelength (<400 nm) (91). To minimize potential phototoxic effects of UV excitation, long-wavelength ratiometric probes such as Zin-naphthopyr-1 (ZNP1), which has a 0.55 nM Zn^{2+} affinity, were designed (**Fig. 2.9**) (93). The ZNP1 probe has dual emission at 545 and 624 nm, where increasing [Zn^{2+}] induces a dramatic increase in the 624-nm emission signal. The diacetate derivative of this probe is membrane permeable and was used to image release of Zn^{2+} from COS-7 cells in real time (93). A NIR-emitting ratiometric probe, DIPCY, is based on a carbocyanine fluorophore scaffold (94) This probe, which has a Zn^{2+} K_d of 98 nM, displays an approximate 50 nm red-shift in its absorbance spectrum upon binding zinc.

Ion selective probes for H^+ are one of the oldest and most studied classes of ion sensors. Their development and use has been vital for investigation of pH changes in the endosomal/lysosomal system. Furthermore, disruption of acid/base homeostasis is associated with the pathophysiology of diseases such as cancer, cystic fibrosis, and immune dysfunction (95–98). Many fluorophores have intrinsic pH sensitivity. For example, fluorescein has a fluorogenic pK_a of approximately 6.4 and has been used as a dual excitation single emission ratiometric probe for intracellular pH (99, 100). However, fluorescein can leak from cells and is used infrequently as a stand-alone probe for intracellular ratiometric pH imaging. New nanomaterials doped with FITC have been developed for ratiometric pH imaging. In one example, fluorescein and rhodamine isothiocyanate fluorophores (the rhodamine is used as a pH insensitive reference) were incorporated into core/shell silica nanoparticles and used for monitoring pH in intracellular compartments of mast cells (101). The dual excitation-single emission pH probe BCECF is one of the most widely used pH probes and is a fluorescein derivative modified with two carboxyethyl groups in the 2′ and 7′ positions of the dye. These additional carboxylate groups significantly improve intracellular retention of the sensor and contribute to an increase in the pH responsive pK_a to 6.97 (102). However, as a result of its dual excitation single-emission response, it is not ideal for imaging with laser microscopes or for flow cytometry experiments. To address this, single excitation dual emission pH responsive fluorophores were developed. In the early 1990s the seminaphthorhodafluor scaffold was designed for ratiometric pH imaging (103). In addition to having a single excitation dual emission response to pH with a pK_a of approximately 7.5, the probe exhibits red-shifted emission between 600 and 640 nm (103). One of these derivatives, carboxy-SNARF-1 has been used for imaging intracellular pH in chicken embryo epithelial cells (104). Probes for sensing pH can also be combined with targeting strategies. A series of pH responsive fluorogenic boron-dipyrromethene (BODIPY) fluorophores with tunable pK_a values between 3.8 and 6.0 were recently reported. These fluorophores can be conjugated to targeting groups such as trastuzumab for use as fluorescence-based switches and are activated by internalization into the endosomal/lysosomal system of cancer cells (105).

NIR fluorescent probes for pH sensing show potential for use in intracellular and in vivo pH measurement. Most current NIR pH probes are based on the carbocyanine scaffold. In one approach, dealkylation of one or both of the indole nitrogens on a non-pH responsive carbocyanine fluorophore renders it sensitive to pH (**Fig. 2.10**) (106–109). In contrast to xanthene based pH sensors, the carbocyanine dyes show an increase in fluorescence emission as the pH decreases. These probes have been used to monitor agonist-induced G protein-coupled receptor internalization into CHO or Hek293 cells (106). One pH responsive dye HCyC-646 with a fluorogenic pK_a of 6.2 and fluorescence emission at 670 nm was paired with pH insensitive Cy7 fluorophores on a bacteriophage particle scaffold for use as an nanoscale NIR ratiometric pH sensor (109). This system was used to monitor

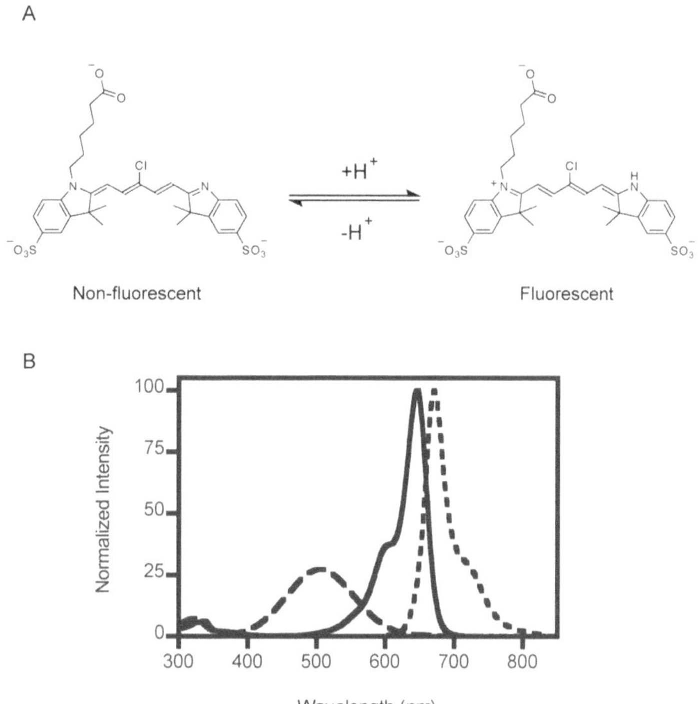

Fig. 2.10. The pH-dependent equilibrium showing activation of HCyC-646 at acidic pH (**panel A**). Absorption traces for HCyC-646 at pH 8 and pH 4 and fluorescence emission at pH 4, *dashed*, *solid*, and *dotted lines*, respectively (**panel B**).

intracellular pH following internalization into RAW cells and its potential for imaging pH in small animal models was demonstrated (109).

Although much recent work has focused on development of ion-selective probes specific for Ca^{2+}, Zn^{2+}, and H^+, many probes for other bio-relevant analytes have been developed. Ratiometric fluorescent indicators based different-sized crown ether ion chelators have been reported for Na^+ (SBFI, K_d = 3.8 mM) (110) and K^+ (PBFI, K_d = 5.1 mM) (110, 111), although they have relatively poor ion selectivity. Due to the potential role of unregulated cellular copper, in various diseases from amyotrophic lateral sclerosis (112) to Alzheimer's disease (113), fluorogenic copper specific probes have been designed (114–116). Additional efforts have focused on preparation of iron selective probes, but only a few turn-on sensors have been reported (117–120) and their ability for live cell imaging remains relatively untested.

3.3. ROS Sensing

There is significant interest in detection of reactive oxygen (ROS) and reactive nitrogen species (RNS) in biology. These reactive compounds are involved in multiple signal transduction and regulatory processes. Furthermore, many of these compounds are strong oxidants and play critical roles in host defense and, when unregulated, in disease progression. One of the key challenges is the development of sensors with a high degree of specificity for a single analyte. This is often difficult since many of these species exhibit similar behavior as oxidants. Without selective sensors, it is very difficult to study the function of a single ROS/RNS since many different species are present simultaneously in the same biological systems. The inability to differentiate between specific reactive species has been the primary flaw of early probes such as 2′,7′-dichlorodihydrofluorescein (DCFH), which shows broad non-specific activation to a variety of oxidant species (121). In addition to this lack of selectivity, DCFH displays marked autoxidation activity when exposed to light.

Nitric oxide (NO) is one of the first reactive species for which selective fluorogenic sensors were developed. There are many approaches to imaging NO, and a more complete summary is given elsewhere (122, 123). The most common strategy today for design of selective NO sensors is based on the *o*-phenylenediamine functional group. In the presence of dioxygen and NO, a selective reaction occurs to convert the *o*-diamine into a triazole derivative (**Fig. 2.11a**). This effectively results in an increase in fluorescence signal since the amine groups of the *o*-phenylenediamine group, which are good photoinduced electron transfer (PET) quenchers, are converted into an electron deficient triazole. This approach has been used for design of a variety of NO sensors using naphthalene (124), fluorescein (125), BODIPY (126), rhodamine (127), and carbocyanine (128) fluorophores spanning the electromagnetic spectrum from the blue to NIR. In general, these probes show excellent selectivity for NO in aerobic environments with little or no observed reactivity to other oxidants such as peroxynitrite ($ONOO^-$), hydrogen peroxide (H_2O_2), or superoxide radical (O_2^-) (125). Although useful in most imaging applications, these sensors do not directly monitor NO. Fluorogenic sensors for NO based on the *o*-phenylenediamine functional group only react with RNS formed by the pre reaction of NO with O_2. Therefore detection is dependent not only on the presence of NO but also local O_2 levels.

A preferred tactic is fluorogenic sensors that are capable of direct reaction with NO. Recently the first probes suitable for live cell imaging based on direct detection of NO were reported (129). These sensors consist of a Cu(II) complex with a modified fluorescein derivative bearing an 8-aminoquinaldine chelating group (130). The paramagnetic properties of the Cu(II) coordinated to the fluorescein probe result in quenched fluorescence emission in the absence of NO. Reaction of this probe, CuFL, with NO results in reduction of the Cu(II) to Cu(I),

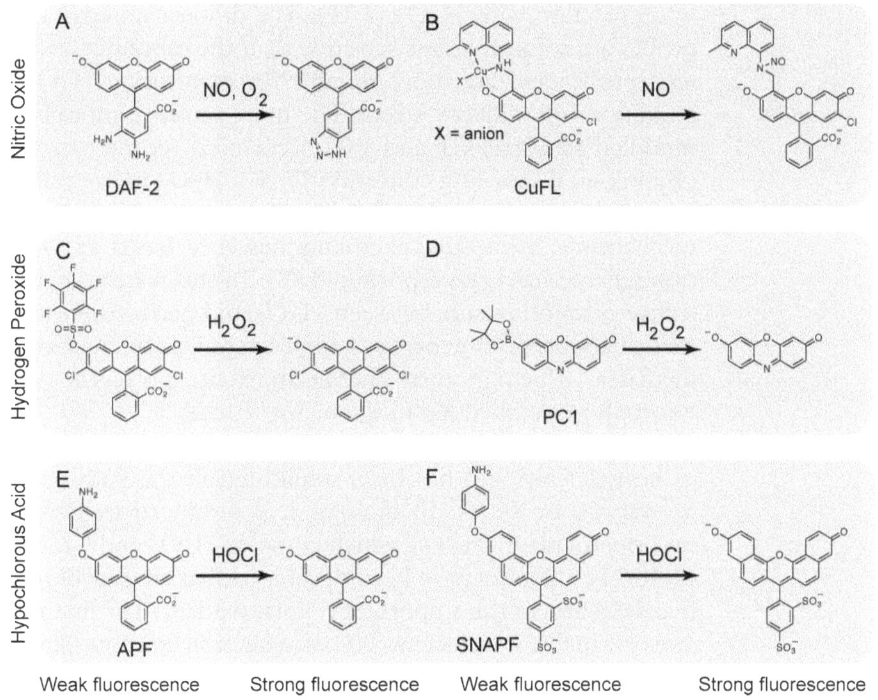

Fig. 2.11. Several examples of intensity based activatable sensors for nitric oxide, hydrogen peroxide, and hypochlorous acid.

release of the Cu(I) ion, formation of a nitrosamine modified fluorophore, and a strong increase in fluorescence emission (**Fig. 2.11b**) (130). CuFL shows a greater than tenfold increase in fluorescence upon combination with NO and like the *o*-phenylenediamine based probes has excellent selectivity, with minimal side reactivity to $ONOO^-$ (130).

There are many approaches for development of selective probes for H_2O_2. One strategy relies on the cleavage of sulfonate-protected fluorescein (**Fig. 2.11c**) (131, 132) or naphthofluorescein (133) scaffolds. However, these probes typically have non-trivial side reactivity with several oxidants such as O_2^-, $OH^{\bullet}$, or hypochlorous acid (HOCl) (131, 133). In an alternative approach, chemospecific probes for H_2O_2 have been reported based on the specific chemical reaction of H_2O_2 with boronate esters to give phenol species (**Fig. 2.11d**) (134). This detection strategy has been used to prepare fluorogenic xanthene and resorufin derivatives with either one or two reactive boronate esters (134–136). As a result of their non-polar composition, the sensors have innate cell permeability and depending on the fluorophore scaffold emit in the blue (135), green (134, 136), or red (135–137). The boronate ester-based H_2O_2 sensors show a dramatic fluorescence response; in the case of the fluorescein-based probe PF1, a >500-fold increase in fluorescence signal is observed upon reaction with H_2O_2 (134). The diboronate ester modified probes are somewhat less sensitive than the monoboronate modified probes and are more suitable for monitoring H_2O_2 under conditions of oxidative stress. The more sensitive monoboronate modified sensors PG1 and PC1 were used recently to visualize physiological signaling concentrations of H_2O_2 in live cells (136). In addition to the more common fluorescence based signaling mechanisms, a selective chemiluminescence based H_2O_2 detection scheme has been reported (138). In this system, a chemoselective reaction occurs between H_2O_2 and peroxylate ester polymeric nanoparticles generating high-energy dioxetanedione intermediates, which in turn chemically excite polycyclic aromatic fluorophores embedded in the nanoparticles.

Hypochlorous acid is an important strong oxidant involved in host defense and has been implicated in the pathogenesis of several disease states. In biology, it is produced by the enzyme myeloperoxidase (MPO), which converts H_2O_2 and Cl^- ions into HOCl. Few probes have been developed that are capable of selective detection of this important cellular oxidant. The first reported sensor capable of efficient HOCl detection is aminophenyl fluorescein (APF) (121). APF shows excellent response to HOCl, generating fluorescein as a final oxidation product and displays several 100-fold fluorescence activation (**Fig. 2.11e**). However, it also shows significant cross reactivity with $ONOO^-$ and $OH^{\cdot}$ (121). This lack of specificity can be overcome when APF is used in conjunction with the chemically related HPF, which only reacts with $ONOO^-$ and $OH^{\cdot}$ (121). The selectivity of this system has been further tuned by careful selection of the fluorophore scaffold. Using the same *p*-aminophenylether reactive group, the sulfonaphthofluorescein based probe (SNAPF) displays enhanced ROS selectivity (**Fig. 2.11f**). SNAPF is activated exclusively by HOCl and has been used to monitor MPO generated HOCl in vitro, in cell culture, and in vivo (139). Although it does react specifically with HOCl, the SNAPF probe is still not ideal as it only shows an approximately tenfold fluorescence response when exposed to HOCl (139). Continued efforts in this field will undoubtedly yield improved probes for selective detection of HOCl.

In addition to efforts for preparation of selective probes for reactive small molecules such as NO, H_2O_2, and HOCl, sensors for several other species have been designed. Of these, there are few fluorogenic probes specific for $OH^{\cdot}$ or O_2^-. However, recently a new class of carbocyanine dyes, the hydrocyanines, has been reported that is sensitive to both superoxide and hydroxy radicals (140). These probes are prepared from conventional carbocyanine fluorophores via selective reduction. When the reduced carbocyanines are exposed to ROS, they are oxidized back to their parent carbocyanine fluorophore. Peroxynitrite responsive probes have also been reported. These sensors only show a seven- to eightfold activation for $ONOO^-$ and have marked side reactivity with other ROS such as OH$^{\bullet}$, with an observed threefold activation (141). Nevertheless, there is promise for development of more selective peroxynitrite sensors based on this activation scheme. Although less active than the invesitgation of ROS or RNS, there is interest the bioimaging of thiols. Imaging of thiols can give insight into local redox status. Furthermore, the presence of thiol containing amino acids such as homocysteine have been associated with a variety of disease states (142, 143). To this end, there have been several recent reports of fluorogenic and ratiometric probes for thiol bioimaging (144–147).

4. Summary

There are many fluorescence-based imaging agents for biological targets, enzymes, and other analytes. Targeted fluorescent labels utilizing affinity groups from antibodies to small molecules have been developed and many activatable sensors for metal ions from calcium to zinc have been designed. In addition, there are several different classes of activatable imaging agents for the detection of enzyme activity and reactive small molecules. As a result of the large pool of potential imaging agent choices, it can often be difficult to select the most appropriate imaging agent for a particular experiment. The ability to identify an effective fluorescent reporter can be facilitated by careful consideration of factors such as the conjugation chemistry, photophysical characteristics, polar ity, cost, and selectivity of the imaging agent.

References

1. Stokes, G. G. (1852) On the change of refrangibility of light. *Phil. Trans. R. Soc. London* **142**, 463–562.
2. Köhler, A. (1904) Mikrophotographische einrichtung: eine für ultraviolettes licht (λ = 275 nm) und damit angestellte untersuchungen organischer gewebe. *Phys. Z.* **5**, 666–673.
3. Stübel, H. (1911) Die fluoreszenz tierischer gewebe in ultraviolettem licht. *Pflug. Arch. Ges. Phys.* **142**, 1–14.
4. Coons, A. H., Creech, H. J., Jones, R. N., and Berliner, E. (1942) The demonstration of pneumococcal antigen in tissues by the use of fluorescent antibody. *J. Immunol.* **45**, 159–170.
5. Coons, A. H., Creech, H. J., and Jones, R. N. (1941) Immunological properties of an antibody containing a fluroescent group. *Proc. Soc. Exp. Biol. Med.* **47**, 200–202.
6. Molecular Probes Inc., Eugene, OR: http://probes.invitrogen.com.
7. Sigma-Aldrich Corp., St. Louis, MO: http://sigmaaldrich.com.
8. ATTO-TEC GmbH, Siegen, Germany: http://atto-tec.com.
9. Dyomics Gmb H, Jena, Germany: http://www.dyomics.com.
10. GE Healthcare Bio-Sciences Corp., Piscataway, NJ: http://www.gelifesciences.com.
11. LI-COR Biosciences, Lincoln, NE: http://www.licor.com.

12. VisEn Medical Inc., Woburn, MA: http://www.visenmedical.com.
13. Chang, P. V., Prescher, J. A., Hangauer, M. J., and Bertozzi, C. R. (2007) Imaging cell surface glycans with bioorthogonal chemical reporters. *J. Am. Chem. Soc.* **129**, 8400–8401.
14. Rostovtsev, V. V., Green, L. G., Fokin, V. V., and Sharpless, K. B. (2002) A stepwise Husigen cycloaddition process: copper(I)-catalyzed regioselective "ligation" of azides and terminal alkynes. *Angew. Chem., Int. Ed.* **41**, 2596–2599.
15. Shao, F., Weissleder, R., and Hilderbrand, S. A. (2008) Monofunctional carbocyanine dyes for bio- and bioorthogonal conjugation. *Bioconjug. Chem.* **19**, 2487–2491.
16. Baskin, J. M., Prescher, J. A., Laughlin, S. T., Agard, N. J., Chang, P. V., Miller, I. A., Lo, A., Codelli, J. A., and Bertozzi, C. R. (2007) Copper-free click chemistry for dynamic in vivo imaging. *Proc. Natl. Acad. Sci. USA* **104**, 16793–16797.
17. Codelli, J. A., Baskin, J. M., Agard, N. J., and Bertozzi, C. R. (2008) Second-generation difluorinated cyclooctynes for copper-free click chemistry. *J. Am. Chem. Soc.* **130**, 11486–11493.
18. Ning, X. H., Guo, J., Wolfert, M. A., and Boons, G. J. (2008) Visualizing metabolically labeled glycoconjugates of living cells by copper-free and fast Husigen cycloadditions. *Angew. Chem. Int. Ed.* **47**, 2253–2255.
19. Laughlin, S. T., Baskin, J. M., Amacher, S. L., and Bertozzi, C. R. (2008) In vivo imaging of membrane-associated glycans in developing zebrafish. *Science* **320**, 664–667.
20. Devaraj, N. K., Weissleder, R., and Hilderbrand, S. A. (2008) Tetrazine-based cycloadditions: applications to pretargeted live cell imaging. *Bioconjug. Chem.* **19**, 2297–2299.
21. Blackman, M. L., Royzen, M., and Fox, J. M. (2008) Tetrazine ligation: fast bioconjugation based on inverse-electron-demand Diels-Alder reactivity. *J. Am. Chem. Soc.* **130**, 13518–13519.
22. Sampath, L., Kwon, S., Ke, S., Wang, W., Schiff, R., Mawad, M. F., and Sevick-Muraca, E. M. (2007) Dual-labeled trastuzumab-based imaging agent for the detection of human epidermal growth factor receptor 2 overexpression in breast cancer. *J. Nucl. Med.* **48**, 1501–1510.
23. Hicke, B. J., Stephens, A. W., Gould, T., Chang, Y.-F., Lynott, C. K., Heil, J., Borkowski, S., Hilger, C.-S., Cook, G., Warren, S., and Schmidt, P. G. (2006) Tumor targeting by an aptamer. *J. Nucl. Med.* **47**, 668–678.
24. Von Wallbrunn, A., Höltke, C., Zühlsdorf, M., Heindel, W., Schäfers, M., and Bremer, C. (2007) In vivo imaging of integrin a_vb_3 expression using fluorescence-mediated tomography. *Eur. J. Nucl. Med. Mol. Imaging* **34**, 745–754.
25. Garanger, E., Boturyn, D., Jin, Z., Dumy, P., Favrot, M.-C., and Coll, J.-L. (2005) New multifunctional molecular conjugate vector for targeting, imaging, and therapy of tumors. *Mol. Ther.* **12**, 1168–1175.
26. Jin, Z.-H., Josserand, V., Foillard, S., Boturyn, D., Dumy, P., Favrot, M.-C., and Coll, J.-L. (2007) In vivo optical imaging of integrin a_vb_3 in mice using multivalent or monovalent cRGD targeting vectors. *Mol. Cancer* **6**, 41.
27. Cheng, Z., Wu, Y., Xiong, Z., Gambhir, S. S., and Chen, X. (2005) Near-infrared fluorescent RGD peptides for optical imaging of integrin a_vb_3 expression in living mice. *Bioconjug. Chem.* **16**, 1433–1441.
28. Ke, S., Wen, X., Gurfinkel, M., Charnsangavej, C., Wallace, S., Sevick-Muraca, E. M., and Li, C. (2003) Near-infrared optical imaging of epidermal growth factor receptor in breast cancer xenografts. *Cancer Res.* **63**, 7870–7875.
29. Tung, C.-H., Lin, Y., Moon, W. K., and Weissleder, R. (2002) A receptor-targeted near-infrared fluorescence probe for in vivo tumor imaging. *Chembiochem* **3**, 784–786.
30. Weissleder, R., Kelly, K. A., Sun, E. Y., Shtatland, T., and Josephson, L. (2005) Cell-specific targeting of nanoparticles by multivalent attachment of small molecules. *Nat. Biotechnol.* **23**, 1418–1423.
31. Jaye, D. L., Geigerman, C. M., Fuller, R. E., Akyildiz, A., and Parkos, C. A. (2004) Direct fluorochrome labeling of phage display library clones for studying binding specificities: applications in flow cytometry and fluorescence microscopy. *J. Immunol. Methods.* **295**, 119–127.
32. Kelly, K. A., Bardeesy, N., Anbazhagan, R., Gurumurthy, S., Berger, J., Alencar, H., DePinho, R. A., Mahmood, U., and Weissleder, R. (2008) Targeted nanoparticles for imaging incipient pancreatic ductal adenocarcinoma. *PLOS Med.* **5**, 657–668.
33. Kelly, K. A., Setlur, S. R., Ross, R., Anbazhagan, R., Waterman, P., Rubin, M. A., and Weissleder, R. (2008) Detection of early prostate cancer using a hepsin-targeted imaging agent. *Cancer Res.* **68**, 2286–2291.
34. Hilderbrand, S. A., Kelly, K. A., Weissleder, R., and Tung, C.-H. (2005) Monofunctional near-infrared fluorochromes for imag-

ing applications. *Bioconjug Chem.* **16**, 1275–1281.

35. Weissleder, R., and Ntziachristos, V. (2003) Shedding light onto live molecular targets. *Nat. Med.* **9**, 123–128.
36. Mitchison, T. J. (1989) Polewards microtubule flux in the mitotic spindle: evidence from photoactivation of fluorescence. *J. Cell. Biol. 109*, 637–652.
37. Theriot, J. A., and Mitchison, T. J. (1991) Actin microfilament dynamics in locomoting cells. *Nature* **352**, 126–131.
38. Zhao, Y., Zheng, Q., Dakin, K., Xu, K., Martinez, M. L., and Li, W.-H. (2004) New caged coumarin fluorophores with extraordinary uncaging cross sections suitable for biological imaging applications. *J. Am. Chem. Soc.* **126**, 4653–4663.
39. Guo, Y.-M., Chen, S., Shetty, P., Zheng, G., Lin, R., and Li, W.-H. (2008) Imaging dynamic cell-cell junctional coupling in vivo using trojan-LAMP. *Nat. Methods* **5**, 835–841.
40. Flanagan, J. H., Jr., Khan, S. H., Menchen, S., Soper, S. A., and Hammer, R. P. (1997) Functionalized tricarbocyanine dyes as near-infrared fluorescent probes for biomolecules. *Bioconjug. Chem.* **8**, 751–756.
41. Narayanan, N., and Patonay, G. (1995) A new method for the synthesis of heptamethine cyanine dyes: synthesis of new near-infrared fluorescent labels. *J. Org. Chem.* **60**, 2391–2395.
42. Galande, A. K., Hilderbrand, S. A., Weissleder, R., and Tung, C.-H. (2006) Enzyme-targeted fluorescent imaging probes on a multiple antigenic peptide core. *J. Med. Chem.* **49**, 4715–4720.
43. Bruchez, M. J., Moronne, M., Gin, P., Weiss, S., and Alivisatos, A. P. (1998) Semiconductor nanocrystals as fluorescent biological labels. *Science* **281**, 2013–2016.
44. Wu, X., Liu, H., Liu, J., Haley, K. N., Treadway, J. A., Larson, J. P., Ge, N., Peale, F., and Bruchez, M. P. (2003) Immunofluorescent labeling of cancer marker Her2 and other cellular targets with semiconductor quantum Dots. *Nat. Biotechnol.* **21**, 41–46.
45. Derfus, A. M., Chan, W. C. W., and Bhatia, S. N. (2004) Probing the cytotoxicity of semiconductor quantum dots. *Nano Lett.* **4**, 11–18.
46. Mancini, M. C., Kairdolf, B. A., Smith, A. M., and Nie, S. (2008) Oxidative quenching and degradation of polymer-encapsulated quantum dots: new insights into the long-term fate and toxicity of nanocrystals in vivo. *J. Am. Chem. Soc.* **130**, 10836–10837.
47. Månsson, A., Sundberg, M., Balaz, M., Bunk, R., Nicholls, I. A., Olming, P., Tågerud, S., and Montelius, L. (2004) In vitro sliding of actin filaments labelled with single quantum dots. *Biochem. Biophys. Res. Commun.* **314**, 529–534.
48. Groc, L., Lafourcade, M., Heine, M., Renner, M., Racine, V., Sibarita, J.-B., Lounis, B., Choquet, D., and Cognet, L. (2007) Surface trafficking of neurotransmitter receptor: comparison between single-molecule/quantum dot strategies. *J. Neurosci.* **27**, 12433–12437.
49. Howarth, M., Liu, W., Puthenveetil, J., Zheng, Y., Marshall, L. F., Schmidt, M. M., Wittrup, K. D., Bawendi, M. G., and Ting, A. Y. (2008) Monovalent, reduced-size quantum dots for imaging receptors on living cells. *Nat. Methods* **5**, 397–399.
50. Miyawaki, A. (2008) Green fluorescent protein glows gold. *Cell* **135**, 987–990.
51. Shaner, N. C., Steinbach, P. A., and Tsien, R. Y. (2005) A guide to choosing fluorescent proteins. *Nat. Methods* **2**, 905–909.
52. Pakhomov, A. A., and Martynov, V. I. (2008) GFP family: structural insights into spectral tuning. *Chem. Biol.* **15**, 755–764.
53. Chalfie, M., Tu, Y., Euskirchen, G., Ward, W. W., and Prasher, D. C. (1994) Green fluorescent protein as a marker for gene expression. *Science* **263**, 802–805.
54. Weiner, O. D., Marganski, W. A., Wu, L. F., Altschuler, S. J., and Kirschner, M. W. (2007) An actin-based wave generator organizes cell motility. *PLOS Biol.* **5**, 2053–2063.
55. Weiner, O. D., Rentel, M. C., Ott, A., Brown, G. E., Jedrychowski, M., Yaffe, M. B., Gygi, S. P., Cantley, L. C., Bourne, H. R., and Kirschner, M. W. (2006) Hem-1 complexes are essential for Rac activation, actin polymerization, and myosin regulation during neutrophil chemotaxis. *PLOS Biol.* **4**, 0186–0199.
56. Sakaushi, S., Nishida, K., Minamikawa, H., Fukada, T., Oka, S., and Sugimoto, K. (2007) Live imaging of spindle pole disorganization in docetaxel-treated multicolor cells. *Biochem. Biophys. Res. Commun.* **357**, 655–660.
57. Fukada, T., Senda-Murata, K., Nishida, K., Sakaushi, S., Minamikawa, H., Dotsu, M., Oka, S., and Sugimoto, K. (2007) A multi-fluorescent MDA435 cell line for mitosis inhibitor studies: simultaneous visualization of chromatin, microtubules, and nuclear envelope in living cells. *Biosci. Biotechnol. Biochem.* **71**, 2603–2605.
58. De Wet, J. R., Wood, K. V., Helinski, D. R., and DeLuca, M. (1985) Cloning of fire-

fly luciferase cDNA and the expression of active luciferase in Escherichia coli. *Proc. Natl. Acad. Sci. USA.* **82**, 7870–7873.
59. Lorenz, W. W., McCann, R. O., Longiaru, M., and Cormier, M. J. (1991) Isolation and expression of a cDNA encoding *Renilla reniformis* luciferase. *Proc. Natl. Acad. Sci. U.S.A.* **88**, 4438–4442.
60. Stolz, U., Velez, S., Wood, K. V., Wood, M., and Feder, J. L. (2003) Darwinian natural selection for orange bioluminescent color in a Jamacian click beetle. *Proc. Natl. Acad. Sci. USA.* **100**, 14955–14959.
61. Wood, K. V., Lam, Y. A., and McElroy, W. D. (1989) Bioluminescent click beetles revisited. *J. Biolumin. Chemilumin.* **4**, 31–39.
62. Cohn, D. H., Mileham, A. J., Simon, M. I., and Nealson, K. H. (1985) Nucleotide sequence of the luxA gene of *Vibrio harveyi* and the complete amino acid sequence of the alpha subunit of bacterial luciferase. *J. Biol. Chem.* **260**, 6139–6146.
63. Johnston, T. C., Thompson, R. B., and Baldwin, T. O. (1986) Nucleotide sequence of the luxA gene of *Vibrio harveyi* and the complete amino acid sequence of the beta subunit of bacterial luciferase. *J. Biol. Chem.* **261**, 4805–4811.
64. Tung, C.-H., Bredow, S., Mahmood, U., and Weissleder, R. (1999) Preparation of a cathepsin D sensitive near-infrared fluorescence probe for imaging. *Bioconjug Chem.* **10**, 892–896.
65. Tung, C.-H., Mahmood, U., Bredow, S., and Weissleder, R. (2000) In vivo imaging of proteolytic enzyme activity using a novel molecular reporter. *Cancer Res.* **60**, 953–4958.
66. Bremer, C., Tung, C.-H., and Weissleder, R. (2001) In vivo molecular target assessment of matrix metalloproteinase inhibition. *Nat. Med.* 7, 743–748.
67. Messerli, S. M., Prabhakar, S., Tang, Y., Shah, K., Cortes, M. L., Murthy, V., Weissleder, R., Breakefield, X. O., and Tung, C.-H. (2004) A novel method for imaging apoptosis using a caspase-1 near-infrared fluorescent probe. *Neoplasia* **6**, 95–105.
68. Jaffer, F. A., Kim, D.-E., Quinti, L., Tung, C.-H., Aikawa, E., Pande, A. N., Kohler, R. H., Shi, G.-P., Libby, P., and Weissleder, R. (2007) Optical visualization of cathepsin K activity in atherosclerosis with a novel, protease-activatable fluorescence sensor. *Circulation* **115**, 2292–2298.
69. Tanabe, H., Kumagai, N., Tsukahara, T., Ishiura, S., Kominami, E., Nishina, H., and Sugita, H. (1991) Changes of lysosomal proteinase activities and their expression in rat cultured keratinocytes during differentiation. *Biochim. Biophys. Acta, Mol. Cell Res.* **1094**, 281–287.
70. Selassie, C. D., Kapur, S., Verma, R. P., and Rosario, M. (2005) Cellular apoptosis and cytotoxicity of phenolic compounds: a quantitative structure-activity relationship study. *J. Med. Chem.* **48**, 7234–7242.
71. Liu, J., Bhlagat, M., Zhang, C., Diwu, Z., Hoyland, B., and Klaubert, D. H. (1999) Fluorescent molecular probes V: a sensitive caspase-3 substrate for fluorometric assays. *Bioorg. Med. Chem. Lett.* **9**, 3231–3236.
72. Tzougraki, C., Noula, C., Geiger, R., and Kokotos, G. (1994) Fluorogenic substrates containing 7-Amino-4-methyl-2-quinolinone for aminopeptidase M, chymotrypsin, elastase and trypsin, determination of enzyme activity. *Liebigs Ann. Chem.*, 365–368.
73. Rotman, B., Zderic, J. A., and Edelstein, M. (1963) Fluorogenic substrates for β-D-dalactosidases and phosphatases derived from fluorescein (3,6-dihydroxyfluoran) and its monomethyl ether. *Proc. Natl. Acad. Sci. U.S.A.* **50**, 1–6.
74. Takahashi, A., Camacho, P., Lechleiter, J. D., and Herman, B. (1999) Measurement of intracellular calcium. *Physiol. Rev.* **79**, 1089–1125.
75. Paredes, R. M., Etzler, J. C., Watts, L. T., Zheng, W., and Lechleiter, J. D. (2008) Chemical calcium indicators. *Methods* **46**, 143–151.
76. Heinonen, E., and Akerman, K. E. (1987) Intracellular free magnesium in synaptosomes measured with entrapped eriochrome blue. *Biochim. Biophys. Acta Biomembr.* **898**, 331–337.
77. Gryzkiewicz, G., Poenie, M., and Tsien, R. Y. (1985) A new generation of Ca^{2+} indicators with greatly improved fluorescence properties. *J. Biol. Chem.* **260**, 3440–3450.
78. Tsien, R. Y. (1980) New calcium indicators and buffers with high selectivity against magnesium and protons: design, synthesis, and properties of prototype structures. *Biochemistry* **19**, 2396–2404.
79. Domaille, D. W., Que, E. L., and Chang, C. J. (2008) Synthetic fluorescent sensors for studying the cell biology of metals. *Nat. Chem. Biol.* **4**, 168–175.
80. Kikuchi, K., Komatsu, K., and Nagano, T. (2004) Sensing for cellular application. *Curr. Opin. Chem. Biol.* **8**, 182–191.
81. Thompson, R. B. (2005) Studying zinc biology with fluorescence: ain't we got fun? *Curr. Opin. Chem. Biol.* **9**, 526–532.
82. Walkup, G. K., Burdette, S. C., and Lippard, S. J. (2000) A new cell-permeable fluorescent

probe for Zn(II). *J. Am. Chem. Soc.* **122**, 5644–5645.
83. Chang, C. J., Nolan, E. M., Jaworski, J., Burdette, S. C., Sheng, M., and Lippard, S. J. (2004) Bright fluorescent chemosensor platforms for imaging endogenous pools of neuronal zinc. *Chem. Biol.* **11**, 203–210.
84. Woodroofe, C. C., Masalha, R., Barnes, K. R., Fredrickson, C. J., and Lippard, S. J. (2004) Membrane-permeable and -impermeable sensors of the zinpyr family and their application to imaging of hippocampal zinc in vivo. *Chem. Biol.* **11**, 1659–1666.
85. Nolan, E. M., Burdette, S. C., Harvey, J. H., Hilderbrand, S. A., and Lippard, S. J. (2004) Synthesis and characterization of zinc sensors based on a monosubstituted fluorescein platform. *Inorg. Chem.* **43**, 2624–2635.
86. Nolan, E. M., Ryu, J. W., Jaworski, J., Feazell, R. P., Sheng, M., and Lippard, S. J. (2006) Zinspy sensors with enhanced dynamic range for imaging neuronal cell zinc uptake and mobilization. *J. Am. Chem. Soc.* **128**, 15517–15528.
87. Hirano, T., Kikuchi, K., Urano, Y., Higuchi, T., and Nagano, T. (2000) Highly zinc-selective fluorescent sensor molecules suitable for biological applications. *J. Am. Chem. Soc.* **122**, 12399–12400.
88. Hirano, T., Kikuchi, K., Urano, Y., and Nagano, T. (2002) Improvement and biological applications of fluorescent probes for ainc, ZnAFs. *J. Am. Chem. Soc.* **124**, 6555–6562.
89. Takeda, A., Nakajima, S., Fuke, S., Sakurada, N., Minami, A., and Oku, N. (2006) Zinc release from schaffer collaterals and its significance. *Brain Res. Bull.* **68**, 442–447.
90. Komatsu, K., Kikuchi, K., Kojima, H., Urano, Y., and Nagano, T. (2005) Selective zinc sensor molecules with various affinities for Zn^{2+}, revealing dynamics and regional distribution of synaptically released Zn^{2+} in hippocampal slices. *J. Am. Chem. Soc.* **127**, 10197–10204.
91. Gee, K. R., Zhou, Z. L., Ton-That, D., Sensi, S. L., and Weiss, J. H. (2002) Measuring zinc in living cells. A new generation of sensitive and selective fluorescent probes. *Cell Calcium* **31**, 245–251.
92. MacDiarmid, C. W., Milanick, M. A., and Eide, D. J. (2003) Induction of the ZRC1 metal tolerance gene in zinc-limited yeast confers resistance to zinc shock. *J. Biol. Chem.* **278**, 15065–15072.
93. Chang, C. J., Jaworski, J., Nolan, E. M., Sheng, M., and Lippard, S. J. (2004) A tautomeric zinc sensor for ratiometric fluorescence imaging: application to nitric oxide-induced release of intracellular zinc. *Proc. Natl. Acad. Sci. U.S.A.* **101**, 1129–1134.
94. Kiyose, K., Kojima, H., Urano, Y., and Nagano, T. (2006) Development of a ratiometric fluorescent zinc ion probe in near-infrared region, based on tricarbocyanine chromophore. *J. Am. Chem. Soc.* **128**, 6548–6549.
95. Kellum, J. A., Song, M., and Li, J. (2004) Extracellular acidosis and the immune response: clinical and physiologic implications. *Crit. Care* **8**, 331–336.
96. Coakley, R. D., Grubb, B. R., Paradiso, A. M., Gatzy, J. T., Johnson, L. G., Kreda, S. M., O'Neal, W. K., and Boucher, R. C. (2003) Abnormal surface liquid pH regulation by cultured cystic fibrosis bronchial epithelium. *Proc. Natl. Acad. Sci. USA* **100**, 16083–16088.
97. Gillies, R. J., Raghunand, N., Garcia-Martin, M. L., and Gatenby, R. A. (2004) pH Imaging. *IEEE Eng. Med. Biol. Mag.* **23**, 57–64.
98. Gillies, R. J., Schornack, P. A., Secomb, T. W., and Raghunand, N. (1999) Causes and effects of heterogenous perfusion in tumors. *Neoplasia* **1**, 197–207.
99. Heiple, J. M., and Taylor, D. L. (1980) Intracellular pH in single motile cells. *J. Cell. Biol.* **86**, 885–890.
100. Khodorov, B., Valkina, O., and Turovetsky, V. (1994) Mechanisms of stimulus-evoked intracellular acidification in frog nerve fibers. *FEBS Lett.* **341**, 125–127.
101. Burns, A., Sengupta, P., Zedayko, T., Baird, B., and Weisner, U. (2006) Core/shell fluorescent silica nanoparticles for chemical sensing: towards single cell particle laboratories. *Small* **2**, 723–726.
102. Rink, T. J., Tsien, R. Y., and Pozzan, T. (1982) Cytoplasmic pH and free Mg^{2+} in lymphocytes. *J. Cell. Biol.* **95**, 189–196.
103. Whitaker, J. E., Haughland, R. P., and Prendergast, F. G. (1991) Spectral and photophysical studies of benz[c]xanthene dyes: dual emission pH sensors. *Anal. Biochem.* **194**, 330–344.
104. Bassnett, S., Reinisch, L., and Bebee, D. C. (1990) Intracellular pH measurement using single excitation dual emission fluorescence ratios. *Am. J. Phys. Cell Physiol.* **258**, 171–178.
105. Urano, Y., Asanuma, D., Hama, Y., Koyama, Y., Barrett, T., Kamiya, M., Nagano, T., Watanabe, T., Hasegawa, A., Choyke, P. L., and Kobayashi, H. (2008) Selective molecular imaging of viable cancer cells with pH-activatable fluorescent probes. *Nat. Med.* **15**, 104–109.

106. Adie, E. J., Kalinka, S., Smith, L., Francis, M. J., Marenghi, A., Cooper, M. E., Briggs, M., Michael, N. P., Milligan, G., and Game, S. (2002) A pH-sensitive fluor, CypHer5, used to monitor agonist-induced G protein-coupled receptor internalization in live cells. *BioTechniques* **33**, 1152–1157.
107. Cooper, M. E., Gregory, S., Adie, E., and Kalinka, S. (2002) pH-Sensitive cyanine dyes for biological applications. *J. Fluoresc.* **12**, 425–429.
108. Hilderbrand, S. A., and Weissleder, R. (2007) Optimized pH-responsive cyanine fluorochromes for detection of acidic environments. *Chem. Commun.*, 2747–2749.
109. Hilderbrand, S. A., Kelly, K. A., Niedre, M., and Weissleder, R. (2008) Near infrared fluorescence-based bacteriophage particles for ratiometric pH imaging. *Bioconjug. Chem.* **19**, 1635–1639.
110. Minta, A., and Tsien, R. Y. (1989) Fluorescent indicators for cytosolic sodium. *J. Biol. Chem.* **264**, 19449–19457.
111. Meuwis, K., Boens, N., De Schryver, F. C., Gallay, J., and Vincent, M. (1995) Photophysics of the fluorescent K^+ indicator PBFI. *Biophys. J.* **68**, 2469–2473.
112. Valentine, J. S., and Hart, P. J. (2003) Misfolded CuZnSOD and amyotrophic lateral sclerosis. *Proc. Natl. Acad. Sci. U.S.A.* **100**, 3617–3622.
113. Barnham, K. J., Masters, C. L., and Bush, A. I. (2004) Neurodegenerative diseases and oxidative stress. *Nat. Rev. Drug Discov.* **3**, 205–214.
114. Yang, L. C., McRae, R., Henary, M. M., Patel, R., Lai, B., Vogt, S., and Fahrni, C. J. (2005) Imaging of the intracellular topography of copper with a fluorescent sensor and by synchrotron X-ray fluorescence microscopy. *Proc. Natl. Acad. Sci. U.S.A.* **102**, 11178–11184.
115. Zeng, L., Miller, E. W., Domaille, D. W., and Chang, C. J. (2006) A selective turn-on fluorescent sensor for imaging copper in living cells. *J. Am. Chem. Soc.* **128**, 10–11.
116. Miller, E. W., Zeng, L., Domaille, D. W., and Chang, C. J. (2006) Preparation and use of coppersensor-1, a synthetic fluorophore for live-cell copper imaging. *Nat. Protoc.* **1**, 824–827.
117. Hua, J., and Wang, Y. G. (2005) A highly selective and sensitive fluorescent chemosensor for Fe(III) in physiological aqueous solution. *Chem. Lett.* **34**, 98–99.
118. Xiang, Y., and Tong, A. (2006) A new rhodamine-based chemosensor exhibiting selective Fe(III)-amplified fluorescence. *Org. Lett.* **8**, 1549–1552.
119. Zhang, M., Gao, Y., Li, M., Yu, M., Li, F., Li, L., Zhu, M., Zhang, J., Yi, T., and Huang, C. (2007) A selective turn-on fluorescent sensor for FeIII and application to bioimaging. *Tet. Lett.* **48**, 3709–3712.
120. Lin, W., Yuan, L., Feng, J., and Cao, X. (2008) A fluorescence-enhanced chemodosimeter for Fe^{3+} based on hydrolysis of a bis(coumarinyl) schiff base. *Eur. J. Org. Chem.*, 2689–2692.
121. Setsukinai, K.-I., Urano, Y., Kakinuma, K., Majima, H. J., and Nagano, T. (2003) Development of novel fluorescence probes that can reliably detect reactive oxygen species and distinguish specific species. *J. Biol. Chem.* **278**, 3170–3175.
122. Lim, M. H., and Lippard, S. J. (2007) Metal-based turn-on fluorescent probes for sensing nitric oxide. *Acc. Chem. Res.* **40**, 41–51.
123. Nagano, T., and Yoshimura, T. (2002) Bioimaging of nitric oxide. *Chem. Rev.* **102**, 1235–1269.
124. Nakatsubo, N., Kojima, H., Sakurai, K., Kikuchi, K., Nagoshi, H., Hirata, Y., Akaike, T., Maeda, H., Urano, Y., Higuchi, T., and Nagano, T. (1998) Improved nitric oxide detection using 2,3-diaminonaphthalene and its application to the evaluation of novel nitric oxide synthase inhibitiors. *Biol. Pharm. Bull.* **21**, 1247–1250.
125. Kojima, H., Nakatsubo, N., Kikuchi, K., Kawahara, S., Kirino, Y., Nagoshi, H., Hirata, Y., and Nagano, T. (1998) Detection and imaging of nitric oxide with novel fluorescent indicators: diaminofluoresceins. *Anal. Chem.* **70**, 2446–2453.
126. Gabe, Y., Urano, Y., Kikuchi, K., Kojima, H., and Nagano, T. (2004) Highly sensitive fluorescence probes for nitric oxide based on boron dipyrromethane chromophore-rational design of potentially useful bioimaging fluorescence probe. *J. Am. Chem. Soc.* **126**, 3357–3367.
127. Kojima, H., Hirotani, M., Nakatsubo, N., Kikuchi, K., Urano, Y., Higuchi, T., Hirata, Y., and Nagano, T. (2001) Bioimaging of nitric oxide with fluorescent indicators based on the rhodamine chromophore. *Anal. Chem.* **73**, 1967–1973.
128. Sasaki, E., Kojima, H., Nishimatsu, H., Urano, Y., Kikuchi, K., Hirata, Y., and Nagano, T. (2005) Highly sensitive near-infrared fluorescent probes for nitric oxide and their application to isolated organs. *J. Am. Chem. Soc.* **127**, 3684–3685.
129. Lim, M. H., Xu, D., and Lippard, S. J. (2006) Visualization of nitric oxide in living cells by a copper-based fluorescent probe. *Nat. Chem. Biol.* **2**, 375–380.

130. Lim, M. H., Wong, B. A., Pitcock, W. H., Mokshagundam, D., Baik, M.-H., and Lippard, S. J. (2006) Direct nitric oxide detection in aqueous solution by copper(II) fluorescein complexes. *J. Am. Chem. Soc.* **128**, 14364–14373.
131. Maeda, H., Fukuyasu, Y., Yoshida, S., Fukuda, M., Saeki, K., Matsuno, H., Yamauchi, Y., Yoshida, K., Hirata, K., and Miyamoto, K. (2004) Fluorescent probes for hydrogen peroxide based on a non-oxidative mechanism. *Angew. Chem., Int. Ed.* **43**, 2389–2391.
132. Maeda, H., Yamamoto, K., Nomura, Y., Kohno, I., Hafsi, L., Ueda, N., Yoshida, S., Fukuda, M., Fukuyasu, Y., Yamauchi, Y., and Itoh, N. (2005) A design of fluorescent probes for superoxide based on a nonredox rechanism. *J. Am. Chem. Soc.* **127**, 68–69.
133. Xu, K., Tang, B., Huang, H., Yang, G., Chen, Z., Li, P., and An, L. (2005) Strong red fluorescent probes suitable for detecting hydrogen peroxide generated by mice peritoneal macrophages. *Chem. Commun.*, 5974–5976.
134. Chang, M. C. Y., Pralle, A., Isacoff, E. Y., and Chang, C. J. (2004) A selective, cell permeable optical probe for hydrogen peroxide in living cells. *J. Am. Chem. Soc.* **126**, 15392–15393.
135. Miller, E. W., Albers, A. E., Pralle, A., Isacoff, E. Y., and Chang, C. J. (2005) Boronate-based fluorescent probes for imaging cellular hydrogen peroxide. *J. Am. Chem. Soc.* **127**, 16652–16659.
136. Miller, E. W., Tulyathan, O., Isacoff, E. Y., and Chang, C. J. (2007) Molecular imaging of hydrogen peroxide produced for cell signaling. *Nat. Chem. Biol.* **3**, 263–267.
137. Albers, A. E., Dickinson, B. C., Miller, E. W., and Chang, C. J. (2008) A red-emitting naphthofluorescein-based fluorescent probe for selective detection of hydrogen peroxide in living cells. *Bioorg. Med. Chem. Lett.* **18**, 5948–5950.
138. Lee, D., Khaja, S., Velasquez-Castano, J. C., Dasari, M., Sun, C., Petros, J., Taylor, W. R., and Murthy, N. (2007) In vivo imaging of hydrogen peroxide with chemiluminescent nanoparticles. *Nat. Mater.* **6**, 765–769.
139. Shepherd, J., Hilderbrand, S. A., Waterman, P., Heinecke, J. W., Weissleder, R., and Libby, P. (2007) A fluorescent probe for the detection of myleoperoxidase activity in atherosclerosis-associated macrophages. *Chem. Biol.* **14**, 1221–1231.
140. Kundu, K., Knight, S. F., Willett, N., Lee, S., Taylor, W. R., and Murthy, N. (2009) Hydrocyanines: a class of fluorescent sensors that can image reactive oxygen species in cell culture, tissue, and in vivo. *Angew. Chem., Int. Ed.* **48**, 299–303.
141. Yang, D., Wang, H.-L., Sun, Z.-N., Chung, N.-W., and Shen, J.-G. (2006) A highly selective fluorescent probe for the detection and imaging of peroxynitrite in living cells. *J. Am. Chem. Soc.* **128**, 6004–6005.
142. Seshadri, S., Beiser, A., Selhub, J., Jacques, P. F., Rosenberg, I. H., D'Agostino, R. B., Wilson, P. W. F., and Wolfe, P. A. (2002) Plasma homocysteine as a risk factor for dementia and alzheimer's disease. *New Engl. J. Med.* **346**, 476–483.
143. Refsum, H., and Ueland, P. M. (1998) Homocysteine and cardiovascular disease. *Annu. Rev. Med.* **49**, 31–62.
144. Duan, L., Xu, Y., Qian, X., Wang, F., Liu, J., and Cheng, T. (2008) Highly selective fluorescent chemosensor with red shift for cysteine in buffer solution and its bioimage: symmetrical naphthalimide aldehyde. *Tet. Lett.* **49**, 6624–6627.
145. Lin, W., Long, L., Yuan, L., Cao, Z., Chen, B., and Tan, W. (2008) A ratiometric fluorescent probe for cysteine and homocysteine displaying a large emission shift. *Org. Lett.* **10**, 5577–5580.
146. Bouffard, J., Kim, Y., Swager, T. M., Weissleder, R., and Hilderbrand, S. A. (2008) A highly selective fluorescent probe for thiol bioimaging. *Org. Lett.* **10**, 37–40.
147. Jiang, W., Fu, W., Fan, H., Ho, J., and Wang, W. (2007) A highly selective fluorescent probe for thiophenols. *Angew. Chem., Int. Ed.* **46**, 8445–8448.

Chapter 3

Live Cell Imaging: An Industrial Perspective

Terry McCann

Abstract

The analysis of live cells using automated fluorescence microscopy systems on an industrial scale is known as high content screening/analysis (HCS/A). Its development has been driven both by the demands of compound screening in the drug discovery industry and by the promise of whole genome functional analyses using siRNA knockouts. This chapter outlines the primary applications of HCS/A within the drug discovery process and in systems cell biology. It discusses specific issues which must be addressed when undertaking HCS/A, such as choice of cells, probes, labels, and assay type. Drawing from information gathered from surveys of key users of HCS/A in industry and academia, it then provides a detailed description of HCS/A user issues and requirements, before concluding with a summary of the imaging instrumentation currently available for live cell HCS/A.

Key words: Live cell imaging, automated microscopy, high content screening, high content analysis.

1. Introduction – An Industrial Perspective to Live Cell Imaging

The term "high-content screening", or "high-content analysis", (HCS/A) has been used since the mid-1990s to describe the processes and technologies that allow the automated imaging and analysis of many samples of cells at greatly increased throughput than had previously been possible. This automation of the labour-intensive processes of microscopy imaging and image analysis enabled the pharmaceutical industry to utilise HCS/A to accelerate aspects of the drug discovery process.

A key aspect of HCS/A is the labelling of cells with multiple fluorescent probes. The combination of multiple probes with automation renders HCS/A incredibly powerful as a tool because

D.B. Papkovsky (ed.), *Live Cell Imaging*, Methods in Molecular Biology 591,
DOI 10.1007/978-1-60761-404-3_3, © Humana Press, a part of Springer Science+Business Media, LLC 2010

it permits the analysis of multiple parameters at the level of individual cells. This, in turn, enables interdependent cellular processes to be investigated simultaneously, with assays being conducted on large populations of cells. With its revolutionising of the way in which cellular behaviour can be investigated, over the last 10 years HCS/A has become an indispensable tool in both academia and the pharmaceutical industry.

This chapter considers the use of HCS/A with an emphasis on live cell applications. The term "industrial" is used here to emphasise the scale of what is possible with HCS/A technologies, rather than the body or organisation undertaking the work. Thus, in this chapter, both academic and commercial labs are regarded as undertaking live cell imaging on an industrial scale. In offering an industrial perspective on live cell imaging, this chapter starts by describing the research needs that have driven the increases in scale for cellular imaging and explains the various applications for HCS/A technology. Following this, some of the requirements and constraints for live cell HCS/A are explored, together with the technological solutions available to meet these requirements. The chapter then highlights three key issues that must be taken into account when using HCS/A for live cell imaging and concludes by summarising the commercial sources of imaging instrumentation.

1.1. Why Increase the Throughput of Live Cell Imaging?

What is the driving force behind the need to increase the throughput of live cell imaging? In short, it is because there is so much to do. Pharmaceutical companies will be spending 10–15 years and in excess of $1 billion to develop a new drug (S Paul, Executive VP, Eli Lilly & Co, 2006). One effective way to accelerate the process and to reduce the cost is to ensure that the drug candidates that enter pre-clinical and clinical development have the best chance of success. Undertaking compound screening and characterisation in cellular systems with high biological relevance, using assays that allow simultaneous elucidation of mechanistic, toxic, and off-target effects, can achieve this. For basic research applications in academia and the pharmaceutical industry, new ways of manipulating the expression of hundreds or thousands of genes simultaneously using RNAi technologies have created the need for instruments that can screen cellular gene expression libraries rapidly.

Two distinct benefits derive from using cell-based assays. First, cells provide a more physiological environment for analyses. Decisions and conclusions based on data from cell-based assays will therefore be more biologically relevant than those based on results from assays that use purified components or cell fragments. Second, and more profoundly, modern techniques of cellular analysis permit the use of multiparametric approaches to assay development. This means that much more complex questions can be addressed in a single assay and that greater insight into the mechanisms and pathways underlying a response can be acquired, and acquired rapidly. This multiparametric approach – HCS/A – is being increasingly adopted, firstly by the pharmaceutical industry as an improved means for selecting compounds with the highest potential to become successful drugs, and secondly by academic institutes when applying systems biology approaches to the investigation of gene function at the cellular level.

The motivation for using live cells in high-content assays is the same as in "low-content" experiments, and is related to the dynamics of the pathway under investigation: rapidly changing cellular behaviour cannot easily be elucidated using fixed endpoint assays. Furthermore, when alterations to the time course of an event are being observed, a live cell assay greatly reduces the number of cells/experiments required, since each time point does not require a separate set of experiments. These benefits of live cell experiments are especially important in drug discovery assay development in which the parameters of a screening assay are being determined, and in functional genomics studies investigating the sequential and/or causally related induction of cellular phenotypes. Another benefit of live cell high-throughput assays is the simpler protocols, with fewer washing steps, that permit fast and robust screening assays. Use of fluorescent proteins and live cell-compatible fluorescent probes has made such assays feasible. Other chapters in this book will describe in more detail the advantages of using live cells in imaging experiments.

The sections below describe the work and needs of different users of HCS/A, beginning with the pharmaceutical industry; and they highlight the requirements necessary for success. Information has been gathered from several sources, including published articles, and through discussions held with key users in the pharma/biotech industry and within academia as part of a number of user surveys. In these surveys, structured telephone interviews were conducted with leading scientists active in the HCS/A field to determine major requirements and issues in various aspects of HCS/A.

2. The Drug Discovery Process

Pharmaceutical companies develop new drugs from compound libraries consisting of many hundreds of thousands of different potential drugs. Most of the cost of developing new drugs occurs after a compound has been nominated as a "candidate drug" and becomes subject to stringent preclinical and clinical testing.

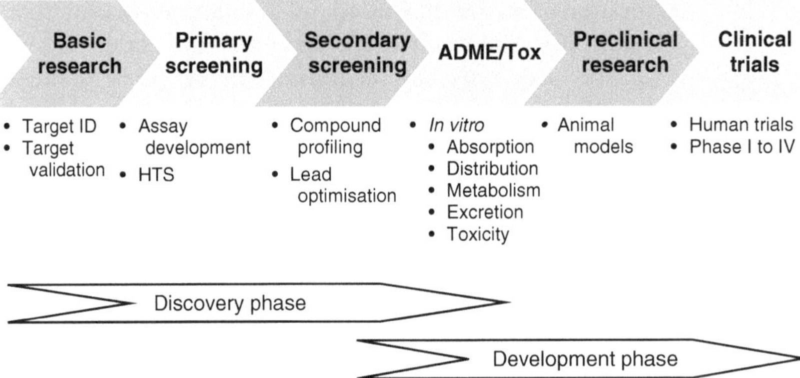

Fig. 3.1. The drug discovery process.

Correctly identifying and rejecting an unsuitable compound before this stage is estimated to save a pharmaceutical company \$1–\$3 million per compound.

Following selection of a drug target, the process of "compound attrition" begins with the primary screen (*see* **Fig. 3.1**). All compounds from a library are screened in an assay to identify those which might have a beneficial impact on the disease target. It is increasingly common to use a cellular assay at this stage, rather than a biochemical one. Output is from populations of cells in wells, in 384- or 1536-well plates. The positive drugs ("hits") are then further investigated in the secondary screening process. In secondary screening, dose-response relationships are determined, functional mechanisms are examined, and newly synthesised, related, compounds are tested for improved drug properties. Efficacy and safety assessments are then initiated in vitro to determine certain pharmacokinetic and toxicity profiles of the lead compounds prior to pre-clinical development in animal models. The purpose of all of these tests is to identify and discard unsuitable compounds, leaving only the most promising candidates to be nominated as candidate drugs.

2.1. Drug Discovery Phases

2.1.1. Target Identification and Validation

A drug discovery programme begins with the search for potential drug targets (usually proteins) that can be manipulated to modify the expression of a disease phenotype. Potential drug targets must then be validated as truly having an influence on the disease (*see* **Section 3.1** for more discussion of the criteria for validation). Basic research in academia underpins much of the work conducted in pharma/biotech's target identification and validation programmes. The outcome of these projects is a well-characterised drug target and the basis for a high-throughput assay to allow primary screening of compound libraries against this target. Live cell HCS/A tools have become an important component of target identification and validation because of their capacity to follow changes in cell behaviour over time.

2.1.2. Primary Screening

Researchers involved in primary screening for new drugs test their company's compound library against selected targets, using an assay that will identify those compounds which affect or modulate target function. Compound libraries are commonly full diversity libraries comprising up to one million compounds, although they may be smaller, targeted libraries of a few tens of thousands of compounds. When HCA is used in primary screening, it is commonly against a targeted library. The need to conduct such large numbers of tests creates many of the key drivers that determine the needs of primary screening laboratories.

The purpose of the primary screen is twofold. First, it should identify all those compounds that change the assay signal readout beyond a certain threshold: these are the positive "hits" that will be taken to the next stage of the process. Failure to detect compounds that do actually affect the assay target gives rise to false negatives and reduced sensitivity. Second, the primary screen should reject all compounds that do not alter the assay signal readout. Failure to reject inactive compounds gives rise to false positives, and the assay lacks specificity. The costs of low assay sensitivity and specificity are considerable.

In addition to being sensitive and specific, primary screening assays should have high throughput in order to screen a full library as rapidly as possible (hence primary screening is also known as high-throughput screening or HTS). High throughput is usually defined as the ability to test 30,000 or 40,000 compounds per day on a single instrument, although higher throughput of up to 100,000 compounds per day is desirable. To facilitate such high numbers of assays, the pharmaceutical industry uses a standardised microtitre plate format that accommodates 96, 384, 1536 or 3456 wells on a plate. Of these, the low density, 96-well plates do not allow sufficiently high throughput for HTS, while 384-well is the most commonly used format. Live cell HCS/A assays are generally not compatible with primary screening because they do not permit the high throughput necessary for this stage of drug discovery. The highest throughput HCS instruments (e.g. PerkinElmer's OPERA, Molecular Devices' ImageXpress instruments, GE's IN Cell 3000, and TTP Labtech's Acumen ^{e}X3 cytometer) have been used for primary screening with end-point assays in fixed cells.

2.1.3. Secondary Screening

Hits identified by the primary screen are subsequently subjected to additional analyses to determine dose-response relationships and to examine functional mechanisms of activity. These hits are also the starting point for the process of designing new, related compounds with improved drug properties. The number of compounds subjected to secondary screening is much lower than in HTS, with only 0.1–1% of all compounds from a full diversity compound library expected to pass through to secondary screening. Thus the throughput required from secondary screening assays is much lower, but more data points are required from multiple tests on each compound. Assays designed for secondary screening should seek to maximise the biological relevance of the assay, which usually means developing a cell-based assay if the primary screen relies on a biochemical approach, or alternatively transferring to a more relevant cell model, as appropriate. These requirements make HCS/A approaches very suitable for aspects of secondary screening, which is a well-developed application for HCS tools (1, 2).

2.1.4. Safety Screening

The final stage of in vitro drug discovery is known as ADME/Tox (for absorption, distribution, metabolism, excretion and toxicology). These investigations are designed to assess the suitability of a compound for development into a drug and to identify potentially toxic actions in standardised metabolic pathways. In the context of drug safety, HCS/A is becoming an increasingly significant tool, as "off-target" effects (i.e. side-effects) of potential drug candidates can be assessed more effectively in a multiparametric cellular assay.

3. Systems Cell Biology

Systems cell biology is predominantly carried out by universities, governmental facilities, and not-for-profit research institutes in order to increase our fundamental understanding of cellular processes. This work often feeds the drug discovery conveyor belt of the pharmaceutical industry and is, to some extent, continued by the industry in its target identification and validation programmes.

3.1. Applications of HCA in Systems Cell Biology

A major driver for applying HCA technologies to basic research is the desire to build a comprehensive understanding of cellular pathways both in the "normal" and diseased states. It is believed that this understanding of how pathways function will provide insight into how disease pathology is expressed and how a pathway might be manipulated to ameliorate or cure disease. To build an as complete as possible understanding of a system, researchers will integrate data and knowledge from several fields.

In the context of drug discovery and industrial applications of HCS/A, academic research plays a major role in target identification and validation (3). Target identification is simply the indication that a gene (and its product) is up- or down-regulated in some way in the disease model, compared with in the control. Target validation requires the demonstration of a functional link between the modulated gene and the onset or expression of the disease phenotype. To be able to establish this link in vitro, the assays used must meet several criteria:

- They must use a cellular system that is a faithful model for the relevant aspects of the disease.
- The assay readouts must accurately reflect the activity of the target and must additionally be related to the disease.
- Methods must exist for modulating the activity or expression level of candidate target genes.

HCA can have a significant impact on each of these criteria. First, it uses a cellular format. Increasingly, the cells being used are not highly transformed, modified cells, but are primary cells – from patients – that accurately reflect the cellular pathology of the disease. Stem cell technology is likely to increase the availability of clinically relevant cells for analysis. Second, HCA utilises multiplexed assays that enable analysis of a complex cellular pathway in a single well. One example is a cellular signalling cascade initiated by a receptor-activated kinase. Here, the activity of a kinase (assay parameter 1) causes translocation of a transcription factor or signalling molecule (assay parameter 2) to the nucleus, which leads to the phenotypic outcome such as apoptosis, mitosis, or differentiation (assay parameter 3, e.g. Ref. 4). Third, in concert with RNAi technologies, the microwell plate format of HCA permits the rapid evaluation of gene silencing experiments on a medium- to high-throughput scale (5–8). It is now possible to target druggable gene families, or even whole genomes, with libraries of small RNA molecules that allow highly effective knock-out experiments, using HCS/A assays that have multiple readouts (9).

4. Needs and Constraints for Live Cell HCS/A

Clearly, HCS/A is among the most promising new technologies for cellular research. However, it is also apparent that obtaining useful and meaningful data requires rigorous attention to detail in experimental design, sample preparation, and choice of assay-, imager-, and analysis-strategies. Some of the specific issues surrounding live cell HCS/A will be considered here, beginning with a description of the experimental process. This is then followed by a more detailed description of user requirements, obtained from recent surveys of users of HCS/A technology.

4.1. HCS/A Workflow

4.1.1. Cells, Probes and Labels

The first requirement for HCS/A is cells. The cell type chosen must clearly be suitable both for the purposes of the assay and for the screening programme. Cell lines are dynamic entities, susceptible to variations that can affect the conclusions drawn from experiments using them. Provision of well-characterised, stable, phenotypically relevant cells for HCS/A requires significant effort and is costly. The pharmaceutical industry has recognised cellular variability as an issue, and has sought to minimise it in several ways: for example, assay-ready cells can be purchased in bulk, saving significant time and resources on maintaining cell line stocks and the associated requirement for quality control. These well-characterised, division-arrested, frozen cells need only to be dispensed into plates prior to assay. However, many projects will require cells with specific characteristics that are not available "off-the-shelf". In this case, automation may be the answer to increasing the quantity of material available. Essentially all cell culture steps, from simple maintenance and passage of cell lines, through plating into 1536-well plates, to clonal expansion, can now be fully automated, bringing greater consistency.

Currently, almost all HCS/A is conducted using fluorescence-emitting probes. These are either chemical-based small molecules or genetically-encoded fluorescent proteins (FPs). The chemical probes are used either covalently linked to a targeting vector, such as an antibody or a peptide ligand or substrate (e.g. Alexa dyes), or as a diffusible chemical that accumulates in certain regions within cells.

The requirement to add a label of some description creates problems for assay design and for data interpretation. The first problem is getting the probe into the cell. For FPs, the cell line must firstly be transfected with a vector expressing the necessary construct(s), and then secondly it must be shown to be stable and to continue to display the required phenotype. Chemical probes soluble in aqueous solution may require special carriers or procedures to enable them to be taken up by cells. Lipid-soluble probes may diffuse to membrane compartments beyond those targeted. In all cases, a second problem associated with fluorescent probes must be considered, namely the potential impact of the probe on both the parameter that it is being used to measure and on cellular function and behaviour in general. These problems may be more acute and pronounced in live cell assays, but the problem is not confined to this group. Issues with probes are dealt with elsewhere in this book.

4.1.2. Fixed Cell or Live Cell Assays

HCS/A users must choose the most appropriate type of assay for their needs. The choice between using an end-point assay, in which the cells are fixed at a particular time point after treatment, or using a live cell assay, in which biological processes can be monitored as they occur over suitable time periods, depends on several factors. Fixed cell assays may be easier to perform, and the data are available more rapidly: these form the majority of HCS/A assays. Specifically, where throughput is an important issue in a compound screening or gene library silencing scenario, end-point assays are often more suitable.

Live cell HCS/A assays are more commonly used in lead optimisation and assay development, and in functional genomics studies. Here, it is more important to follow the biology of the system, in order to determine drug action on multiple pathways, and to ensure that an assay truly reflects the requirements of the study. As well as taking a longer time (and therefore having lower throughput), live cell assays are much more demanding technically. Cells are relatively delicate things; they do not like temperature variations (to within 0.5°C); they do not like being shaken around (so robotic plate-moving systems must be slow and gentle); and they need to be both provided with the right nutrients and also to be maintained at the correct pH. All of these considerations have to be built into the design of the imaging system used for live cell assays. One additional complication associated with live cell assays is the need to manipulate or activate cells whilst they are being measured. Some imagers are able to add reagents to cells whilst they are being imaged: *see* **Table 3.1** for examples.

4.1.3. Imaging Instrumentation

Imagers provide the data acquisition core of HCS/A experiments. The first systems were developed by Cellomics in the mid-1990s, and Cellomics were largely responsible for the creation of the field.

Imagers must be able to acquire images, at 4× to >40× magnification, of cells in SBS-standard microwell plates (most commonly 96-well or 384-well plates) and on standard (i.e. 25 × 75 mm) microscope slides or coverslips. It must be possible to acquire images at several (at least three) wavelengths compatible with commonly used probes and to overlay these images to produce a composite image of the field of view. Users sometimes need to be able to acquire several images from one well and to combine them together to create a composite image of the well. A system should provide for the automated movement of the sample carrier so that images can be taken from each well (or cell array spot) in turn, and the plate- and well-identification data can be stored with the images. In addition, there should be a means to integrate the imager with an external robot that can supply and retrieve microwell plates to and from the imager, before and after image acquisition. For imagers used in live cell assays, the environmental needs of cells must also be accommodated, as described above. HCS/A imagers create very large volumes of data and

Table 3.1.
Summary of HCS imaging instruments

Manufacturer	Instrument	Confocal?	Detector	Light source (s)	# λ_{ex}	# λ_{em}	Environmental control	Liquid-handling	Optimised for live cell analysis	Transmitted light	Robotics integration compatible	Auto-focus method	Objectives	Plates and slides
BD Bio-sciences	Pathway-855	Yes – true confocal	1× CCD	White light	16	8	Yes	Yes	Yes	Yes	Yes	Laser	Olympus – 4× to 60× air	Both
	Pathway-435	Yes – true confocal	1× CCD	White light	8	8	No	No	No	Yes	Yes	Laser	Olympus – 4× to 60× air	Both
Compucyte	iCyte – laser-scanning-cytome-ter	No	4× PMT	4× laser	3	6	No	No	No	Yes (scatter)	Yes	NA	10×, 20×, 40× air	Both
GE Healt-hcare	IN Cell 1000	Optional – structured light	1× CCD	White light	6	6	Optional upgrade	Optional upgrade	With upgrades	Optional	Yes	Laser	Nikon – 4× to 40× air	Plates. Upgrade for slides
	IN Cell 3000	Yes – true confocal	3× CCD	2× lasers	3	3	Yes	Yes	Yes	Optional	Yes	"Dynamic"	?	Plates
Molecular Devices	ImageXpress Ultra	Yes – true confocal	PMTs	2 to 4× laser	4	?	No	No	No	No	Yes	Laser	Nikon – 4× to 100× oil	Both

	ImageXpress Micro	No	1× CCD	White light	5	5	Optional upgrade	Optional upgrade	With upgrades	Optional	Yes	Image, laser option	Up to 4 Nikon – 4× to 100× oil	Both
Olympus Europa	Scan[R]	No	1× CCD	White light	8	6	Optional upgrade	No	With upgrades	Yes	Yes	Image, laser option	Olympus – 2× to 100× air	Both
PerkinElmer	Opera	Yes – true confocal	3× CCD	4× laser	4	4	Optional upgrade	Optional upgrade	With upgrades	No	Yes	Laser	10× to 60× water	Plates
	Opera LX	Yes – true confocal	1× CCD	3× lasers	3	3	Optional upgrade	Optional upgrade	With upgrades	No	Yes	Laser	10× to 60× water	Plates
Thermo-Fisher Cellomics	ArrayScan VTI	Optional – structured light	1× CCD	White light	10	5	Optional upgrade	Optional upgrade	With upgrades	Optional	Yes	?	Zeiss – 2.5× to 40× air	Both
With Applied Precision	cellWoRx	Digital deconvolution	1× CCD	White light	4	4	No	No	No	No	Yes	Image	10× or 20×	Both
TTP Labtech	Acumen ^eX3 – laser scanning cytometer	No	4× PMT	3× lasers	3	4	No	No	No	Yes	Yes	NA	Equivalent to 20× maximum	Both

All data from manufacturers' websites, January 2009. NA: not applicable; ?: no data available.

many images: 1 terabyte per month is an acquisition rate commonly cited by users. The data must be stored on non-volatile media, be readily accessible from the analysis programme user interface, and be easily and securely identified with the experiment and with the plate and the well from which they were acquired.

Three basic types of instrument are in common use for HCS/A assays. Two of these are based on the imaging capabilities of fluorescence microscopy, and they generally rely on CCD cameras for image acquisition. One type uses standard wide-field optics, and another uses high resolution confocal optics. A third type of instrument creates images using a scanning laser beam, in a manner analogous with laser scanning confocal microscopes, but does not use confocal optics. This technology is not commonly used for live cell imaging and will not be described further here. Other approaches to instrument design are discussed below and instrument features are summarised in **Table 3.1**.

The most common type of instrumentation in use is the wide-field CCD imager: this might be described as the default system. The wide-field CCD imager provides sufficient resolution for the majority of HCS/A assays but, where precise colocalisation of probes is required, the image depth of field is too great to provide sufficient accuracy. The relatively thick depth of field does mean that, except for very large cells or 3D aggregates, all of the cellular fluorescence is collected and contributes to the recorded signal. However, there is also the possibility to collect fluorescence signals from the medium bathing the cells, especially if the compounds being screened for activity in the assay are, themselves, fluorescent. This phenomenon of drug-compound fluorescence can significantly both decrease the assay signal-to-noise ratio and increase the between-well variability. Wide-field systems are less expensive than confocal systems, and some manufacturers have designed optional "confocal-like" upgrade components: for example, Cellomics offers the Zeiss Apotome structured light image enhancement system as an upgrade. Such "confocal" upgrades generally both significantly increase image acquisition time and decrease overall system throughput.

True confocal HCS/A imagers provide the highest resolution available for cellular analysis and are able to take "optical slices" from the sample. Some of these systems are compatible with water-immersion objectives (to 60×, PerkinElmer Opera) or oil-immersion objectives (to 100×, Molecular Devices ImageXpress^{ULTRA}). As suggested above, there are two primary advantages to the confocal format. First, precise colocalisation (or separation) of probes is possible, and this can enhance the performance of certain types of assay. Second, almost all of the fluorescence collected comprises signal emitted from the probes because the optical sectioning rejects any background fluorescence emitted by screening compounds. This means that the signal-to-noise ratio is enhanced, even though the signal itself is somewhat attenuated. Of the confocal imagers currently available, two use multiple cameras to capture images at different wavelengths simultaneously (PerkinElmer's OPERA and GE's IN Cell 3000). This feature means that these systems have among the fastest imaging speeds and highest throughput of all HCS/A imagers. The OPERA uses Yokogawa's Nipkow disc-based multibeam scanner to enable high-speed imaging, whilst the IN Cell 3000 uses a slit-scanner system. The other CCD-based confocal imager (BD's Pathway855) also uses a Nipkow disc, but uses a white light source and is without microlenses. The Pathway855 has only one CCD and does not match the acquisition speed of the OPERA and IN Cell 3000. Molecular Devices'' ImageXpress^{ULTRA} confocal HCS/A imager uses a conventional point-scanning confocal beam and is also marketed as a high-throughput system for primary screening.

4.1.4. Image Analysis and Data Management

Critical to effective utilisation of HCS/A are both the software tools for extracting information from images and the informatics and data storage solutions used to process and store the images. Usually, several images are collected from each well of a 96- or 384-well plate, and each is processed with software routines that recognise features and extract data regarding those features. These objective, numerical data are then used to quantify and compare the effects of compounds on the biological assay output. Such detailed measurements of cellular behaviour define the phenotype expressed by the cell, and describe the changes elicited by compound addition or by RNAi-mediated gene silencing.

The complex process of describing quantitatively the phenotype of a cell requires image analysis algorithms optimised for the biology under study (cell type, assay format, etc.). Also necessary are sufficient computing power to ensure fast processing of many images; fast and efficient data storage and retrieval systems; and an intuitive, easy-to-use interface that allows biologists to interact with the system to extract useful and meaningful information from it. HCS/A users require several key elements to implement an effective image analysis and data management programme. These include the following: fully validated algorithms that accurately describe the biology; access to new algorithms; and the ability to develop new (or adapt existing) ones themselves. Data storage solutions must be scalable to meet future needs (current data accumulation rates can be 1 terabyte per month) and must allow for image storage for 5 years or more. A commonly cited (but seemingly unmet) need is for effective image databasing tools that allow full integration between the images, their associated metadata, and complementary data such as genomic or proteomic data sets.

5. Users' Requirements and Issues

Information on users' requirements was obtained from discussions with experts active in the field of HCS/A, from both academia and the drug discovery industry. **Table 3.2** summarises the key attributes for HCS/A and the most commonly expressed requirements associated with each attribute.

There is a clear consensus from discussions with HCS/A users that the areas of image analysis, data mining, and data management provide the greatest frustration for users and that vendors have so far failed to meet users' needs in these areas.

5.1. Users' Issues

The technical issues raised in discussions with users cover many aspects of HCS/A workflow. These discussions suggest that successful implementation of HCS/A requires developments in several areas, including improvements to instruments, to increase productivity. These technical issues are summarised in **Table 3.3**, and selected points are discussed in more detail below.

5.1.1. Maintaining Cell Viability

For some kinetic applications, cells need to be maintained in a healthy and functioning state whilst being illuminated in the presence of organic fluorophores and/or fluorescent proteins.

Table 3.2
Key user requirements for HCS/A

Attribute	Requirement
Sample volume/throughput	At least 5000 data points/day and, for many users, >10,000 data points/day is preferable. No difference in needs between basic research and drug discovery.
Resolution	All users want to be able to resolve subcellular structures at high resolution. Most also want the flexibility to use lower resolution if assay permits it.
Number of wavelengths/probes	At least four detection channels are strongly desired, with significant emphasis on flexibility of wavelength selection.
Image quality	Absolutely critical to have excellent image quality, as all analysis depends upon it.
Live cell imaging	Currently around 25% of HCS/A experiments use live cells.
Kinetic analysis	Essential element of live cell analysis, mostly in the "minutes" and "hours" time frames.
Image analysis algorithms	Image analysis algorithms should be both simple to use and customisable.
Data management	A source of great frustration for users is inadequate data management systems. Currently there is great scope for improvement on existing systems.

Table 3.3
Technical issues identified in user discussions

Issue	Description	Potential Solution(s)
Speed: image acquisition	Kinetic analysis requires the acquisition of several images per second to gain an accurate view of cellular responses. It should be possible to add drugs while images are being taken.	Acquisition of multiwavelength images requires multiple detectors *and* on-board dispensing technology to add compounds whilst imaging.
Speed: throughput	The throughput demands of both drug discovery and basic research continue to rise. Throughput is particularly an issue for kinetic experiments, where time courses must be acquired from each well.	Consider parallelisation of imaging – but this would add significant cost.
Image analysis	Many issues to address: image formats; throughput; availability of new algorithms; ease of development of new algorithms; lack of effective data-mining tools; availability and analysis of metadata.	Requires consensus on image formats – OME appears to be only contender. Multiprocessor computers are needed. Several vendors are addressing these issues.
Image quality	Poor image quality seriously affects image analysis capabilities. The main cause of low image quality is variable, imprecise autofocus. Other causes cited were as follows: cell biology (uneven monolayers and cell clumping); plate flatness; and poor discrimination of fluorescence channels.	Laser-based autofocus is the most effective: it is increasingly available on instrumentation. Stringent quality control is required on plates and cells prior to assay. Laser-scanning systems could implement spectral separation such as that seen on point-scanning confocals.

These applications enable detailed analysis of cell proliferation, differentiation, and cytotoxicity. However, these applications also demand mechanisms for assaying cell plates over long time periods (hours and days), with full environmental control. In addition to the environmental conditions that must be maintained to keep cells alive, the cells must also be protected from excessive illumination by the fluorescence excitation source. There is considerable stress imposed on live cells by the light used for illumination, since fluorophores are prone to photodegradation and, once destroyed, are lost for imaging within a cellular sample. Furthermore, the breakdown products of photodegradation can seriously damage cellular processes. These "photobleaching" and "phototoxicity" effects can affect reproducible imaging for screening quite critically, and must be minimised.

Two aspects of imaging must be considered in minimising the unwanted effects of illumination. First, determining the correct focus position in the *z*-axis can increase phototoxic and photobleaching effects. There are two methods for automated focussing. Image-based focussing uses a z-stack of images taken through the focal plane to determine the z-position of maximum intensity or contrast. This approach requires many subsequent exposures in every position to be imaged, and it is not well suited for live cell imaging because it takes considerable time and bleaches the sample. The preferred option for live cell screening is to employ an infrared laser as a light source, and a detector to observe reflections from the interfaces between sample carrier and medium. The second consideration is the mode of excitation during image acquisition: the more the sample is illuminated, the greater the extent of phototoxic effects and loss of signal. These effects are cumulative and can be a significant problem in long-term live cell assays. The optimum mode of illumination for maintaining cell viability is to use high-frequency pulses (~300 Hz to ~1 kHz) of relatively low intensity illumination, rather than continuous illumination (wide-field microscopy) or low frequency (1–10 Hz) illumination of high intensity (point scanning confocal, *see* Ref. 10). In practical terms, the optimum illumination mode for long-term imaging is a multi-beam confocal configuration, such as provided by Nipkow disc systems including the PerkinElmer OPERA and the BD Biosciences Pathway. Whichever mode is selected, imaging systems for live cell HCS/A should utilise high NA optics and sensitive detectors, to minimise the illumination required. Intelligent scheduling of imaging steps is also important, to ensure that cells are exposed to the illumination beam for the shortest possible time.

5.1.2. Improving Speed

There are three main technical requirements for high-speed imaging in HCS/A. First, it is important that data from different emission wavelengths can be collected simultaneously. Second, the detectors should have high sensitivity to allow them to detect signal in the very short time available for image acquisition. Finally, the signal should be well discriminated from the background noise. It follows that a high-speed HCS/A imager should have multiple detectors and that these should be of sufficient sensitivity for simultaneous acquisition of images at several wavelengths. Confocal optics will improve the signal-to-noise ratio by rejecting out-of-focus light and have been shown to improve the performance of high-speed imagers, such as GE Lifesciences' IN Cell 3000 and PerkinElmer's Opera.

Further increase of the throughput of HCS/A, especially of live cell HCS/A, can probably only come through the parallelisation of detection channels, i.e. using more objectives. However, this is likely to be a very expensive solution, and it is not clear whether there is sufficient demand for it at present.

5.1.3. Image Quality

HCS/A users who participated in the surveys discussed here identified several factors contributing to poor image quality. Some of these factors were quality control issues relating to cell biology and consumables, such as the clumping of cells and the design of microplates with insufficiently flat well bottoms. Two issues relevant to instrument design were problems with autofocus and the poor discrimination of fluorescence emission. Inconsistent performance of autofocus image-based systems was the most frequently cited cause of poor image quality. Manufacturers appear to be responding to this complaint, with more recently introduced systems benefiting from higher performance laser-based autofocus, or offering this as an optional upgrade. A second problem identified by the user surveys is poor discrimination of wavelengths. This is usually a problem for fluorescence emission but, where white light is used as an excitation source, proper discrimination must also be implemented in the excitation path. Emission wavelength selection in HCS/A instrumentation is universally made with high-quality band-pass filters. In principle, laser-scanning instruments have the option of adopting spectral separation of emitted light, as seen in many point-scanning research confocal microscopes. Using spectral separation should provide better discrimination of emitted fluorescence signals than is currently possible. For camera-based systems the options are more limited, and selecting alternative probes with larger Stokes' shifts may be the best route to improving wavelength discrimination.

6. Instrumentation

Since the inception of HCS/A, the predominant imaging format for instrumentation has been based on the wide-field fluorescence microscope light path, using a CCD camera detector, and filters and dichroic mirrors to select and discriminate wavelengths. When coupled with automated microplate-handling hardware and suitable acquisition and analysis software, this arrangement meets effectively the basic requirements for automated high-content image acquisition and analysis.

Recent product launches have seen the wider availability of confocal (or confocal-like) technology, either as upgrades to existing systems or as newly developed instruments. Cellomics has incorporated Carl Zeiss's Apotome structured light image enhancement system into its ArrayScan system and has also partnered with Applied Precision to develop a high-resolution cell analyser using deconvolution. GE has introduced an optical sectioning module as an accessory for the IN Cell 1000. This module also uses a structured light system to increase resolution. BD Biosciences has launched lower cost instruments, which have spinning disk confocal capability. Arguably the most innovative development in confocal HCS/A imaging is Molecular Devices' ImageXpress^{ULTRA} which, paradoxically, is based on standard, single beam, point-scanning confocal technology that was first commercialised in the 1980s. This instrument uses a galvanometer for scanning in the *x* co-ordinate and the movement of the stage to scan in the *y*. Like standard research beam-scanning confocals, it uses multiple PMT detectors to perform simultaneous detection of light from several different wavelengths: this provides the potential for high-speed imaging.

There is a range of suppliers and instruments available for HCS. They are summarised in **Table 3.1**, and a brief description of those suitable for live cell imaging follows.

6.1. BD Biosciences

BD markets two instruments for high-content imaging, both of which utilise single CCD camera detection and a Nipkow disc-based confocal system. The Pathway 855 system is designed for all types of analysis, including live cell kinetic experiments. Features include the ability to make compound additions to the plate as it is being imaged, along with full environmental control. The Pathway 435 is a bench-top instrument designed primarily for endpoint assays with fixed samples.

6.2. GE Healthcare

GE's IN Cell 1000 and 3000 instruments incorporate options to support live cell imaging, including environmental control and integrated liquid-handling. The IN Cell 1000 may also have a structured light image enhancement module added on, to provide confocal-like images. GE's other HCS/A system is the IN Cell 3000: this uses multiple CCD camera detectors and a confocal slit scanner to provide high-throughput automated imaging.

6.3. Molecular Devices

The ImageXpress product range comprises a CCD camera/white light illumination system (ImageXpress^{MICRO}) and a point scanning confocal system with up to four laser lines (ImageXpress^{ULTRA}). The ImageXpress^{MICRO} has several upgrade options for live cell imaging, including a more sensitive camera, full environmental control, and single-channel pipetting for kinetic experiments. These options are not yet available for the ImageXpress^{ULTRA}. Both instruments are compatible with high-NA oil-immersion objectives up to 100× magnification, providing high resolution.

6.4. Olympus

Olympus' Scan^{R} system was developed by EMBL in Heidelberg and is based on Olympus microscopy components (automated microscope and white light source) and Hamamatsu CCD cameras. It has environmental control options for live cell imaging, but it is available only in certain countries.

6.5. PerkinElmer

The Opera system is a multi-beam confocal system with four CCD detectors and laser illumination. Multiple detectors confer high acquisition speeds for multiparametric assays; optional environmental control and dispensing accessories provide facilities for automated live cell imaging applications. Fluorescence lifetime imaging and FRET analyses are also a possibility with this system. The Opera LX system is very similar but has only one detector, so multiple wavelength images are acquired sequentially. The LX has the same live cell imaging capabilities as the Opera. The systems are compatible with high-NA water-immersion objectives to 60× magnification.

6.6. ThermoFisher Cellomics

The ArrayScan V^{TI} HCS reader is the fifth generation of Cellomics' instrumentation. It is a white-light, single-CCD detector system compatible with objectives up to 40× magnification. For live cell applications, upgrade modules are available for environmental control and liquid handling.

7. Summary

Technologies are commercially available that enable live cell imaging on an unprecedented scale, permitting the imaging and analysis of thousands of samples in a single day. Such HCS/A instruments have been in use in the pharmaceutical industry in one form or another for a decade; they are now extremely well-proven. HCS/A has additionally found applications in academic institutions, often used in concert with RNAi gene-silencing libraries, for functional genomics analyses.

Several manufacturers produce instruments, with associated accessories and consumables, mostly designed around a standard epifluorescence light-path with a CCD detector. Some systems include laser scanning confocal optics and multiple CCDs or PMTs. Further increases in the scale and consistency of live cell imaging are expected to come from the automation of both cell line maintenance and the preparation of cells for HCS/A assays.

Acknowledgements

I would like to acknowledge the help of G. Gradl (PerkinElmer), M. Sjaastad (Molecular Devices), M. Collins (ThermoFisher Cellomics), and J. McCann in the preparation of this article.

References

1. Zanella, F., Rosado, A., Blanco, F., Henderson, B.R., Carnero, A., and Link, L. (2007) An HTS approach to screen for antagonists of the nuclear export machinery using high content cell-based assays. *ASSAY Drug Dev. Technol.* **5**, 333–342.
2. Haasen, D., Merk, S., Seither, P., Martyres, D., Hobbie, S., and Heilker, R. (2008) Pharmacological profiling of chemokine receptor-directed compounds using high-content screening. *J. Biomol. Screen* **13**, 40–53.
3. Krausz, E. and Korn, K. (2007) Academic contribution to high-content screening for functional and chemical genomics. *Eur. Pharm. Rev.* **12**, 52–58.
4. Nelson, D.E., Ihekwaba, A.E., Elliott, M. et al. (2004) Oscillations in NF-kappaB signalling control the dynamics of gene expression. *Science* **306**, 704–708.
5. Kittler, R., Putz, G., Pelletier, L., Poser, I., Heninger, A.K., Drechsel, D., Fischer, S., Konstantinova, I., Habermann, B., Grabner, H., Yaspo, M.L., Himmelbauer, H., Korn, B., Neugebauer, K., Pisabarro, M.T., and Buchholz, F. (2004) An endoribonuclease-prepared siRNA screen in human cells identifies genes essential for cell division. *Nature* **432**, 1036–1040.
6. Pelkmans, L., Fava, E., Grabner, H., Hannus, M., Habermann, B., Krausz, E., and Zerial, M. (2005) Genome-wide analysis of human kinases in clathrin- and caveolae/raft-mediated endocytosis. *Nature* **436**, 78–86.
7. Collins, C.S., Hong, J., Sapinoso, L., Zhou, Y., Liu, Z., Micklash, K., Schultz, P.G., and Hampton, G.M. (2006) A small interfering RNA screen for modulators of tumor cell motility identifies MAP4K4 as a promigratory kinase. *Proc. Natl. Acad. Sci. USA* **103**, 3775–3780.
8. Laketa, V., Simpson, J.C., Bechtel, S., Wiemann, S., and Pepperkok, R. (2007) High-content microscopy identifies new neurite outgrowth regulators. *Mol. Biol. Cell* **18**, 242–252.
9. Neumann, B., Held, M., Liebel, U., Erfle, H., Rogers, P., Pepperkok, R., and Ellenberg, J. (2006) High-throughput RNAi screening by time-lapse imaging of live human cells. *Nat. Methods* **3**, 385–390.
10. Wang, E., Babbey, C.M., and Dunn, K.W. (2005) Performance comparison between the high-speed Yokogawa spinning disk confocal system and single-point scanning systems. *J. Microsc* **218**, 148–159.

Part II

Imaging Techniques, Probes, and Applications

Chapter 4

Design of Fluorescent Fusion Protein Probes

Elizabeth Pham and Kevin Truong

Abstract

Many fluorescent probes depend on the fluorescence resonance energy transfer (FRET) between fluorescent protein pairs. The efficiency of energy transfer becomes altered by conformational changes of a fused sensory protein in response to a cellular event. A structure-based approach can be taken to design probes better with improved dynamic ranges by computationally modeling conformational changes and predicting FRET efficiency changes of candidate biosensor constructs. FRET biosensors consist of at least three domains fused together: the donor protein, the sensory domain, and the acceptor protein. To more efficiently subclone fusion proteins containing multiple domains, a cassette-based system can be used. Generating a cassette library of commonly used domains facilitates the rapid subcloning of future fusion biosensor proteins. FRET biosensors can then be used with fluorescence microscopy for real-time monitoring of cellular events within live cells by tracking changes in FRET efficiency. Stimulants can be used to trigger a range of cellular events including Ca^{2+} signaling, apoptosis, and subcellular translocations.

Key words: FRET biosensors, fusion proteins, computational modeling, cell imaging, structure-based design.

1. Introduction

Biosensors relying on the fluorescence resonance energy transfer (FRET) between fluorescent proteins have been used extensively, including for live-cell imaging of cellular events such as caspase activation, protein phosphorylation, and calcium ion (Ca^{2+}) signaling (1–7). FRET is the natural phenomenon of energy transfer via resonance between two fluorophores with a spectral overlap between the donor emission and the acceptor excitation. The efficiency of this energy transfer depends on the relative distance and orientation between the donor and acceptor (8). In FRET

D.B. Papkovsky (ed.), *Live Cell Imaging*, Methods in Molecular Biology 591,
DOI 10.1007/978-1-60761-404-3_4,

protein biosensors, natural sensory proteins for the desired cellular events are inter- or intramolecularly fused with a pair of fluorescent proteins of suitable spectral overlap, such as cyan fluorescent protein (CFP) as the donor and yellow fluorescent protein (YFP) as the acceptor (2, 9). The efficiency of energy transfer between the donor and acceptor fluorescent proteins becomes altered by conformational changes of a fused sensory protein caused by a cellular event. Hence, a change in FRET efficiency of a biosensor can be correlated with the cellular event.

In the case of FRET Ca^{2+} biosensors, changes in FRET efficiency can be correlated with Ca^{2+} concentrations (10–13). CFP and YFP have been genetically fused with calmodulin (CaM), a cytoplasmic Ca^{2+}-sensitive protein. Upon binding Ca^{2+} ions, CaM undergoes a conformational change from an extended to a compact conformation by wrapping around a fused CaM-binding peptide (14, 15), which alters the relative distance and orientation of the FRET pair. As a result, an increase in FRET efficiency corresponds to a higher ion concentration. Different configurations of fusion proteins can be used to monitor different cellular events. A fusion construct of a peptide fused between two fluorescent proteins can be used to observe proteolytic cleavage, where a decrease in FRET efficiency would correspond to a cleavage event (16). In the presence of protease proteins, the biosensor is cleaved, separating its fluorescent pair and causing a loss in FRET efficiency.

1.1. Structure-Based Computational Modeling

The design of fluorescent probes is often dependent on the structure of the sensory domain chosen. With available structural information, a computational model can be constructed to assist in the design process. This structure-based design of fluorescent probes allows existing probes to be better designed to have improved dynamic ranges. The NMR structure of CaM bound to a CaM-binding peptide from CaM-dependent kinase kinase was used to improve existing CaM-based biosensors. A computational model of CaM bound to its peptide showed that it would be possible to splice the peptide within the CaM structure to improve the dynamic range attainable (17). Similarly, the structure of caspase-3 bound to its inhibitor peptide was used to computationally design an improved caspase-3 activation biosensor (18, 19).

A computational modeling tool can thus be developed to estimate FRET efficiency changes. We previously developed a computational tool called FPMOD (Fusion Protein MODeler) to predict FRET efficiency changes of a range of biosensor constructs. FPMOD can be used to generate fusion protein models from PDB (protein databank) files containing the three-dimensional structures of proteins and other biological macromolecules. By defining regions of flexible linkers between different domains and then rotating the domains around these flexible linkers to produce random conformations, FPMOD samples the conformational space of a biosensor design and provides average predicted FRET efficiency changes. These predicted values can then be used to evaluate potential biosensor designs (13, 20).

1.2. Cassette System

To create fusion proteins, subcloning techniques are employed to insert PCR products of individual domains into an expression vector at restriction enzyme cut sites. The availability of these sites limits the number and configuration of the biosensor constructs possible. Future fusions of other domains into existing vectors are not always possible due to available restriction sites either exhausted in previous subcloning steps or incompatible. To more efficiently construct fusion proteins containing multiple domains, a cassette-based system can be used. Cassettes will have a standard vector structure based on specific restriction endonuclease sites that can be used to fuse domains in any configuration and number of times. If properly designed, this cassette vector can also be used to simplify the process of screening successful recombination of insertion fragments using fluorescence (21).

Specifically for FRET biosensors, often at least three domains will need to be fused together: the donor protein, the sensory domain, and the acceptor protein. Generating a cassette library of commonly used domains (e.g., CFP and YFP) facilitates the rapid subcloning of fusion biosensor proteins that can be recombined any number of times irrespective of order while maintaining the same simple management of restriction sites.

1.3. Live-Cell Imaging Using Fluorescence Microscopy

Fluorescence microscopy allows the real-time monitoring of cellular events within live cells by tracking the change in FRET efficiency. Stimulants can be used to trigger cellular events such as Ca^{2+} signaling, apoptosis, and subcellular translocations (16, 22–24). Simultaneous signaling processes can also be observed using a co-culture of cells transfected with different FRET biosensors (22). An advantage of live-cell imaging with fluorescence is the real-time monitoring of molecular signaling pathways using FRET biosensors and any corresponding morphological changes (16).

2. Materials

2.1. Structure-Based Computational Modeling

1. Software used to develop FPMOD (individual.utoronto.ca/ktruong/software.htm): C++ Development Environment – Bloodshed Dev-C++ IDE (www.bloodshed.net/devcpp.html), Perl language platform – ActivePerl (www.activestate.com/Products/activeperl/index.mhtml),

gGraphical uUser iInterface (GUI) – wxWindows toolkit (sourceforge.net/projects/wxwindows).

2. Structural resources: Protein 3D structure – protein databank (PDB) files (www.rcsb.org), 3D structure viewing – Swiss PDB viewer (www.expasy.ch/spdbv/text/getpc.htm), protein structure manipulation and rendering – PyMOL (pymol.sourceforge.net).

2.2. Cassette System

1. Base vectors: pTriEx 1.1 – Hygro (Novagen, Madison, WI, USA), designed primers (Invitrogen, Carlsbad, CA, USA).
2. Subcloning enzymes (*see* **Note 1**): restriction enzymes *Nco*I, *Spe*I, *Bam*HI, *Stu*I, *Bgl*II, *Sma*I, *Nhe*I, *Pme*I, *Xho*I (New England Biolabs, Ipswich, MA, USA), ligation enzyme T4 DNA ligase (New England Biolabs), and *Pfu* DNA polymerase (Fermentas, Burlington, ON, Canada) with 5 mM deoxyribonucleotide (dNTP) mixture (Fermentas).
3. Subcloning equipment: PCR amplification Mastercycler Personal system (Eppendorf, Mississauga, ON, Canada), Mini Electrophoresis system for DNA electrophoresis (VWR, Mississauga, ON, Canada).
4. DNA purification kits: PureLink Quick Plasmid Miniprep kit, PureLink PCR Purification kit, and PureLink Gel Extraction kit with 1% UltraPure agarose gel in Tris-acetate–EDTA buffer (0.5× TAE bBuffer) (Invitrogen). Ethidium bromide (Sigma-Aldrich, Oakville, ON, Canada), O'GeneRuler DNA Ladder Mix (Fermentas) and Electronic UV transilluminator (Ultra Lum, Inc., Claremont, CA).
5. Bacterial transformation and growth reagents: *Escherichia coli* DH5α strain competent cells (Subcloning Efficiency, Invitrogen) grown in Luria broth (LB) (Sigma-Aldrich) with 100 μg/mL ampicillin (Sigma-Aldrich) and plated on LB agar plates (Sigma-Aldrich) with 100 μg/mL ampicillin (Sigma-Aldrich) grown in a shaking incubator (Barnstead, Dubuque, IA).
6. Fluorescence screening of bacterial cells: Lighttools Illuminatool Tunable Lighting System LT-9500 (Lighttools Research, Encinitas, CA) equipped with 535 and 470 nm viewing filters and 488/10 and 440/10 nm filter cups.
7. Protein purification: Ni-NTA 6% agarose beads charged with Ni^{2+} ions (Qiagen, Valencia, CA).

2.3. Live-Cell Imaging Using Fluorescence Microscopy

1. Cell culture reagents (*see* **Note 2**): Cos-7 cells, Dulbecco's Modified Eagle Medium (DMEM) with high glucose, L-glutamine and sodium pyruvate (Invitrogen), fetal bovine serum (FBS) (Sigma-Aldrich), trypsin–EDTA (Sigma-Aldrich), cell freezing medium dimethyl sulfoxide

(DMSO) (Sigma-Aldrich), Dulbecco's phosphate-buffered saline (PBS) without Ca^{2+}, Mg^{2+}, or phenol red (Invitrogen), PBS with Ca^{2+} (Invitrogen), and Lipofectin Transfection Reagent (Invitrogen) (stored at 4°C).

2. Materials for cell culture: T-25 flasks (Sarstedt, Montreal, QC, Canada) and 35-mm glass-bottom dishes (MatTek, Ashland, MA).
3. Stimulants (*see* **Note 1**): Ionomycin (1 μM in PBS) (Sigma-Aldrich), ATP (10 μM in PBS) (Fermentas), and Staurosporine (STS) (5 μM in PBS) (Sigma-Aldrich).
4. Microscope: Inverted microscope IX81 with Lambda DG4 Xenon lamp source and CCD camera, objective (10×), oil-immersion objectives (20×, 40×, 60×, 100×) (Olympus, Markham, ON, Canada) (*see* **Note 3**), filter sets (BRIGHTLINE CFP FILTER SET EX:438/24, 458DM, EM:483/32, BRIGHTLINE YFP FILTER SET EX:500/24, 520DM, EM:542/27) (Semrock, Rochester, NY), 2-channel filter for simultaneous dual-band imaging of CFP and YFP (Roper Biosciences SpecEM, Tucson, AZ).
5. Imaging software packages: QEDInVivo and ImagePro-Plus (MediaCybernetics, Bethesda, MD).

3. Methods

3.1. Structure-Based Computational Modeling

The prediction of FRET efficiency changes of potential candidate biosensor constructs prior to subcloning helps to determine which constructs will likely have an appropriate dynamic range for the specific application. A computational modeling tool such as FPMOD (13) can be developed to construct fusion proteins based on determined atomic structural information from PDB files. The modeling tool should allow the sampling of a biosensor's conformational space and estimate the FRET efficiency change in response to a stimulus. This requires that structures are available for domains and proteins before and after the desired cellular event. Estimated values provided should include the distance factor, orientation factor, and FRET efficiency for each candidate biosensor. These values are sufficient to compare the candidate biosensor constructs and to select the construct that should be subcloned in vitro and tested further.

FRET efficiency (E%) is the percentage of energy transferred between a donor–acceptor fluorophore pair. This efficiency is a function of the Forster distance factor, R_o, the distance between the fluorophores, R, and the orientation factor, κ^2. In turn, the orientation factor depends on the angle between donor or acceptor fluorophore dipoles and the joining vector (θ_A and θ_D, respectively), as well as the angle between fluorophore pair planes (α). A common assumed constant for κ^2 is $2/3$ but this does not apply here because the linkers within the biosensors are not in isotropic motion upon Ca^{2+} binding (25, 26). Several relevant parameters are constants defined for the donor–acceptor pair used in the biosensor. For CFP and YFP, these constants are quantum yield ($Q_D = 0.42$), refractive index ($n = 1.4$), and overlap integral ($J = 1.46e\text{-}9$). For each conformation, the dipoles and related angles were determined from the PDB files and used in Equations [1], [2], and [3] to determine R, κ^2, and $E\%$.

$$E\% = \frac{R_0^6}{R_0^6 + R^6} \quad [1]$$

$$R_0 = 9.78 \times 10^3 \times \left(Q_d \kappa^2 n^{-4} J\right)^{1/6} \text{Å} \quad [2]$$

$$\kappa^2 = [\sin(\theta_D)\sin(\theta_A)\cos(\phi) - 2\cos(\theta_D)\cos(\theta_A)]^2 \quad [3]$$

The orientation of transition dipoles is defined with respect to PDB atom coordinates of the HETATM for CFP and YFP, from atom N15 to C4 and N3 to CZ, respectively. This assumed direction is kept consistent for all constructs simulated. For other donor–acceptor fluorescent pairs, the transition dipoles will need to be determined from their PDB files.

A graphical user interface (GUI) should be developed to simplify use of the modeling tool. It should include at least the following features: custom dialog boxes to ensure that the user will enter the necessary arguments and an output window for viewing results. Such a GUI can be developed using the wxWindows toolkit (wxWidgets Open Source Software).

3.1.1. High-Level Organization of FRET Biosensor Modeling Tool

1. The solved atomic structure of each domain of a FRET biosensor construct must be available. Most importantly, the structure of the sensory domain before and after the desired cellular event must be available as separate structural files.
2. Domains are treated as rigid bodies while linkers fusing the domains together are considered flexible.
3. To generate the conformational space of a biosensor construct, a sufficient number of models must be generated where rigid-body domains are rotated around the flexible linkers. For each residue in a linker, there are three torsional or dihedral angles: ψ, Φ, and ω. During a random rotation step, all linker residues are randomly rotated such that all atoms of a linker residue preceding the N atom along the N–C_α bond are rotated by the torsional angle Φ. Next, all atoms after the C atom along the C_α–C bond are rotated by

angle ψ. While the angles ψ and Φ do not have any restriction so that they range from $-180°$ to $180°$, ω is fixed at $180°$. These dihedral angles are given random values, representing a random rotation of each peptide bond within the user-defined flexible linkers.

4. Valid conformations must be screened from the collection of randomly generated conformations. This involves selecting only conformations whose atoms do not sterically collide after random rotations of the linker regions. Each randomly generated model is then saved as individual PDB files.
5. This linker rotation procedure is repeated until a representative number of models are generated to sufficiently span the conformational space (*see* **Note 4**).
6. From each PDB file generated previously, the distance factor, orientation factor, and FRET efficiency estimations can be tabulated.

3.1.2. Structure-Based Computational Design Process

1. Determine availability of solved atomic structures for all domains and proteins used in fusion biosensor protein. For the sensory domain, structural information for both before and after a cellular event must be available.
2. Use modeling tool to construct the biosensor construct, defining domains as rigid bodies and linkers as flexible sequences with no secondary structure. Generate a sufficient collection of valid conformations. Determine the distance factor, orientation factor, and FRET efficiency value for each generated model. Tabulate values and average to determine a FRET efficiency value representing the conformational space of each candidate biosensor construct. Determine the conformational space and FRET efficiency value for the FRET biosensor before and after the desired cellular event.
3. Use the change in FRET efficiency to determine appropriateness of using the proposed biosensor construct for in vitro and cell imaging studies.

3.1.3. Example of the Use of the Computational Modeling Tool FPMOD to Design a New Class of Ca^{2+} Biosensor According to 3.1.2

1. Solved atomic structures for epithelial cadherin, CFP, and YFP were downloaded from the Protein Databank: 1MYW (Venus (27), a variant of YFP), 1OXD (CFP), 1EDH with Ca^{2+} removed and linkers defined as flexible (epithelial cadherin domain in the absence of Ca^{2+} ions), 1EDH (epithelial cadherin domain in the presence of Ca^{2+} ions).
2. Constructs were created in FPMOD (13) for CEcadY12 (**Fig. 4.1**) before and after Ca^{2+} binding. Average predicted values of the distance factor, orientation factor, and FRET efficiency were determined for 130 conformations each of

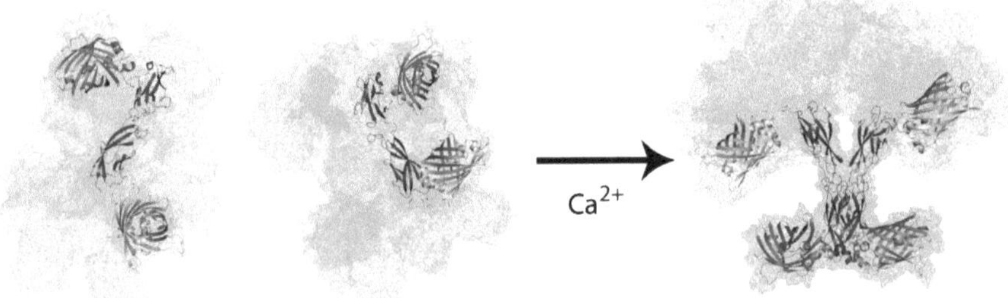

Fig. 4.1. Superposition of multiple conformations of the CEcadY12 biosensor in the presence and absence of Ca^{2+} binding. The conformational space conferred by generated models of possible conformations are shown for CEcadY12 before (*left*) and after (*right*) Ca^{2+} binding. After Ca^{2+} binding, biosensors homodimerize, forcing the CFP/YFP pair to move closer and increasing the FRET efficiency.

the unbound case (91 Å, 0.59, 3.1%) and bound case (85 Å, 0.39, 4.5%), respectively.

3. For CEcadY12, the FRET efficiency change predicted was an increase of 3.6% upon Ca^{2+} binding. FPMOD demonstrated that CEcadY12, which is a new class of Ca^{2+} biosensors, showed a sufficient predicted change in FRET efficiency to warrant further study. In vitro FRET efficiency measurements showed an increase of 14% upon binding of Ca^{2+}. The developed FRET biosensor can be used further in live-cell imaging studies (data not shown). Other relevant characteristics of a FRET biosensor can be discerned from the predicted values if desired (*see* **Note 5**).

3.2. Cassette System

3.2.1. Components of a Standard Cassette Vector

1. The standard cassette vector we created follows the scheme illustrated in **Fig. 4.2**. Restriction cut sites 2a and 2b should be sequences for different restriction enzymes that produce blunt or compatible cohesive ends such as *Spe*I/*Nhe*I, *Bam*HI/*Bgl*II, and *Stu*I/*Sma*I (more than one pair of compatible restriction enzymes can be included at this multiple cloning site). When these compatible cohesive ends are ligated together, the ligation product becomes unrecognizable by either restriction enzyme. Cut sites 1 and 3, however, should be sequences that produce unique cohesive ends on the vector such as *Nco*I and *Xho*I.
2. The standard cassette vector should also contain a gene for a fluorescent protein flanked by blunt-end restriction cut sites 4a and 4b such as *Pme*I.
3. The presence of the stop codon upstream of the fluorescent protein gene allows for quick, reliable screening of successful subclones and recombination by fluorescence. The vectors do not fluoresce when expressed because of the presence of the stop codon. However, when the stop codon is replaced

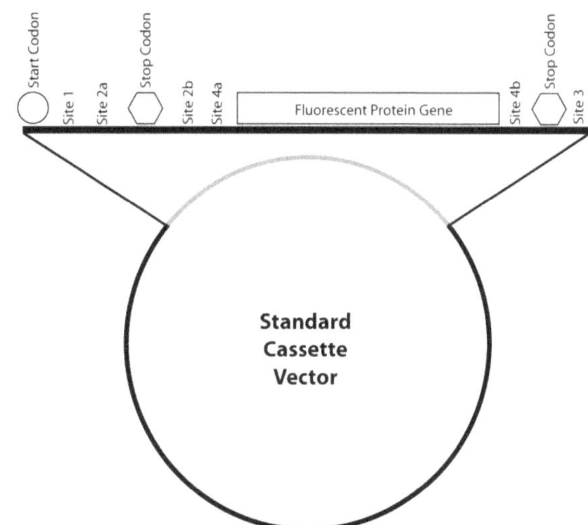

Fig. 4.2. Schematic diagram of a standard cassette vector. Restriction sites 1 and 3 should produce unique cohesive ends, such as *Nco*I and *Xho*I. Sites 2a and 2b should produce either blunt ends or compatible cohesive ends such as *Spe*I/*Nhe*I, *Bam*HI/*Bgl*II, and *Stu* I/*Sma* I. Multiple compatible restriction sites can be included in the vector at these sites. Finally, sites 4a and 4b should produce blunt ends to allow the removal of the fluorescent protein gene.

by a gene of interest, the fluorescent protein gene will be expressed, indicating a successful subclone.

4. If a non-fluorescent cassette is desired, the fluorescent protein gene can be easily removed by cutting at blunt-end sites 4a and 4b followed by a self ligation.

3.2.2. Construction of a Standard Cassette Vector – pCfvtx

1. Our standard expression vector, pCfvtx (21), was created from pTriEx 1.1 – Hygro and pVenus (27) (*see* **Notes 6** and 7).
2. Primers were used to PCR amplify the Venus gene with the restriction enzyme sequences for *Nco*I, *Spe*I, *Nhe*I, and *Pme*I upstream of Venus (5′-CATGCCATGGG-CCTGACTAGTAGGCCTGCTAGCCTGTTTAAACTGG-TGAGCAAGGGCGAGGAGCTG-3′) and *Pme*I and *Xho*I downstream of Venus (5′-CCGCTCGAGTTACAGTTT-AAACAGGGCGGCGGTCACGAACTCCA-3′). The PCR fragment was subcloned into pTriEx 1.1 – Hygro at *Nco*I and *Xho*I, by screening for a fluorescent colony to form an intermediate vector.
3. Multiple cloning sites were subcloned into the intermediate vector to create pCfvtx by using PCR fragments containing *Spe*I, *Bam*HI, *Stu*I and *Bgl*II, *Sma*I, *Nhe*I sandwiching a stop codon using 5′-end phosphorylated primers (sense: 5′-CTAGTGGATCCAGGCCTTAAAGATCTCCCGGGG-3′ and anti-sense: 5′-CTAGCCCCGGGAGATCTTTAAGGC-

CTGGATCCA-3′). This PCR fragment was subcloned into the intermediate vector at *Spe*I and *Nhe*I. A non-fluorescent colony was selected.

3.2.3. Generating a Cassette Library

Fusion proteins of greater complexity and multiple domains can be easily constructed following a consistent set of protocols (**Fig. 4.3**). Each new cassette becomes part of a growing cassette library.

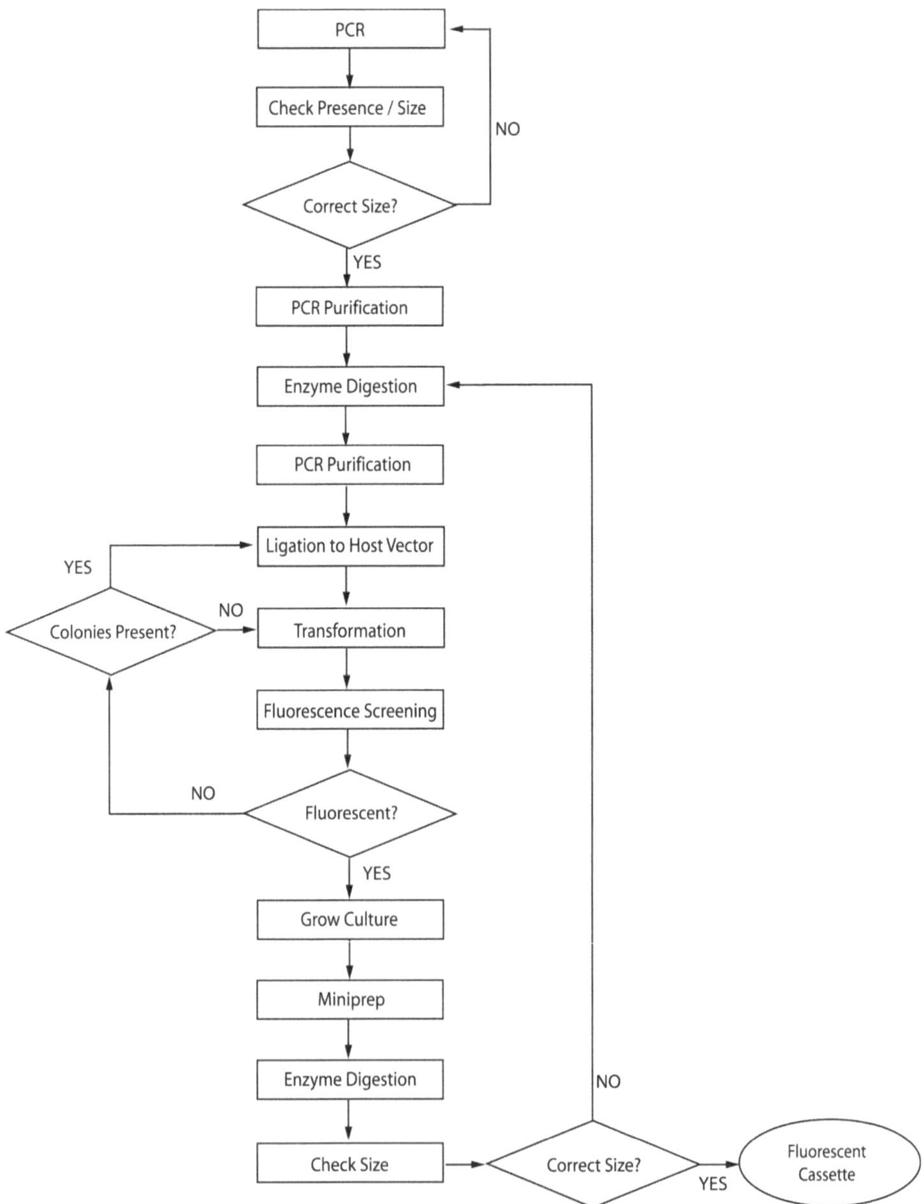

Fig. 4.3. Insertion of a gene into the standard cassette vector. See text for subcloning details.

Insertion of a New Gene of Interest

1. Polymerase chain reaction (PCR): Primers are designed to contain the necessary restriction sites, matching those on pCfvtx, to facilitate insertion into the standard host vector. Annealing temperature, length, and cDNA or plasmid source depend on the gene of interest being amplified. Gene fragments are amplified using the Eppendorf Mastercycler Personal and *Pfu* DNA polymerase. PCR reaction buffer contains 5 μL of 10× PCR buffer with $MgSO_4$ (provided with enzyme), 5 μL each of the 5′–3′ and 3′–5′ primers, 5 μL of 5 mM dNTP, 1 μL of cDNA or plasmid source, 28.5 μL water, and 0.5 μL *Pfu* DNA polymerase.
2. Purification of PCR fragment: The Invitrogen PureLink PCR Purification kit was used to purify PCR fragments. The kit can be used to purify fragments ranging in length from 100 to 12,000 bp. It is highly recommended that a DNA gel (1% agarose gel in 0.5× TAE buffer, 25 min at 100 V) be run to check for the presence of the gene of interest, as well as to assess the specificity of the PCR amplification. The gene of interest should appear as a clear band at the correct size when compared to a DNA ladder. If a blur appears or no band is visible, the PCR conditions should be modified to better amplify the gene.
3. Enzyme digestion: pCfvtx (host vector) and the PCR fragment are digested at corresponding restriction cut sites using restriction enzymes. The reaction buffer consists of 1 μL each of the restriction enzymes, 3 μL of 10× buffer (provided with enzyme), 1 μL 30× BSA buffer (provided with enzyme), and 24 μL water. Digestions are incubated at 37°C for 3 h using the Eppendorf Mastercycler Personal to keep conditions constant.
4. Purification of digested products: The Invitrogen PureLink PCR Purification kit can be used to purify the digested PCR fragment and host vector.
5. Ligation: T4 DNA Ligase is used to ligate the host vector and PCR fragment. The reaction is carried out with 2 μL of ligase buffer (provided with enzyme) and 1 μL T4 DNA ligase at 16°C for 1 h using the Eppendorf Mastercycler Personal to keep conditions constant.
6. Transformation of competent *E. coli* cells: 20 μL aliquots of DH5α *E. coli* cells are stored at –80°C. Prior to transformation, an aliquot is left to slowly thaw on ice (4°C). One microliter of the ligation product is mixed into the competent cell aliquot. Transformation is achieved by heat-shocking the cells, first quickly subjecting the aliquot to

42°C for 40 s and then placing it back on ice for another 5 min.

7. *E. coli* cell culturing: Transformed cells are grown in 1 mL of LB containing 1 μL of 100 μg/mL ampicillin overnight at 37°C in a shaking incubator at 200 rpm.
8. Fluorescence screening: Cell growths are diluted and plated on agar plates. The diluted cell solution is spread onto a pre-warmed plate. The plate is then incubated at 37°C overnight to form colonies. Colonies containing successful subclones are screened by fluorescence using an Illumatool Tunable Lighting System LT-9500. Colonies from a transformation will form both fluorescent and non-fluorescent colonies. Only successful subclones will fluoresce. A fluorescent colony can thus be picked off the plate and grown in 3 μL LB overnight at 37°C.
9. Plasmid purification: The Invitrogen PureLink Quick Plasmid Miniprep kit is used to extract plasmid DNA from *E. coli* cells. Extracted plasmids are stored at –20°C.
10. Size and sequence checking: A purified plasmid from a new insertion subclone can be size-checked by digesting the plasmid at available restriction cut sites on the standard cassette vector for 1.5 h at 37°C. The digested product can be run through a DNA gel and the size of the insertion product verified with a DNA ladder. If the size of the insertion is correct, a final check should be performed by sending the plasmid out for sequencing to ensure there were no frameshifts or other misligations during the subcloning process.

Removing the Fluorescent Protein Gene

1. Enzyme digestion: The fluorescent protein gene can be easily removed (**Fig. 4.4**) by digesting the plasmid at the blunt-end (e.g., *Pme*I) restriction cut site.
2. Purification: The Invitrogen PureLink PCR Purification kit can be used to purify the digested plasmid.
3. Ligation and transformation: The same procedure can be used as in **Section 3.2.3.1** (steps 5–7) to ligate and transform the digested plasmid.
4. Screening: Since the fluorescent protein gene is removed, non-fluorescent colonies should be picked off the plate and grown overnight. The purification of the new, non-fluorescent plasmid remains the same as in **Section 3.2.3.1**. This new plasmid can be added to the growing cassette library.

Recombination of Cassette Plasmids

The availability of compatible cohesive ends allows cassettes to be combined and recombined irrespective of order, each time resulting in a fusion cassette with the same standard restriction

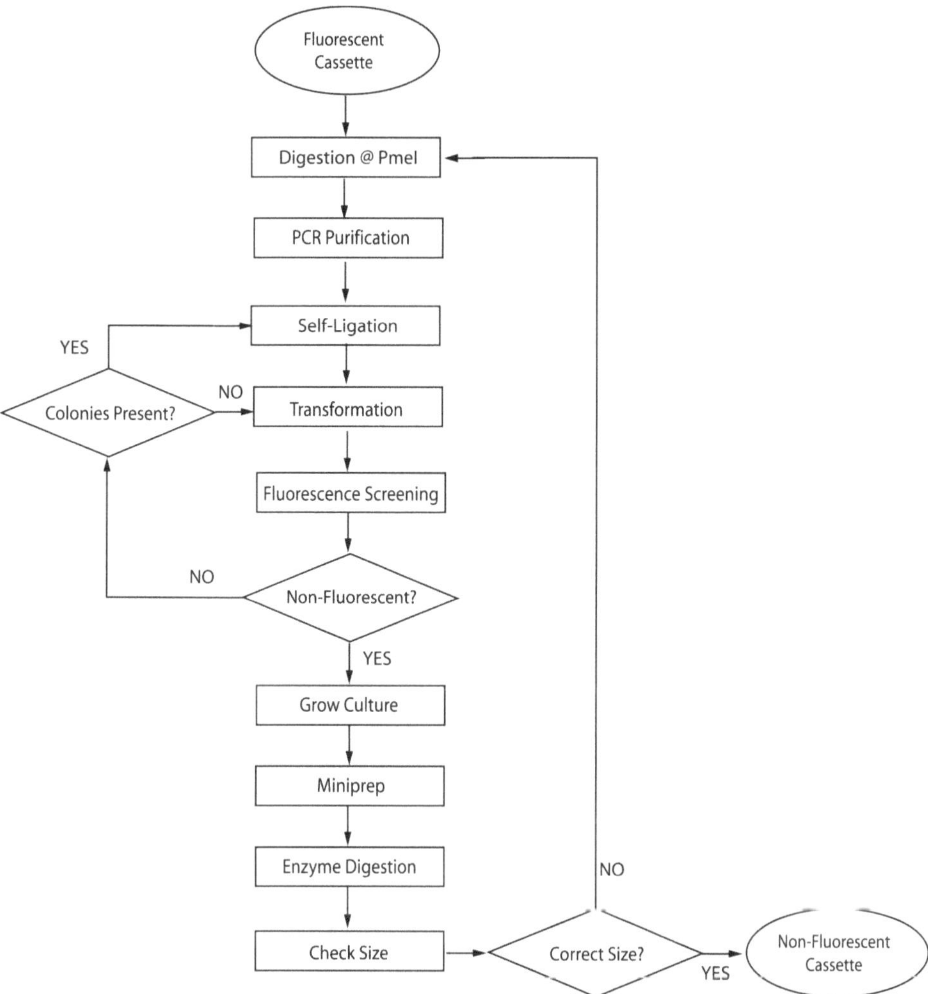

Fig. 4.4. Removal of the fluorescent protein gene. See text for subcloning details.

cut site scheme (**Fig. 4.5a**). There are two ways to create an AB fusion from cassette A and B: ligate the "insert" from cassette B (digested at 2a and 3) to the "host" cassette A (digested at 2b and 3), or ligate the "insert" from cassette A (digested at 1 and 2b) to the "host" cassette B (digested at 1 and 2a) (**Fig. 4.5b**). Creating a BA fusion from cassette A and B follows a similar scheme (**Fig. 4.5c**). Both the resulting AB and BA fusion cassettes will have the same standard vector structure and can be added to the cassette library.

To facilitate easy fluorescence-based screening, the "insert" should be fluorescent, while the "host" should be non-fluorescent (**Fig. 4.6**). Alternatively, both the cassettes can be fluorescent as long as two spectrally different fluorescent genes are used.

1. Digestion of "host" plasmid: The "host" plasmid is digested at restriction cut sites 1 and 2a (the "insert" will be inserted at the 5′-end) or cut sites 2b and 3 (the "insert" will be

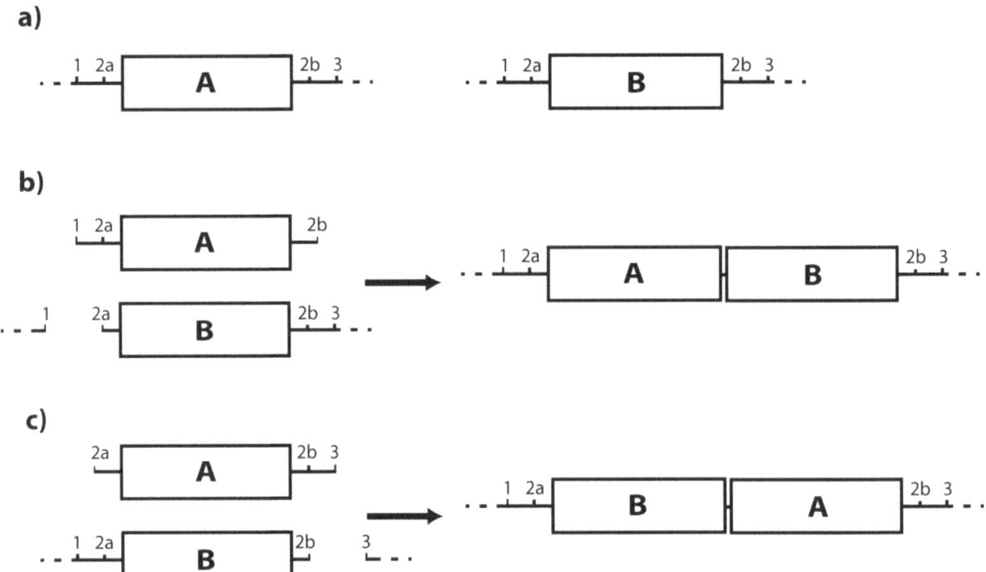

Fig. 4.5. Construction of fusion cassettes. There are two ways to create an AB fusion from available cassettes A and B (**a**). One method is to ligate the insert from cassette A digested at sites 1 and 2b into the host cassette B digested at 1 and 2a (**b**). Ligating the compatible cohesive ends at sites 2a and 2b will result in a sequence unrecognized by either restriction enzyme. A similar method can be used to form a BA fusion (**c**). Any cassette formed from the standard cassette vector will contain the same structure after subcloning.

inserted at the 3′-end). Follow steps 3 and 4 of **Section 3.2.3.1**.

2. Digestion of "insert" plasmid: The insertion fragments should be fluorescent and digested at two restriction cut sites flanking both the gene of interest and the fluorescent protein gene (sites 2a and 3 or 1 and 2b). Step 3 of **Section 3.2.3.1** can be followed to digest the plasmid; however, this digested fragment should be purified differently. The digestion product for this "insert" plasmid will result in both cut insert fragment and the rest of the plasmid with same protruding cohesive ends. To isolate just the insertion fragment, a DNA gel electrophoresis can be run and the insertion fragment excised from the gel. The Invitrogen PureLink Quick Gel Extraction kit can be used to purify the excised gel piece.
3. Ligation and transformation: Follow steps 5–10 of **Section 3.2.3.1**.

3.2.4. Example of Creation of CEcadY12 Ca^{2+} Biosensor

Subcloning CEcadY12 from Cassette Plasmids

1. Following **Section 3.2.3.1**, the following plasmids were previously created and added to the cassette library: pCFPtx (21) contains the gene coding for Cerulean (a variant of CFP) (28); pEcad12vtx (13) contains domains 1 and 2 of epithelial cadherin upstream of Venus.
2. Following **Section 3.2.3.3**, pCFPtx was digested at *Nco*I and *Nhe*I while pEcad12vtx was digested at *Nco*I and

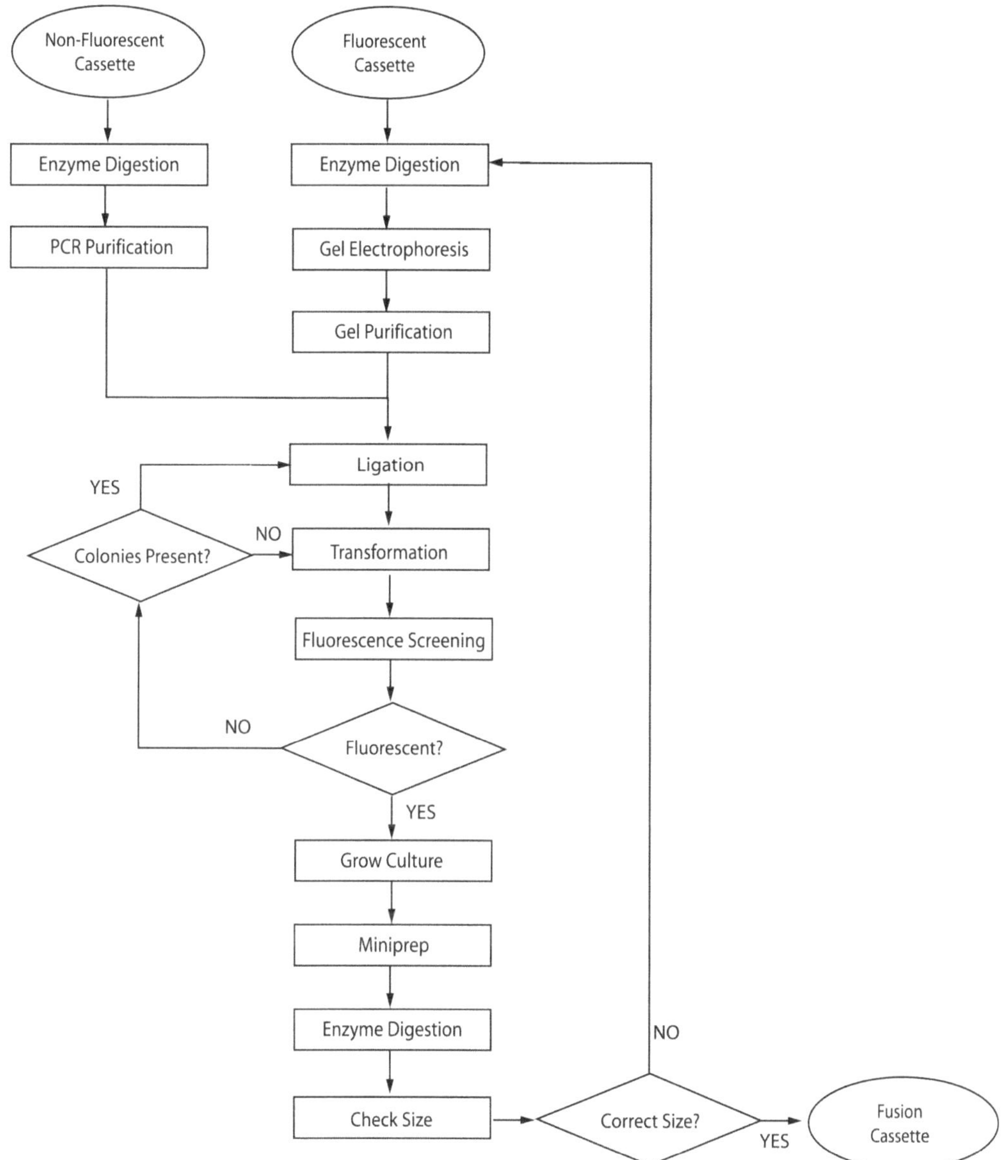

Fig. 4.6. Recombination of existing cassette plasmids into new fusion proteins. See text for subcloning details.

*Spe*I. (Recall: *Spe*I and *Nhe*I produce compatible cohesive ends.)

3. *E. coli* cells were transformed with the ligated plasmid (named pC12Ytx). Colonies with both cyan and yellow fluorescence were screened and selected. pC12Ytx: Cerulean-Ecad12-Venus.

Tagging of Fusion Proteins for Purification (e.g., with Histidine, His, or Glutathione-S-Transferase, GST)

1. The following plasmids were previously created and added to the cassette library: pC12Ytx (13) and pHistx (21), which contains a 6x-His tag.
2. Following **Section 3.2.3.3**, pC12Ytx was digested at *Spe*I and *Xho*I while pHistx was digested at *Nhe*I and *Xho*I.
3. *E. coli* cells were transformed with the ligated plasmid (named pHisC12Ytx). Colonies with both cyan and yellow fluorescence were screened and selected. pHisC12Ytx contains a 6xHis affinity tag that can be used to purify the CEcadY12 biosensor protein using Ni-NTA 6% agarose beads charged with Ni^{2+} ions.

Tagging of Fusion Proteins for Subcellular Localization

1. The following plasmids were previously created and added to the cassette library: pC12Ytx (13) and pTattx (21), which contains a peptide that localizes to the nucleolus and nucleus.
2. Following **Section 3.2.3.3**, pC12Ytx was digested at *Spe*I and *Xho*I while pTattx was digested at *Nhe*I and *Xho*I.
3. *E. coli* cells were transformed with the ligated plasmid (named pTatC12Ytx). Colonies with both cyan and yellow fluorescence were screened and selected. pTatC12Ytx contains a peptide derived from the HIV TAT protein transduction domain that can be used to target the CEcadY12 Ca^{2+} biosensor protein to the nucleolus and nucleus.

3.3. Live-Cell Imaging Using Fluorescence Microscopy

3.3.1. Cell Culture Procedures

These procedures are used for maintaining Cos-7 cells. Dilutions and growth conditions may vary for different cell lines (*see* **Note 8**).

1. Growing cells from stock: Cos-7 cells are stored at –80°C in DMSO. Prior to growing cells in T-25 flasks, thaw cells in a water-bath at 37°C and pre-heat DMEM with 10% FBS growth medium at 37°C. Add 5 mL DMEM with 10% FBS growth medium to T-25 flask. Gently mix thawed cell solution. Slowly add cell solution (10^6–10^7 cells/mL) to T-25 flask dropwise. Gently rock the flask to distribute cells and incubate overnight at 37°C at 5% CO_2. Cells should be about 80–100% confluent by the next day.
2. Passaging cells: Pre-heat DMEM with 10% FBS growth medium, PBS without Ca^{2+}, and trypsin–EDTA at 37°C. Add 5 mL DMEM with 10% FBS growth medium to a new sterile T-25 flask. Remove old growth medium from cell T-25 flask. Gently add 5 mL of PBS to wash cells and remove PBS solution. Add 1 mL of trypsin–EDTA to the flask. Incubate cells for 3 min at 37°C. Rock the flask and tap gently to detach cells from bottom surface. Remove trypsinized cells and place into an eppendorf tube. Centrifuge at 500*g* for

3 min. Remove supernatant, being careful to avoid disturbing the cell pellet. Resuspend cells in 100 μL DMEM with 10% FBS. Add cells at desired dilution ratio to new T-25 flask.

3. Stocking cells: Cells should be at 80–100% confluency. Preheat DMEM with 10% FBS growth medium, PBS without Ca^{2+}, 1 mL of trypsin–EDTA, and 0.5 mL DMSO solution. Remove old growth medium from T-25 flask. Gently add 5 mL of PBS to wash cells and remove PBS solution. Mix 1 mL of trypsin–EDTA solution and add to the flask. Incubate cells for 3 min at 37°C. Rock the flask and tap gently to detach cells from bottom surface. Remove trypsinized cells and place into an eppendorf tube. Centrifuge at 500*g* for 3 min. Remove supernatant, being careful to avoid disturbing the cell pellet. Resuspend cells in 100 μL DMEM with 10% FBS. Remove 50 μL of cell solution and add to a storage tube. Mix DMSO solution and add dropwise to storage tube containing cells. Very gently mix the solution. Freeze storage tube in a styrofoam freezer box at –20°C overnight. The following day, move Styrofoam box to –80°C for long-term storage. Follow step 1 in this section to re-grow cells from stock.

3.3.2. Preparing Cells for Transfection and Imaging

1. Growing cells in glass-bottom wells: When cells are at 80–100% confluency, passage cells as in **Section 3.3.1**. Pre-heat DMEM with 10% FBS growth medium, PBS without Ca^{2+} (Invitrogen), and 1 mL of trypsin–EDTA at 37°C. Add 2 mL DMEM with 10% FBS growth medium to a new sterile glass-bottom well. Remove old growth medium from T-25 flask. Gently add 5 mL of PBS to wash cells and remove PBS solution. Add 1 mL of trypsin–EDTA to flask. Incubate cells for 3 min at 37°C. Rock the flask and tap gently to detach cells from bottom surface. Remove trypsinized cells and place into an eppendorf tube. Centrifuge at 500*g* for 3 min. Remove supernatant, being careful to avoid disturbing the cell pellet.
2. Resuspend cells in 100 μL DMEM with 10% FBS. Add cells at desired dilution ratio to the glass-bottom well. For next day transfection, dilute Cos-7 cells at 1:10.

3.3.3. Transfection

1. Preparing transfection solution: When cells in glass-bottom wells are at 50% confluency, prepare transfection solution using Lipofectin. Mix 2 μg of plasmid with 100 μL of pre-heated DMEM without FBS. In a separate eppendorf tube, mix 10 μL of Lipofectin solution with 100 μL of DMEM without FBS. Incubate for 30–45 min at room temperature. Mix the two solutions gently and incubate for 15–30 min,

allowing plasmid DNA to complex with the Lipofectin solution.

2. Preparing cells for transfection: Carefully remove old growth medium from glass-bottom wells. Wash cells with 2 mL of DMEM without FBS and remove the solution.
3. Adding transfection solution to washed cells: Gently mix 0.8 mL of DMEM without FBS into incubated transfection solution containing Lipofectin and plasmid DNA. Add this solution gently to the glass-bottom well. Incubate at 37°C with 5% CO_2 for 5–8 h.
4. Growing transfected cells: Remove Lipofectin-containing medium and replace with 2 mL of DMEM with 10% FBS growth medium. Grow transfected cells at 37°C with 5% CO_2 for 12–48 h.

3.3.4. Imaging Experiment

1. Imaging experiment: Set up fluorescence microscope with appropriate filters and objectives (see **Section 2.3**, step 4 for microscope specifications). Prepare software settings for fluorescence measurements as appropriate for the imaging experiment to be performed.
2. Preparing cells for imaging (*see* **Note 9**): Remove old growth medium and wash cells with 1 mL PBS. Add 1.8 mL PBS solution to prevent cells from drying out.
3. Preparing stimulus solution: Dilute stimulus stock solution to 10× desired concentration. For example, if a concentration of 1 μM is required to stimulate cells during imaging, prepare a concentrated solution of 10 μM.
4. During imaging experiment: Add 200 μL stimulant dropwise to glass-bottom well, allowing stimulant solution to diffuse across cells (*see* **Note 10**).
5. Fluorescence measurement: Define regions of interest and monitor fluorescence changes for both donor and acceptor emission channels.

3.3.5. Example of Use of FRET Biosensor to Monitor Ca^{2+} Response

1. Ca^{2+} biosensor construct: A troponin-based Ca^{2+} biosensor named C-TNXL-V was amplified from TN-XL (29) and subcloned into our cassette system. In the presence of Ca^{2+}, C-TNXL-V shows an increase in FRET efficiency; this corresponds to a decrease in CFP emission with a corresponding increase in YFP emission.
2. 10 μM ATP or 1 μM ionomycin can be used to stimulate cells during imaging (**Fig. 4.7a**). A plot of the changes in fluorescence intensity for the CFP and YFP channels is shown in **Fig. 4.7b, c**.

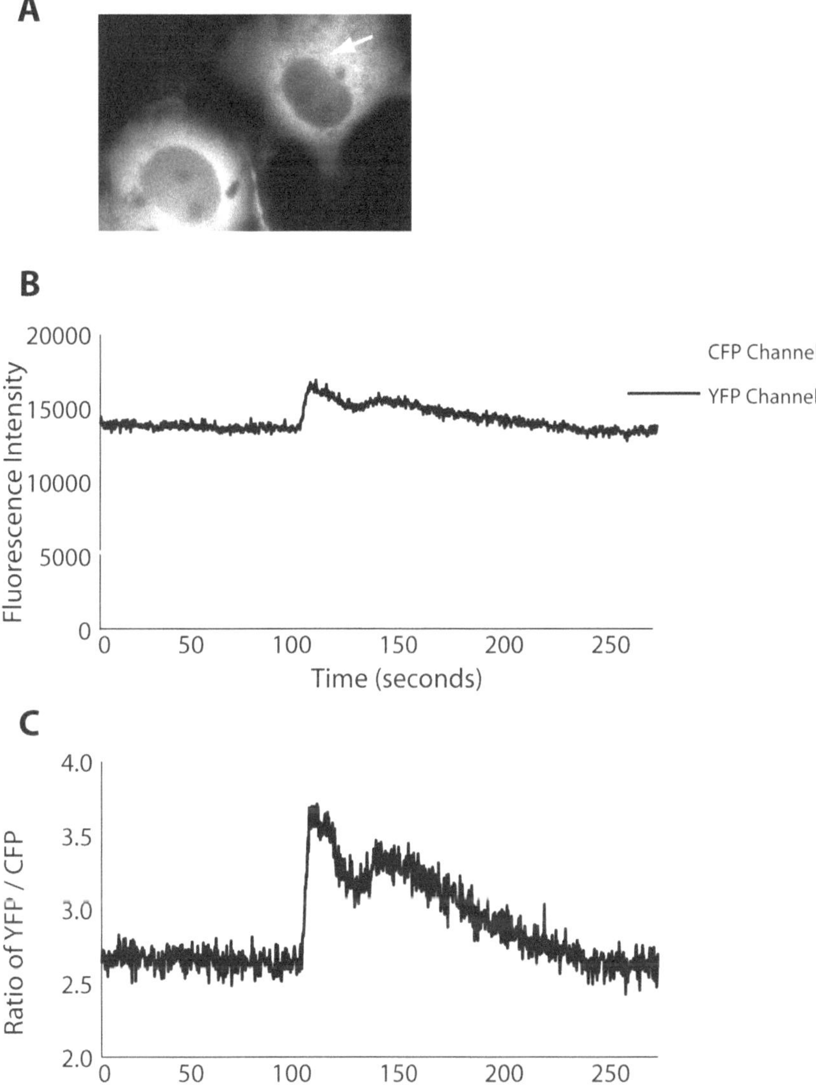

Fig. 4.7. Response of C-TNXL-V biosensor to ATP stimulation. Cos-7 cells were transfected with the C-TNXL-V Ca^{2+} biosensor (**a**). Fluorescence intensities for both the YFP and CFP channels were measured where indicated by an arrow. Ten micromolars of ATP was added to cause a Ca^{2+} response at 100 s (**b**). YFP/CFP ratio changes are also shown (**c**).

3.3.6. Example of Use of FRET Biosensor to Monitor Apoptosis

1. vDEVDc is a caspase-7 FRET biosensor that consists of a CFP and a YFP sandwiching a caspase recognition peptide, Asp-Glu-Val-Asp (DEVD) (16, 18).
2. 5 mM of STS is used to induce apoptosis in Cos-7 cells. During apoptosis, caspase-7, an executioner protease, is activated. In the presence of endogeneous caspase-7, the vDEVDc biosensor is cleaved, separating the fluorescent pair and causing a decrease in FRET efficiency. **Figure 4.8b, c** shows

an increase in CFP emission with a corresponding decrease in YFP emission upon cleavage.

3. Corresponding morphological changes can also be observed, confirming apoptotic changes within the cell (16, 22) (**Fig. 4.8a**).

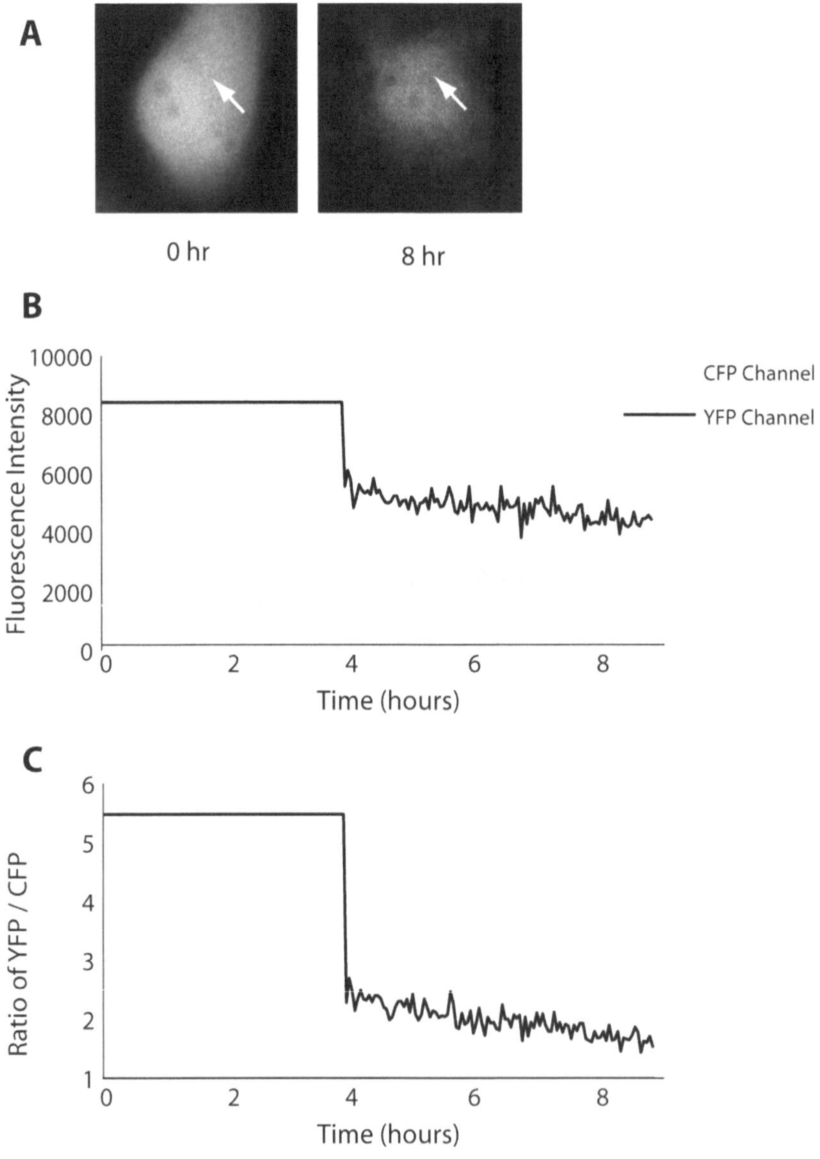

Fig. 4.8. Response of vDEVDc biosensor to STS-induced apoptosis. Cos-7 cells were transfected with the vDEVDc caspase biosensor. Cell images are taken before (*left*) and after (*right*) STS-induced apoptosis (**a**). Fluorescence intensities for both the YFP and CFP channels were measured where indicated by an *arrow* (**b**). YFP/CFP ratio changes are also shown (**c**).

4. Notes

1. Store all enzymes at –20°C. To prevent cross-contamination, we often aliquot the solutions into smaller eppendorf tubes that are clearly labeled. This prevents the contamination of larger volumes.
2. Unless otherwise stated, store growth medium (DMEM with or without FBS) at 4°C and all other cell culture solutions at –20°C. To prevent repeated cycles of thawing and freezing, aliquot these solutions into smaller tubes, for instance, trypsin–EDTA is aliquoted into 1 mL aliquots since this is the amount needed for each passage.
3. Make sure to immediately clean off unused oil from objectives using lens paper sprayed with ethanol. The buildup of oil is much harder to clean off and will affect cell images taken.
4. The number of models generated for each biosensor was deemed sufficient when doubling the number of generated models did not change the overall average obtained (13).
5. Developing a computational tool that outputs the distance and orientation factors in addition to the FRET efficiency allows other biosensor characteristics to be studied. For instance, the conformational change of CEcadY12 involves homodimerization in the presence of Ca^{2+}. Closer inspection of the Ca^{2+}-bound models revealed a shorter, more restrained distance between the CFP and the YFP (91 Å versus 85 Å). This restrained distance is caused by limited rotational freedom of the linkers in the presence of Ca^{2+} as observed from the change in orientation factor from 0.59 to 0.39 (13).
6. We often use Venus instead of other YFP variants because it matures (folds) faster and is a generally brighter variant.
7. Another standard vector that was created is pCfcerutx3, which has the same structure as pCfvtx with the Venus gene replaced by a gene coding for Cerulean, a variant to CFP (28). Instead of the base vector pTriEx 1.1 – Hygro, pTriEx 3 – Hygro (Novagen) was used.
8. Perform all cell-culture steps in a laminar hood to prevent contamination.
9. Leave cells in growth medium until ready to image. Removal of growth medium too early will cause cells to detach from the glass surface. Similarly, it is important to add and remove solutions very slowly and gently to prevent detaching cells.

10. When adding stimulants, be careful not to disturb the surface of the PBS solution, as this will cause a noticeable change in measured fluorescence intensity levels. Similarly, be careful not to touch any part of the microscope setup, as measurements are also sensitive to small vibrations.

Acknowledgments

This work was supported by a fellowship to EP from National Science and Engineering Research Council (NSERC) and grants to KT from the Canadian Foundation of Innovation (#10296), Canadian Institutes of Health Research (#81262), Heart and Stroke Foundation (#NA6241), and the National Science and Engineering Research Council (#283170).

References

1. Griesbeck, O. (2004) Fluorescent proteins as sensors for cellular functions. *Curr Opin Neurobiol.* **14**, 636–41.
2. Li, I. T., Pham, E., and Truong, K. (2006) Protein biosensors based on the principle of fluorescence resonance energy transfer for monitoring cellular dynamics. *Biotechnol Lett.* **28**, 1971–82.
3. Luo, K. Q., Yu, V. C., Pu, Y., and Chang, D. C. (2001) Application of the fluorescence resonance energy transfer method for studying the dynamics of caspase-3 activation during UV-induced apoptosis in living HeLa cells. *Biochem Biophys Res Commun.* **283**, 1054–60.
4. Miyawaki, A. (2003) Visualization of the spatial and temporal dynamics of intracellular signaling. *Dev Cell.* **4**, 295–305.
5. Pozzan, T., Mongillo, M., and Rudolf, R. (2003) The Theodore Bucher lecture. Investigating signal transduction with genetically encoded fluorescent probes. *Eur J Biochem.* **270**, 2343–52.
6. Truong, K., and Ikura, M. (2001) The use of FRET imaging microscopy to detect protein-protein interactions and protein conformational changes in vivo. *Curr Opin Struct Biol.* **11**, 573–8.
7. Kurokawa, K., Mochizuki, N., Ohba, Y., Mizuno, H., Miyawaki, A., and Matsuda, M. (2001) A pair of fluorescent resonance energy transfer-based probes for tyrosine phosphorylation of the CrkII adaptor protein in vivo. *J Biol Chem.* **276**, 31305–10.
8. Clegg, R. M. (1995) Fluorescence resonance energy transfer. *Curr Opin Biotechnol.* **6**, 103–10.
9. Tsien, R. Y. (1998) The green fluorescent protein. *Annu Rev Biochem* **67**, 509–44.
10. Berridge, M. J., Lipp, P., and Bootman, M. D. (2000) The versatility and universality of calcium signalling. *Nat Rev Mol Cell Biol.* **1**, 11–21.
11. Miyawaki, A., Griesbeck, O., Heim, R., and Tsien, R. Y. (1999) Dynamic and quantitative Ca^{2+} measurements using improved cameleons. *Proc Natl Acad Sci USA* **96**, 2135–40.
12. Miyawaki, A., Llopis, J., Heim, R., McCaffery, J. M., Adams, J. A., Ikura, M., and Tsien, R. Y. (1997) Fluorescent indicators for Ca^{2+} based on green fluorescent proteins and calmodulin. *Nature* **388**, 882–7.
13. Pham, E., Chiang, J., Li, I., Shum, W., and Truong, K. (2007) A computational tool for designing FRET protein biosensors by rigid-body sampling of their conformational space. *Structure* **15**, 515–23.
14. Kuboniwa, H., Tjandra, N., Grzesiek, S., Ren, H., Klee, C. B., and Bax, A. (1995) Solution structure of Ca^{2+}-free calmodulin. *Nat Struct Biol.* **2**, 768–76.
15. Wriggers, W., Mehler, E., Pitici, F., Weinstein, H., and Schulten, K. (1998) Structure and dynamics of calmodulin in solution. *Biophys J.* **74**, 1622–39.
16. Li, I. T., Pham, E., Chiang, J. J., and Truong, K. (2008) FRET evidence that an isoform of caspase-7 binds but does not cleave its sub-

strate. *Biochem Biophys Res Commun.* **373**, 325–9.

17. Truong, K., Sawano, A., Mizuno, H., Hama, H., Tong, K. I., Mal, T. K., Miyawaki, A., and Ikura, M. (2001) FRET-based in vivo Ca^{2+} imaging by a new calmodulin-GFP fusion molecule. *Nat Struct Biol.* **8**, 1069–73.
18. Chiang, J. J., and Truong, K. (2006) Computational modeling of a new fluorescent biosensor for caspase proteolytic activity improves dynamic range. *IEEE Trans Nanobiosci* **5**, 41–5.
19. Xu, X., Gerard, A. L., Huang, B. C., Anderson, D. C., Payan, D. G., and Luo, Y. (1998) Detection of programmed cell death using fluorescence energy transfer. *Nucleic Acids Res.* **26**, 2034–5.
20. Winters, D. L., Autry, J. M., Svensson, B., and Thomas, D. D. (2008) Interdomain fluorescence resonance energy transfer in SERCA probed by cyan-fluorescent protein fused to the actuator domain. *Biochemistry* **47**, 4246–56.
21. Truong, K., Khorchid, A., and Ikura, M. (2003) A fluorescent cassette-based strategy for engineering multiple domain fusion proteins *BMC Biotechnol.* **3**, 8.
22. Chiang, J. J., and Truong, K. (2005) Using co-cultures expressing fluorescence resonance energy transfer based protein biosensors to simultaneously image caspase-3 and Ca^{2+} signaling. *Biotechnol Lett.* **27**, 1219–27.
23. Violin, J. D., Zhang, J., Tsien, R. Y., and Newton, A. C. (2003) A genetically encoded fluorescent reporter reveals oscillatory phosphorylation by protein kinase C *J Cell Biol.* **161**, 899–909.
24. Presley, J. F., Cole, N. B., Schroer, T. A., Hirschberg, K., Zaal, K. J., and Lippincott-Schwartz, J. (1997) ER-to-Golgi transport visualized in living cells. *Nature* **389**, 81–5.
25. Dale, R. E., Eisinger, J., and Blumberg, W. E. (1979) The orientational freedom of molecular probes. The orientation factor in intramolecular energy transfer. *Biophys J.* **26**, 161–93.
26. Hillel, Z., and Wu, C. W. (1976) Statistical interpretation of fluorescence energy transfer measurements in macromolecular systems. *Biochemistry* **15**, 2105–13.
27. Nagai, T., Ibata, K., Park, E. S., Kubota, M., Mikoshiba, K., and Miyawaki, A. (2002) A variant of yellow fluorescent protein with fast and efficient maturation for cell-biological applications. *Nat Biotechnol.* **20**, 87–90.
28. Rizzo, M. A., Springer, G. H., Granada, B., and Piston, D. W. (2004) An improved cyan fluorescent protein variant useful for FRET. *Nat Biotechnol.* **22**, 445–9.
29. Mank, M., Reiff, D. F., Heim, N., Friedrich, M. W., Borst, A., and Griesbeck, O. (2006) A FRET-based Ca^{2+} biosensor with fast signal kinetics and high fluorescence change. *Biophys J.* **90**, 1790–6.

Chapter 5

Synthetic Fluorescent Probes for Imaging of Peroxynitrite and Hypochlorous Acid in Living Cells

Dan Yang, Zhen-Ning Sun, Tao Peng, Hua-Li Wang, Jian-Gang Shen, Yan Chen, and Paul Kwong-Hang Tam

Abstract

Peroxynitrite ($ONOO^-$) and hypochlorous acid (HOCl) are two highly reactive oxygen species generated in biological systems. The overproduction of peroxynitrite or hypochlorous acid is implicated in a broad array of human pathologies including vascular, immunological, and neurodegenerative diseases. However, unambiguous detection of these reactive oxygen species has been relatively difficult due to their short biological half-lives and multiple reaction pathways. Based on their specific chemical reactions, we have developed fluorescent probes **HKGreen-1** and **HKOCl-1** for highly sensitive detection of peroxynitrite and hypochlorous acid, respectively. Both probes have been demonstrated to be able to discriminate corresponding reactive species from other reactive oxygen and nitrogen species (ROS and RNS) in not only chemical systems but also biological systems. The endogenous production of peroxynitrite in neuronal cells under oxygen-glucose deprivation (OGD) conditions has been visualized for the first time by utilizing **HKGreen-1** probe, whilst the endogenous production of hypochlorous acid in macrophage cells upon stimulation with LPS, IFN-γ, and PMA has been imaged by utilizing **HKOCl-1** probe.

Key words: Fluorescent probe, **HKGreen-1**, **HKOCl-1**, peroxynitrite, hypochlorous acid, reactive oxygen species, reactive nitrogen species, neuronal cell, oxygen-glucose deprivation, macrophage cell.

1. Introduction

Peroxynitrite ($ONOO^-$), a short-lived highly reactive species, is formed in vivo by the diffusion-controlled reaction between overproduced nitric oxide ($^{\bullet}NO$) and superoxide ($O_2^{\bullet -}$) radicals (1–4). Accumulating evidence has suggested that peroxynitrite

D.B. Papkovsky (ed.), *Live Cell Imaging*, Methods in Molecular Biology 591,
DOI 10.1007/978-1-60761-404-3_5, © Humana Press, a part of Springer Science+Business Media, LLC 2010

contributes to the pathogenesis of many human diseases including ischemic reperfusion injury, circulatory shock, diabetes, chronic heart failure, stroke, myocardial infarction, cancer, chronic inflammatory diseases, and neurodegenerative disorders (5, 6). Unfortunately, comprehensive investigations and understandings of the pathological and physiological roles of peroxynitrite have been hampered by the difficulty in achieving specific detection of intracellular peroxynitrite for its short biological half-life and multiple reaction pathways (7). Despite numerous efforts in the literature to develop new methods for the detection of peroxynitrite, none of them turns out to be specific and highly sensitive (7–9).

On the basis of a specific chemical reaction (10–12), we have developed a novel fluorescent probe **HKGreen-1** (**Fig. 5.1**) for highly sensitive detection of peroxynitrite (12). The probe **HKGreen-1** has been proven to be applicable for the detection and imaging of peroxynitrite in living cells (12, 13). Probe-loaded primary cultured neuronal cells displayed strong fluorescence signals after treatment with peroxynitrite donor 3-morpholinosydnonimine hydrochloride (SIN-1), whereas negligible fluorescence signals were observed in the cells upon treatment with nitric oxide donor *S*-nitroso-*N*-acetyl-*DL*-penicillamine (SNAP) or superoxide donor (xanthine/xanthine oxidase). Furthermore, by utilizing **HKGreen-1** probe, visualization of endogenous peroxynitrite production in neuronal cells was achieved for the first time under oxygen-glucose deprivation (OGD) conditions (14, 15).

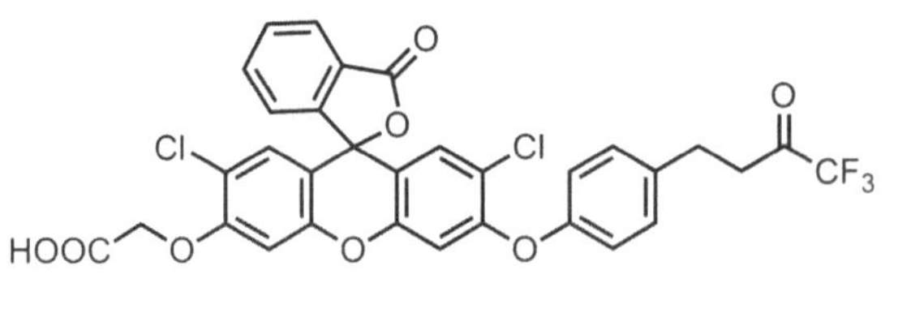

HOOC O O O O Cl Cl O O O CF_3

HKGreen-1

OMe HO N B N F F Et_2NOC $CONEt_2$

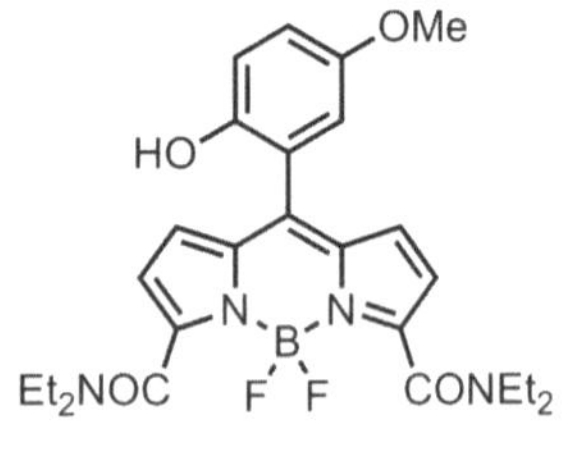

HKOCl-1

Fig. 5.1. Chemical structures of fluorescent probes **HKGreen-1** (12) and **HKOCl-1** (24).

Hypochlorous acid (HOCl, pK_a = 7.5) or its basic form, hypochlorite (OCl^-), is another reactive species present in cells. It is generated predominantly in activated leukocytes, including neutrophils, macrophages, and monocytes (16), by myeloperoxidase (MPO)-catalyzed oxidation of chloride ions in the presence of hydrogen peroxide. In biological systems, hypochlorous acid functions mainly in the prevention of microbial invasion as neither bacteria nor mammalian cells can neutralize its toxic effects due to a lack of enzymes for catalytic detoxification (17). Many studies have pointed to the role of hypochlorous acid in the pathogenesis of human diseases including cardiovasculopathies, osteoarthritis, and neurodegeneration (16, 18–21). Two red fluorescent probes for hypochlorous acid were reported recently (22, 23), though issues concerning their sensitivity remain to be addressed. We have developed a green fluorescent probe **HKOCl-1** (**Fig. 5.1**) for highly sensitive and selective detection of hypochlorous acid based on its specific reaction with *p*-methoxyphenol (24). By utilizing **HKOCl-1** probe, endogenous production of hypochlorous acid in macrophage cells was visualized after exposure to stimuli such as lipopolysaccharide (LPS), interferon-γ (IFN-γ), and phorbol 12-myristate 13-acetate (PMA) (24).

2. Materials

2.1. Cell Culture

1. Pregnant Sprague-Dawley rats (Harlan).
2. Neuronal cell growth medium: Neurobasal medium (Gibco) supplemented with 2% B27 (Gibco), 0.5 mM glutamine, 100 units/mL penicillin, and 100 μg/mL streptomycin (all Sigma-Aldrich).
3. Neurobasal medium (Gibco) supplemented with 2% B27 minus antioxidants (B27 minus AO; Gibco).
4. Oxygen-glucose deprivation (OGD) medium: Dulbecco's modified Eagle's medium (DMEM) without glucose (Gibco) gassed with 94% N_2/1% O_2/5% CO_2 for 15 min.
5. Macrophage cell growth medium: Dulbecco's modified Eagle's medium (DMEM; Gibco) containing 10% fetal bovine serum (FBS; Gibco) supplemented with 100 units/mL penicillin (Sigma-Aldrich) and 100 μg/mL streptomycin (Sigma-Aldrich).
6. RAW 264.7 macrophage cells (ATCC).
7. T-75 culture flasks (Corning).
8. Teflon cell scrapers (Thermo Fisher).
9. Poly-L-lysine-coated 6-well culture plates (BD Biosciences).
10. 35 mm cover-slip dishes (MatTek).
11. Phosphate-buffered saline (PBS; 1×, Gibco).

2.2. Imaging of Peroxynitrite Production with Probe HKGreen-1

1. Acetonitrile (CH_3CN; ≥99.93%, biotech. grade, Sigma-Aldrich).
2. Dimethyl sulfoxide (DMSO; ≥99.9%, for molecular biology, Sigma-Aldrich).

3. **HKGreen-1** is synthesized according to published procedures (12). Prepare a 2 mM stock solution of **HKGreen-1** (F.W. = 659.39) in CH_3CN by dissolving 2.0 mg of the probe crystals in 1.517 mL of anhydrous acetonitrile (*see* **Note 1**).
4. Dihydroethidium (HE; Invitrogen). Prepare a 4 mM stock solution of HE (F.W. = 315.41) in DMSO by dissolving 1.3 mg of HE in 1.030 mL of anhydrous DMSO (*see* **Note 2**).
5. 3-Morpholinosydnonimine hydrochloride (SIN-1; Sigma-Aldrich), dissolved at 1 mM in water immediately before the experiment.
6. *S*-Nitroso-*N*-acetyl-*DL*-penicillamine (SNAP; Sigma-Aldrich), dissolved at 1 mM in DMSO immediately before the experiment.
7. Xanthine and xanthine oxidase (X/XO; Sigma-Aldrich).

2.3. Imaging of Hypochlorous Acid Production with Probe HKOCl-1

1. *N,N*-dimethylformamide (DMF; ≥99.93%, biotech. grade, Sigma-Aldrich).
2. **HKOCl-1** is synthesized according to published procedures (24). Prepare a 2 mM stock solution of **HKOCl-1** (F.W. = 512.24) in DMF by dissolving 1.0 mg of the probe crystals in 0.976 mL of anhydrous DMF (*see* **Note 3**).
3. Lipopolysaccharide (LPS; purified by phenol extraction, Sigma-Aldrich), dissolved at 1 mg/mL in 1× PBS immediately before the experiment.
4. Interferon-γ (IFN-γ; ≥98%, Sigma-Aldrich), dissolved at 50 μg/mL in 1× PBS immediately before the experiment.
5. Phorbol 12-myristate 13-acetate (PMA; ≥99%, molecular biology grade, Sigma-Aldrich), dissolved at 20 μg/mL in DMSO immediately before the experiment.

3. Methods

3.1. Imaging of Exogenous Peroxynitrite with HKGreen-1 in Primary Cultured Neuronal Cells

1. Sprague-Dawley rats were used with permission from local health authorities and maintained in compliance with the principles set forth in the "Guide for Care and Use of Laboratory Animals." Primary cortex neurons from embryonic day 15 Sprague-Dawley rats were isolated as described in the literature (14, 25).
2. Seed the suspension of dispersed neuronal cells at a density of 2 × 10^6 cells/well on poly-L-lysine coated 6-well

culture plates or 35 mm cover-slip dishes with neuronal cell growth medium. Maintain the cells in a humidified incubator at 37°C, in 5% CO_2-95% air. After 4 days, replace half of the growth medium with fresh medium. Change growth medium twice a week. On day 9, change growth medium to neurobasal medium supplemented with B27 minus AO. On day 10, use cultured cortical cells for imaging experiments.

3. Retrieve plates or dishes from the incubator. Discard the old medium. Add fresh medium (1.980 mL; neurobasal medium supplemented with 2% B27 minus AO) and incubate the plates or dishes for at least 1 h.
4. Add 20 μL of **HKGreen-1** stock solution to each well to give a final concentration of 20 μM in the medium. Incubate for 15 min at 37°C to load cells with the probe (*see* **Note 4**).
5. Take out the cells from the incubator and wash them with 1× PBS three times. Treat the cells with 10 μM SIN-1 or 10 μM SNAP or 100 μM of xanthine plus 0.1 IU of xanthine oxidase (X/XO) in 2 mL PBS for 15 min (*see* **Notes 5** and **6**).
6. Wash the cells again with 1× PBS. After washing, add 2 mL of 1× PBS to each sample (*see* **Note 5**).
7. Observe cells under a fluorescent microscope. Use bright-field illumination to obtain phase-contrast images of cells after focusing on them. Record fluorescence images by using excitation at 488 nm. Examples of the fluorescence images are shown in **Fig. 5.2** (*see* **Notes 7** and **8**).

3.2. Imaging of Endogenous Peroxynitrite Production with HKGreen-1 in Primary Cultured Neuronal Cells Under Hypoxia Conditions

1. Prepare neuronal cells on imaging plates or dishes (*see* **Section 3.1**). Change to fresh medium (1.978 mL) 1 h before the experiments.
2. Add 20 μL of **HKGreen-1** and 2 μL of HE stock solutions to each well to give final concentrations of 20 and 4 μM, respectively. Incubate for 15 min at 37°C to load cells with the probes.
3. Take out the cells from the incubator and gently wash them with 1× PBS twice. Then incubate the cells in OGD medium (2 mL) at 37°C for indicated time (2 h or 12 h) in a humidified hypoxia chamber equilibrated with 94% N_2/1% O_2/5% CO_2.
4. At the end of OGD, wash the cells twice with 1× PBS and observe them under a fluorescent microscope. Use bright-field illumination to obtain phase-contrast images of cells after focusing on them. Record fluorescence images by using excitation at 488 nm. Detect green fluorescence at 510 nm produced by **HKGreen-1** oxidation and the red

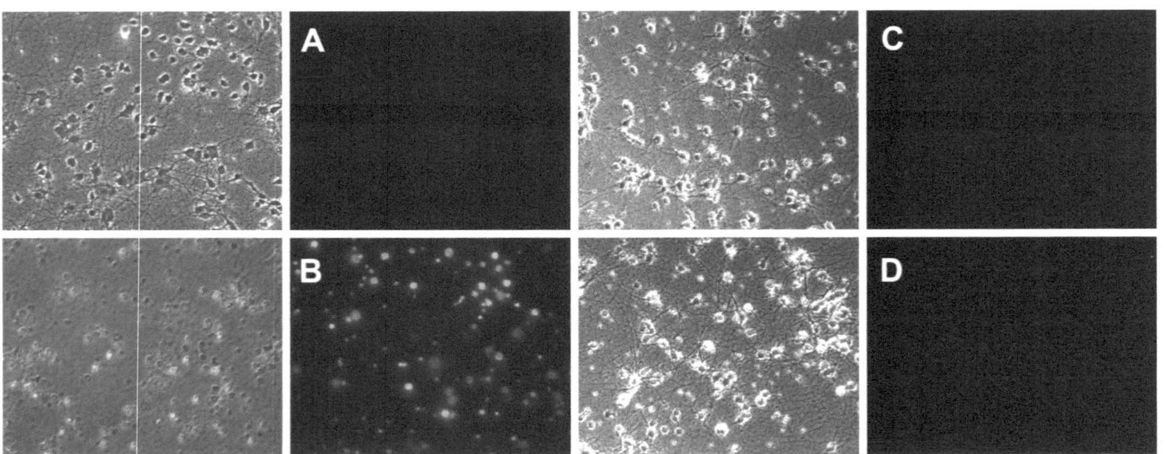

Fig. 5.2. Imaging of primary cultured neuronal cells: phase contrast (*left*) and fluorescence (*right*). Neuronal cells were incubated with **HKGreen-1** (20 μM) for 15 min and then subjected to different treatments: (**a**) Control; (**b**) 10 μM of SIN-1; (**c**) 10 μM of SNAP; (**d**) 100 μM of xanthine plus 0.1 IU of xanthine oxidase. (Reprinted with permission from (12), Copyright 2006 American Chemical Society.).

fluorescence at 590 nm produced by HE oxidation. Examples of fluorescence images are shown in **Fig. 5.3** (*see* **Notes 7**, **8**, and **9**).

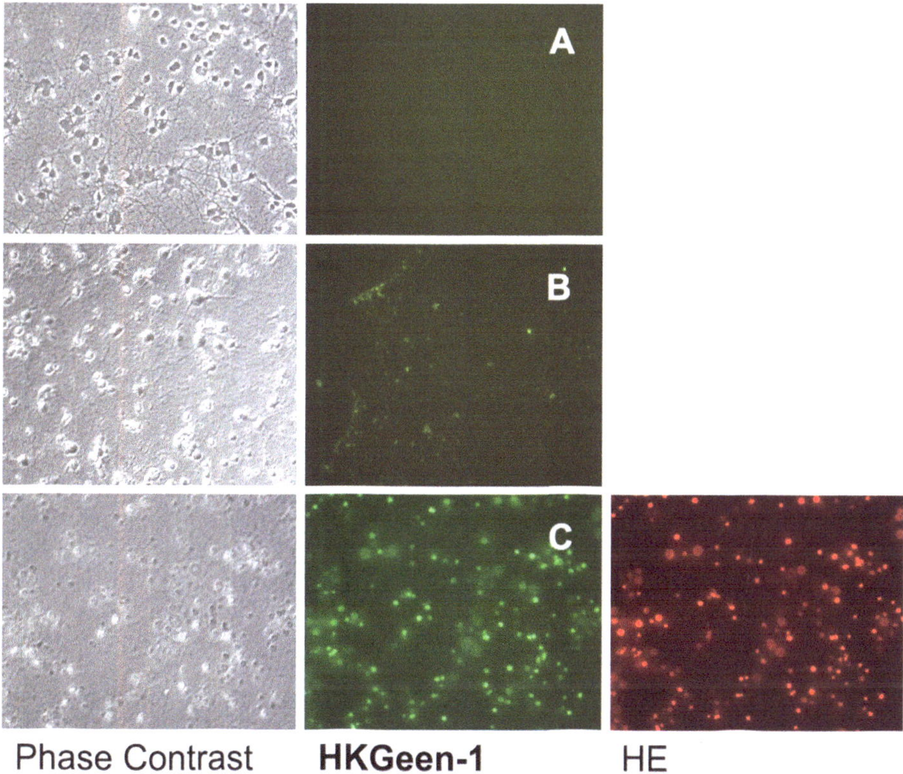

Fig. 5.3. Imaging of primary cultured neuronal cells: phase contrast (*left*); fluorescence via green channel (*middle*); fluorescence via red channel (*right*). Neuronal cells were incubated with **HKGreen-1** and HE (20 μM and 4 μM, respectively) for 15 min and then subjected to OGD condition: (**a**) Control; (**b**) OGD for 2 h; (**c**) OGD for 12 h.

3.3. Imaging of Endogenous Hypochlorous Acid Production with HKOCl-1 in Macrophage Cells

1. Culture RAW 264.7 macrophage cells in DMEM containing 10% FBS supplemented with 100 U/mL of penicillin and 100 μg/mL streptomycin at 37°C in 5% CO_2. Subculture cells by scraping and seeding them in T-75 flasks. Change growth medium every two or three days. Grow cells to confluence for imaging experiments.
2. Seed macrophage cells at a density of 2×10^6 cells/well on 35 mm cover-slip dishes with serum-free DMEM. After 24 h, change to fresh serum-free DMEM (1.996 mL) and incubate the cells for 1 h.
3. Take out the cells from the incubator. Add 2 μL of LPS stock solution and 2 μL of IFN-γ stock solution (final concentrations of 1 μg/mL and 50 ng/mL in medium, respectively) and incubate cells for further 4 h (*see* **Note 10**).

4. Take out the cells from the incubator. Add 1 μL of PMA stock solution (final concentration 20 μg/mL) and incubate for 30 min (*see* **Note 11**).
5. Add 20 μL of **HKOCl-1** stock solution to each well to give a final concentration of 20 μM in the medium and incubate for 1 h to load cells with the probe (*see* **Note 12**).
6. Take out the cells from the incubator. Wash them gently with 1× PBS twice and then add 1 mL of 1× PBS.
7. Observe the cells under a fluorescent microscope. Use bright-field illumination to obtain phase-contrast images of cells after focusing on them. Record green fluorescence images by using excitation at 488 nm. Examples of fluorescence images are shown in **Fig. 5.4** (*see* **Note 7**).

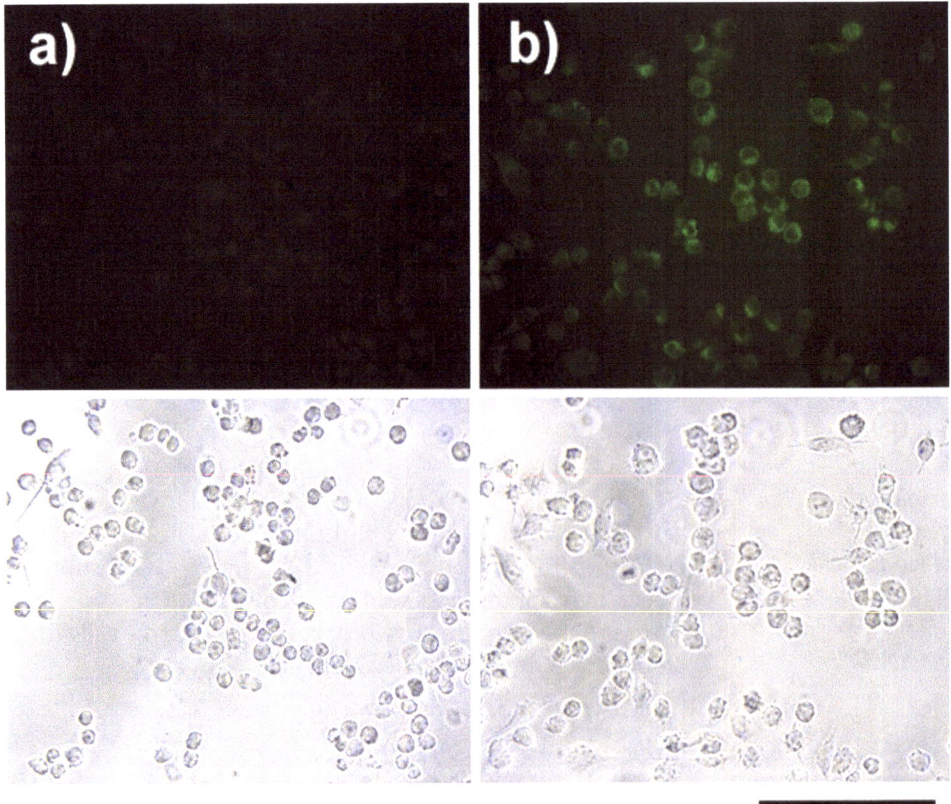

Fig. 5.4. Images of RAW 264.7 macrophages: phase contrast (*lower*); fluorescence (*upper*). Cells were treated with various stimulants and then incubated with **HKOCl-1** (20 μM) for 1 h. (**a**) Control; (**b**) LPS (1 μg/mL) and IFN-γ (50 ng/mL) for 4 h, then PMA (10 nM) for 0.5 h. (Reprinted with permission from (24), Copyright 2008 American Chemical Society.)

4. Notes

1. Crystalline form of probe **HKGreen-1** can be kept in a glass vial wrapped with aluminum foil for several months at –20°C. Use CH_3CN of high purity to make stock solutions, which can be stored at –20°C in the dark for no more than 3 weeks. Note that CH_3CN is harmful and highly flammable.
2. Use DMSO of high purity to make HE stock solutions, which can be stored at –20°C in the dark for no more than 3 weeks (26).
3. Crystalline form of probe **HKOCl-1** can be kept in a glass vial wrapped with aluminum foil for several months at –20°C. Use DMF of high purity to make stock solutions, which can be stored at –20°C in the dark for no more than 3 weeks. Note that DMF is harmful.
4. Incubation time and concentration of **HKGreen-1** need to be carefully optimized, since optimal conditions may vary for different cell types.
5. Warm PBS solution to 37°C before use and wash the cells gently.
6. Use freshly prepared solutions of SIN-1, SNAP, xanthine, and xanthine oxidase. SIN-1 solid should be kept at –20°C. Purchase SIN-1 in small packages as its quality declines after opening. The duration of incubation of cells with SIN-1 need to be optimized since simultaneous generation of superoxide and nitric oxide from SIN-1 is slow.
7. Image cells immediately after washing them with PBS. Choose healthy cells based on their morphology for imaging.
8. Adjust illumination intensity and exposure time for the sample to minimize photobleaching of **HKGreen-1**.
9. Duration of light exposure and image acquisition of HE probe should be minimized to prevent HE photo-oxidation, which can generate fluorescent products (26).
10. Stock solutions of LPS can be stored at 4°C in the dark for no more than 2 months. Stock solutions of IFN-γ can be stored at –20°C in the dark for no more than 2 months.
11. Stock solutions of PMA can be stored at –20°C in the dark for no more than 2 months.
12. Incubation time and concentration of **HKOCl-1** need to be carefully optimized, since optimal conditions may vary for different cell types.

Acknowledgments

The authors would like to thank The University of Hong Kong, Hong Kong Research Grants Council (HKU7060/05P), the Area of Excellence Scheme (AoE/P-10-01) established under the University Grants Committee (HKSAR), and Morningside Foundation for financial support.

References

1. Beckman, J. S., Beckman, T. W., Chen, J., Marshall, P. A. and Freeman, B. A. (1990) Apparent hydroxyl radical production by peroxynitrite: Implications for endothelial injury from nitric oxide and superoxide. *Proc Natl Acad Sci USA* **87**, 1620–1624.
2. Gryglewski, R. J., Palmer, R. M. J. and Moncada, S. (1986) Superoxide anion is involved in the breakdown of endothelium-derived vascular relaxing factor. *Nature* **320**, 454–456.
3. Huie, R. E. and Padmaja, S. (1993) The reaction of NO with superoxide. *Free Radic Res* **18**, 195–199.
4. Pryor, W. A. and Squadrito, G. L. (1995) The chemistry of peroxynitrite: A product from the reaction of nitric oxide with superoxide. *Am J Physiol Lung Cell Mol Physiol* **268**, 699–722.
5. Szabo, C., Ischiropoulos, H. and Radi, R. (2007) Peroxynitrite: Biochemistry, pathophysiology and development of therapeutics. *Nat Rev Drug Discov* **6**, 662–680.
6. Pacher, P., Beckman, J. S. and Liaudet, L. (2007) Nitric oxide and peroxynitrite in health and disease. *Physiol Rev* **87**, 315–424.
7. Radi, R., Peluffo, G., Alvarez, M. N., Naviliat, M. and Cayota, A. (2001) Unraveling peroxynitrite formation in biological systems. *Free Radic Biol Med* **30**, 463–488.
8. Gomes, A., Fernandes, E. and Lima, J. L. F. C. (2006) Use of fluorescence probes for detection of reactive nitrogen species: A review. *J Fluoresc* **16**, 119–139.
9. Wardman, P. (2007) Fluorescent and luminescent probes for measurement of oxidative and nitrosative species in cells and tissues: Progress, pitfalls, and prospects. *Free Radic Biol Med* **43**, 995–1022.
10. Yang, D., Tang, Y. C., Chen, J., Wang, X. C., Bartberger, M. D., et al. (1999) Ketone-catalyzed decomposition of peroxynitrite via dioxirane intermediates. *J Am Chem Soc* **121**, 11976–11983.
11. Yang, D., Wong, M. K. and Yan, Z. (2000) Regioselective intramolecular oxidation of phenols and anisoles by dioxiranes generated in situ. *J Org Chem* **65**, 4179–4184.
12. Yang, D., Wang, H. L., Sun, Z. N., Chung, N. W. and Shen, J. G. (2006) A highly selective fluorescent probe for the detection and imaging of peroxynitrite in living cells. *J Am Chem Soc* **128**, 6004–6005.
13. Zhou, G., Li, X., Hein, D. W., Xiang, X., Marshall, J. P., et al. (2008) Metallothionein suppresses angiotensin II-induced nicotinamide adenine dinucleotide phosphate oxidase activation, nitrosative stress, apoptosis, and pathological remodeling in the diabetic heart. *J Am Coll Cardiol* **52**, 655–666.
14. Furuichi, T., Liu, W., Shi, H., Miyake, M. and Liu, K. J. (2005) Generation of hydrogen peroxide during brief oxygen-glucose deprivation induces preconditioning neuronal protection in primary cultured neurons. *J Neurosci Res* **79**, 816–824.
15. Yang, D., Wang, H. L. and Shen, J. G. (2008) Unpublished Results in Dan Yang's Research Group
16. Yap, Y. W., Whiteman, M. and Cheung, N. S. (2007) Chlorinative stress: An under appreciated mediator of neurodegeneration?. *Cell Signal* **19**, 219–228.
17. Lapenna, D. and Cuccurullo, F. (1996) Hypochlorous acid and its pharmacological antagonism: An update picture. *Gen Pharmacol* **27**, 1145–1147.
18. Steinbeck, M. J., Nesti, L. J., Sharkey, P. F. and Parvizi, J. (2007) Myeloperoxidase and chlorinated peptides in osteoarthritis: Potential biomarkers of the disease. *J Orthop Res* **25**, 1128–1135.
19. Sugiyama, S., Kugiyama, K., Aikawa, M., Nakamura, S., Ogawa, H., et al. (2004) Hypochlorous acid, a macrophage product, induces endothelial apoptosis

and tissue factor expression: Involvement of myeloperoxidase-mediated oxidant in plaque erosion and thrombogenesis. *Arterioscler Thromb Vasc Biol* **24**, 1309–1314.

20. Sugiyama, S., Okada, Y., Sukhova, G. K., Virmani, R., Heinecke, J. W., et al. (2001) Macrophage myeloperoxidase regulation by granulocyte macrophage colony-stimulating factor in human atherosclerosis and implications in acute coronary syndromes. *Am J Pathol* **158**, 879–891.
21. Hoy, A., Leininger-Muller, B., Kutter, D., Siest, G. and Visvikis, S. (2002) Growing significance of myeloperoxidase in non-infectious diseases. *Clin Chem Lab Med* **40**, 2–8.
22. Kenmoku, S., Urano, Y., Kojima, H. and Nagano, T. (2007) Development of a highly specific rhodamine-based fluorescence probe for hypochlorous acid and its application to real-time imaging of phagocytosis. *J Am Chem Soc* **129**, 7313–7318.
23. Shepherd, J., Hilderbrand, S. A., Waterman, P., Heinecke, J. W., Weissleder, R., et al. (2007) A fluorescent probe for the detection of myeloperoxidase activity in atherosclerosis-associated macrophages. *Chem Biol* **14**, 1221–1231.
24. Sun, Z. N., Liu, F. Q., Chen, Y., Tam, P. K. H. and Yang, D. (2008) A highly specific BODIPY-based fluorescent probe for the detection of hypochlorous acid. *Org Lett* **10**, 2171–2174.
25. Brewer, G. J. (1995) Serum-free B27/neurobasal medium supports differentiated growth of neurons from the striatum, substantia nigra, septum, cerebral cortex, cerebellum, and dentate gyrus. *J Neurosci Res* **42**, 674–683.
26. Robinson, K. M., Janes, M. S. and Beckman, J. S. (2008) The selective detection of mitochondrial superoxide by live cell imaging. *Nat Protocols* **3**, 941–947.

Chapter 6

Photo-Activatable Probes for the Analysis of Receptor Function in Living Cells

Wen-Hong Li

Abstract

Photo-activatable (caged) probes are powerful research tools for biological investigation. The superb maneuverability of a light beam allows researchers to activate caged probes with pinpoint accuracy. Recent developments in caging chemistry and two-photon excitation technique further enhance our capability to perform photo-uncaging with even higher spatial and temporal resolution, offering new photonic approaches to study cell signaling dynamics in greater detail. Here we present a sample method that combines the techniques of photo-activation and digital fluorescence microscopy to assay an important class of intracellular receptors for the second messenger D-*myo*-inositol 1,4,5-trisphosphate (Ins(1,4,5)P_3, or IP_3). The imaging assay is performed in fully intact living cells using a caged and cell membrane permeable ester derivative of IP_3, cm-IP_3/PM.

Key words: IP_3, IP_3 receptors, IP_3R, cell permeable ester of IP_3, caged IP_3, two-photon uncaging, calcium signaling, calcium imaging.

1. Introduction

Caged probes are molecules whose biological, biochemical, or physicochemical activities are masked by light-sensitive protecting groups. The parent molecule of a caged compound usually contains functional groups (carboxylate, phosphate, amine, hydroxy, phenoxy, amide, etc.) that are crucial for its function. Caging these functionalities reduces or eliminates the activity of the parent molecule, yet its activity can be abruptly restored with a flash of light (typically ultraviolet or UV light). Because light beams can be precisely guided to targeted areas at the time of our choice,

D.B. Papkovsky (ed.), *Live Cell Imaging*, Methods in Molecular Biology 591,
DOI 10.1007/978-1-60761-404-3_6, © Humana Press, a part of Springer Science+Business Media, LLC 2010

photo-uncaging offers the advantage of high spatiotemporal definition for controlling dynamics of cellular biochemistry. Since the pioneering work of Kaplan and Hoffman who first invented caged ATP to study cellular ATPase (1), a large number of caged probes have been developed and applied to biological and biochemical research (2, 3).

To assay receptor functions in cells, caged ligands for both cell surface and intracellular receptors have been developed. For example, caged glutamate (4) is widely used to control the activation of glutamate receptors on the plasma membrane of neurons; caged D-*myo*-inositol 1,4,5-trisphosphate (IP_3) (5, 6) and caged cyclic ADP ribose (7) were developed to study Ca^{2+} release from intracellular Ca^{2+} stores gated by the IP_3 receptors (IP_3Rs) or ryanodine receptors, respectively. Ideally, caged probes for the analysis of receptor functions in living cells need to meet a number of requirements: (1) be inert, i.e., neither activating nor inactivating cellular receptors or proteins; (2) be reasonably soluble in aqueous solution and biocompatible (non-toxic); (3) have high photolysis efficiency by UV or two-photon excitation so that only a small dose of light is needed for photo-activation, thus minimizing potential photo-damage to live cells; (4) have favorable photolysis kinetics so that the parent molecule can be rapidly released to fully activate its receptor prior to inducing desensitization; (5) non-invasive delivery of caged compounds into cells if their targets are located intracellularly.

cm-IP_3/PM (*see* **Fig. 6.1**), a caged and cell membrane permeable ester of IP_3, meets these requirements and is ideally suitable for studying dynamics and functions of Ca^{2+} release from intracellular Ca^{2+} stores gated by IP_3Rs. This compound is neutral, with three phosphates protected by six propionyloxymethyl (PM) esters. The PM ester masks the negative charge of phosphate and conveys lipophilicity to the molecule to allow cm-IP_3/PM diffuse passively across cell membranes. Once inside cells, the PM ester is hydrolyzed by ubiquitous cellular esterases to produce cm-IP_3, a caged IP_3 analogue that remains trapped inside cells. In cm-IP_3, the 6-hydroxy of *myo*-inositol is caged by 4,5-dimethoxy-2-nitrobenzyl group (DMNB). Because 6-hydroxy plays a crucial role in the interaction between IP_3 and its receptors, cm-IP_3 has negligible binding to IP_3Rs. Photolysis of the DMNB group frees 6-hydroxy and generates m-IP_3, a highly potent IP_3 analogue with 2- and 3-hydroxies protected by a methoxymethylene group (**Fig. 6.1**). M-IP_3 binds to IP_3Rs with an affinity about 75% of that of IP_3 (6), and it is rapidly metabolized in cells at a rate very close to that of natural IP_3 (8). These properties make m-IP_3 and its caged precursor cm-IP_3 ideal pharmacological reagents for controlling the activity of IP_3Rs and for studying the regulation and function of IP_3-Ca^{2+} signaling pathway in living cells.

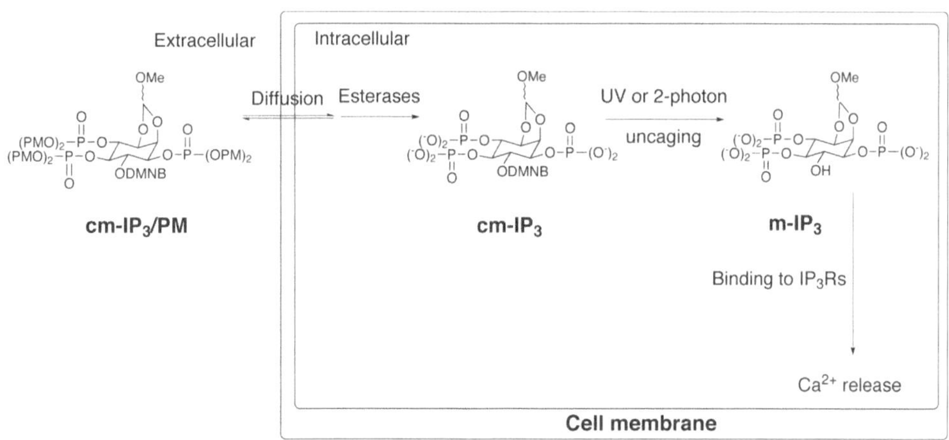

Fig. 6.1. Chemical structures of m-IP_3, cm-IP_3 (caged m-IP_3), and cm-IP_3/PM (neutral PM ester of cm-IP_3) and their mode of actions in living cells.

Intracellular Ca^{2+} activities ($[Ca^{2+}]_i$) control a variety of essential biological processes, and IP_3 is an important and ubiquitous second messenger that regulates Ca^{2+} homeostasis. IP_3 is hydrolyzed from phosphatidylinositol 4,5-bisphosphate (PIP_2) by phospholipase C, which is activated by cell surface receptors upon ligand stimulation. PIP_2 also releases diacylglycerol that activates protein kinase C. Since methods relying on endogenous mechanisms to produce IP_3 also activate many branching signaling events, it is difficult to dissect the function of IP_3-Ca^{2+} signaling branch. Moreover, to examine the spatial heterogeneity of IP_3 sensitivity or Ca^{2+} release activity of IP_3Rs, it would require elevating IP_3 concentration selectively in subcellular areas.

Since two-photon excitation only occurs at the focal point of a focusing objective, two-photon uncaging achieves very high spatial selectivity of photo-activation with resolution of about 1 μm^3 (9). This technique is particularly useful for studying cell signaling dynamics in three dimensions because areas above or below the focus are not excited. Biological preparations including dissected tissues, organotypic cell cultures, or living model organisms are ideal subjects for the application of this technique. In addition, two-photon uncaging offers a unique and powerful experimental approach to the analysis of subcellular heterogeneity of receptor distribution, receptor activation, and functional consequences of localized signaling events.

To monitor $[Ca^{2+}]_i$ in living cells, a number of fluorescent indicators are available for Ca^{2+} imaging. These sensors vary in their chemical composition (small synthetic dyes and genetically encoded protein sensors), Ca^{2+} affinity (from sub-micromolar to over tens of micromolar), excitation and emission wavelengths (UV excitable to red emitting), dynamic range of signal change (from less than twofold to more than 100-fold), and cellular localization (10, 11). A suitable Ca^{2+} indicator should be chosen based on the biological question to be addressed and the biological preparation to be used. In this protocol, I describe how to apply cm-IP_3/PM to activate IP_3Rs in living cells by photo-uncaging, using cultured HeLa cells as a model system. $[Ca^{2+}]_i$ is monitored with Fluo-3, a long wavelength and high affinity Ca^{2+} sensor ($K_d = 0.39\ \mu M$) that exhibits a more than 100-fold fluorescence enhancement upon binding Ca^{2+}. Detailed procedures are provided for cell loading, imaging of Fluo-3 by digital wide-field and confocal laser scanning microscopy, UV and two-photon uncaging, data processing, and measuring the amount of released m-IP_3 using the IP_3 binding assay after photo-uncaging.

2. Materials

2.1. Reagents and Solutions

1. cm-IP_3/PM (synthesized from *myo*-inositol according to (6)) prepared as stock solution (1 or 2 mM) in dimethyl sulfoxide (DMSO). The stock solution is stable for at least 4 months if stored at –20°C. High purity anhydrous DMSO (Sigma-Aldrich) should be used to minimize the hydrolysis of the PM ester during storage. Vials containing the compound should be wrapped in aluminum foil to avoid light exposure, and kept on ice during usage.
2. Fluo-3/AM (Invitrogen), a membrane-permeable fluorescent calcium indicator with the excitation maximum near 490 nm, prepared as 1 mM stock solution in dry DMSO (stored at –20°C, stable for at least 6 months).
3. DMSO stock (10% w/v) of Pluronic F127 (BASF, cat. no. 583106). It is stable for at least several months if stored at room temperature.
4. Histamine (Sigma) (5 mM stock solution in water), a compound which stimulates Ca^{2+} release in HeLa cells by activating phospholipase C to produce endogenous IP_3, is used it as a control to check that cells are functional in IP_3-Ca^{2+} signaling.
5. Ionomycin (LC Laboratories, Woburn, MA), a Ca^{2+} ionophore which raises cellular Ca^{2+} to a high level ($\sim 1\ \mu M$), is used to check Ca^{2+} responsiveness of Fluo-3 at the end of an imaging experiment.
6. 2-Aminoethoxydiphenyl borate (Aldrich) and Xestospongin C (Sigma) are membrane permeable inhibitors of IP_3Rs.

7. Hank's balanced salt solution (HBS) (Gibco, cat. no. 14065, diluted ten times with water) supplemented with 20 mM HEPES buffer, adjusted to pH 7.3 with NaOH or HCl and filtered through a 0.2-μm sterile filter.
8. Ca^{2+}-free DPBS buffer (Gibco cat no. 14190) supplemented with 5.5 mM glucose, 1 mM $MgCl_2$, and 20 mM HEPES, pH 7.3, and filtered through a 0.2-μm sterile filter.
9. Amersham IP_3 [^{3}H] Biotrak Assay kit (GE Healthcare Life Sciences).
10. 1 M aqueous solution of trichloroacetic acid (Aldrich).
11. Mixture of trichlorotrifluoroethane and trioctylamine (3 vol: 1 vol, both from Aldrich).

2.2. Cell Culture

1. HeLa cells (American Tissue Culture Collection) cultured in Dulbecco's modified Eagle's medium (DMEM, Gibco) supplemented with 10% fetal bovine serum (FBS, Gemini Bio-Products) and 1% penicillin/streptomycin (Sigma-Aldrich). Cells are passed regularly when they reach high density (>80%). During passage, cells are detached from culture dishes by trypsinization (0.25% trypsin and 1 mM EDTA, Gibco/BRL) for several minutes at 37°C.
2. For imaging, cells are seeded on glass-bottom imaging dishes (MatTek, cat. no. P35G-0-10-C, 3.5 cm diameter) at low to medium density (< 50%).

2.3. Imaging and Uncaging Equipment

1. Axiovert 200 M inverted fluorescence microscope (Carl Zeiss) for wide-field Ca^{2+} imaging and UV uncaging. The scope is equipped with a cooled CCD camera (ORCA-ER, Hamamatsu), a 40× oil-immersion objective (Fluar, 1.3 NA, Carl Zeiss), and an excitation source for rapid wavelength switching (Lambda DG-4, with a 175 W Xenon lamp, Sutter Instrument). Equivalent imaging platforms and hardware from other manufacturers may also be used.
2. Optical filters for imaging Fluo-3: 480 nm ± 20 nm (excitation filter), 535 nm ± 25 nm (emission filter), and a long-pass 505 nm beamsplitter coated with UV reflection material to facilitate UV uncaging (Chroma Technology or Omega Optical).
3. Open-Lab imaging software (http://www.improvision.com/products/openlab/) for controlling image acquisition, uncaging, and post-acquisition analysis. Similar softwares from other vendors can be used.
4. Hand-held UV lamp B-100 AP (UVP, Upland, CA) for global uncaging of all cells in a culture dish.

5. IL 1700 Research Radiometer with SED033 detector (International Light Inc., Newburyport, MA) for measuring light intensity of the hand-held UV lamp.
6. Electronic timed mechanical shutter (Uniblitz, Model VMM-T1; Vincient Associates, Rochester, NY) for gating UV exposure.
7. Zeiss LSM510 imaging system (Carl Zeiss) equipped with a 30-mW Argon laser, a Chameleon-XR laser (Coherent), and a 40× oil-immersion objective (Fluar, 1.3 NA, Carl Zeiss) for laser scanning confocal imaging and two-photon uncaging. Other equivalent imaging platforms can be used.
8. Power meter with PM 10 sensor (FieldMate, Coherent) for measuring the average power of femtosecond pulsed laser for two-photon excitation.

3. Methods

All caged compounds including cm-IP_3/PM and its cellular hydrolysis product cm-IP_3 should be protected from room light throughout an experiment, and never be exposed to day light. Use a red safety light (available from local hardware stores) to provide illumination when handling caged compounds.

All the experiments in this protocol are carried out at room temperature (~25°C) unless specified otherwise.

3.1. Cellular Loading of cm-IP_3/PM and Fluo-3/AM

Since it takes 15 steps to synthesize cm-IP_3/PM, consumption of this valuable material should be minimized. The following procedure uses approximately 0.4 nmol or 0.5 μg of cm-IP_3/PM per loading, and it works well for a number of cultured cell lines, including HeLa, HEK293 human embryonic kidney cells, NIH3T3 fibroblasts, and 1321N1 astrocytoma cells. Optimal loading conditions should be empirically determined for each cell type.

1. To prepare loading solution, mix DMSO stock solutions of cm-IP_3/PM (1 mM, 0.4 μL), Fluo-3/AM (1 mM, 0.3 μL), and pluronic (10%, 0.3 μL) in a 0.6-mL centrifuge tube. Then add 0.1 mL of HBS solution and vortex it briefly (*see* **Notes 1, 2, 3**).
2. Remove culture medium from an imaging dish containing cultured cells using a disposable Pasteur pipette. Gently rinse cells twice with 1× HBS solution. After the second rinse, remove residual HBS solution from the dish and carefully wipe off cells from plastic surface with a piece of folded

Kimwipe, without touching cells on the central part of the dish. This leaves cells only on the glass surface (*see* **Note 4**).

3. Gently add 0.1 mL of HBS on top of the glass surface to cover the cells, then add 0.1 mL of loading solution from step 1. Cover the dish with a lid to minimize water evaporation (*see* **Note 5**).
4. Incubate cells in the loading solution in dark for 20–45 min. Incubation time should be adjusted based on the cell type and confluence: longer incubation loads more probe into cells, and higher cell confluence requires longer incubation time.
5. Remove the loading solution with a Pasteur pipette and rinse cells once with HBS. Add 1.5 mL HBS to the dish and incubate cells in the dark for another 10 min to allow complete hydrolysis of the AM and PM esters in cells.

3.2. Wide-Field Imaging of Fluo-3 and Global UV Uncaging of cm-IP$_3$

Fluo-3 and other Ca^{2+} indicators may gradually lose their Ca^{2+} responsiveness upon intense, prolonged excitation. When attempting Ca^{2+} imaging for the first time, acquisition parameters such as intensity of excitation light, exposure time, acquisition frequency, and the duration of imaging experiments should be optimized. To check the Ca^{2+} responsiveness of Ca^{2+} indicators, Ca^{2+} ionophores such as ionomycin (2–10 μM) are used to raise $[Ca^{2+}]_i$ to fairly high levels. If the indicator remains sensitive to Ca^{2+} at the end of an experiment, it should respond to ionomycin similarly as a freshly loaded indicator. However, if it loses Ca^{2+} sensitivity after excessive excitation, the amplitude of ionomycin-stimulated Ca^{2+} signal is seen to be reduced.

1. Place an imaging dish on the microscope stage. Bring cells into focus by observing cellular Fluo-3 signal. If loading is successful, fluorescence should be dim but visible, and uniformly distributed inside cells.
2. To perform global UV uncaging through a 40× objective, choose a field containing 5–15 cells. Adjust the exposure time of the CCD camera to obtain a Fluo-3 image. In resting or unstimulated cells, $[Ca^{2+}]_i$ is low and Fluo-3 signal is fairly weak. In our set-up, we typically set the exposure time at 50–200 ms so that the average cellular Fluo-3 intensity is 2–3 times above the background (signal in cells without Fluo-3). Since excess illumination of Fluo-3 diminishes its Ca^{2+} responsiveness, we do not recommend using long exposure time or high intensity excitation light to bring up image intensity.
3. Start image acquisition by acquiring Fluo-3 signal every 5–10 s for about a minute. Baseline signal should be stable. Just prior to uncaging, increase acquisition frequency

to $\geq$ 1 image every 2 s. Quickly switch the filter to UV excitation to photolyze cm-IP_3 then back to image acquisition immediately. A successful uncaging should produce enough m-IP_3 to stimulate Ca^{2+} release that is detectable by Fluo-3. The optimal uncaging duration can only be determined empirically, since it depends on a number of variables including UV light output from the excitation source, UV transmission efficiency of the imaging system, amount of cm-IP_3 loaded into cells, cellular expression level of IP_3Rs, IP_3 sensitivity of different IP_3R isoforms, etc. In our set-up, the uncaging duration ranges from tens of milliseconds to over a second depending on the magnitude of Ca^{2+} increase which we aim to generate (*see* **Note 6**).

4. Continue imaging $[Ca^{2+}]_i$ fluctuations until Fluo-3 signal drops back to its basal level. M-IP_3, once being generated from cm-IP_3 by photolysis, is rapidly metabolized in cells by cellular phosphatases (8), so it typically induces Ca^{2+} transients that last less than a minute. When $[Ca^{2+}]_i$ returns to the resting level, decrease the frame rate to $\leq$ 1 image every 5 s to minimize Fluo-3 excitation.
5. Steps 3 and 4 can be repeated multiple times to produce many Ca^{2+} spikes mimicking natural Ca^{2+} oscillations (6, 8). Higher doses of UV light should be used for subsequent episodes of uncaging to compensate for the consumption of cm-IP_3 in previous photolysis. To reliably generate repetitive Ca^{2+} spikes in the same cells, it is necessary to control the extent of UV photolysis and avoid producing too much m-IP_3 in any single uncaging event.
6. cm-IP_3 loaded into cells is metabolically stable for at least 8 h at the room temperature (6). This allows multiple uncaging experiments to be conducted on the same dish of loaded cells, by moving each time to a different imaging field. We usually use cells that have been kept in HBS for less than 3 h. To check the health of these cells and to confirm that they still maintain robust IP_3-Ca^{2+} signaling at the end of an experiment, a cell surface receptor agonist (histamine, for example) can be added to activate phospholipase C to produce endogenous IP_3 which raise $[Ca^{2+}]_i$. Healthy cells should respond to agonist stimulation like freshly loaded cells.
7. To confirm that photo-released m-IP_3 induces Ca^{2+} release from intracellular stores, replace HBS with Ca^{2+}-free DPBS solution. Similar Ca^{2+} transients should be observed in Ca^{2+}-free solutions after photolyzing cm-IP_3.
8. To confirm that m-IP_3 induced Ca^{2+} release is from intracellular stores gated by IP_3Rs, IP_3R antagonists including

heparin, 2-aminoethoxydiphenyl borate, and Xestospongin C can be applied. These compounds are expected to block the effect of m-IP_3 (12). The later two drugs are membrane permeable, and their cellular application is straightforward, whereas heparin requires microinjection. Since these drugs affect multiple cellular targets including ryanodine receptors, sarco/endoplasmic reticulum Ca^{2+}-ATPase (SERCA) and various ion channels of plasma membranes, none of them should be considered as specific inhibitors of IP_3Rs.

3.3. Localized Two-Photon Uncaging of cm-IP_3 and $[Ca^{2+}]_i$ Imaging by Laser Scanning Microscopy

Caged probes based on the traditional 2-nitrobenzyl caging group typically are not very sensitive to two-photon photolysis, and there have been a fair amount of efforts devoted to developing new caging chemistries suitable for two-photon photolysis (13–18). In cm-IP_3/PM, the caging group (DMNB) has a relatively low two-photon uncaging cross-section, on the order of 0.01 Goeppert-Mayer (GM, 1 GM = 10^{-50} cm^4s/photon) (19). However, since cm-IP_3 is loaded into cells at sub-millimolar concentrations, and because IP_3 binds to IP_3Rs with nanomolar affinity (20, 21), only a small fraction (~0.1%) of loaded cm-IP_3 needs to be photolyzed within the two-photon excitation volume to locally activate IP_3Rs. This is an important consideration that eliminates the need of using high doses of laser light to uncage, thus avoiding cell damage. Even though two-photon excitation is restricted to the focal area, femtosecond laser pulses can still cause substantial photobleaching and cell injury at high power levels (22, 23).

For clarity and simplicity, we illustrate with cultured cells the procedure of two-photon uncaging of cm-IP_3 and confocal imaging of Fluo-3, though the protocol is applicable to studying IP_3-Ca^{2+} signaling in tissues or other biological preparations exhibiting three-dimensional architecture. In addition to confocal imaging, two-photon laser scanning microscopy can be combined with two-photon uncaging in the same experiment (*see* **Note 7**).

1. Load cells with cm-IP_3 and Fluo-3 using the procedure described in **Section 3.1**. Place the imaging dish on the microscope stage and bring the cells into focus by observing cellular Fluo-3 signal.
2. When using Zeiss LSM510 system, configure it as shown in **Fig. 6.2** and start with the settings given in **Table 6.1**. During two-photon uncaging of cm-IP_3, set laser power at the specimen at ~10 mW (determined with a power meter placed just above the objective).
3. Acquire a z-stack of confocal images of Fluo-3 using 488 nm excitation. Like in the wide-field imaging, baseline signal of Fluo-3 at resting $[Ca^{2+}]_i$ is fairly weak. Do not use high laser power to enhance image intensity.

4. Choose a cell to perform two-photon uncaging. Adjust the objective focus to a z-plane that cuts across the middle of the cell. Take a single confocal image at this height and use this image to define the area for two-photon uncaging.
5. Open the "Bleach Control" module in the LSM510 imaging software. Define the uncaging area using the "Define Region" function. Depending on the goal of experiments, the dimension of the uncaging area can be set from less than 1 μm^2 to larger than tens of μm^2. The uncaging area shown in **Fig. 6.3** corresponds to a circle of about 3.5 μm in diameter.
6. Using "Laser Control" function, set the wavelength of the Chameleon XR laser to 730 nm. Under the "Bleach Control," set the uncaging wavelength to 730 nm. During two-photon uncaging, the laser repeatedly scans through the defined uncaging area. The scanning repetition can be defined using the "Iterations" function. We typically set this value between 10 and 20 when the average power of laser input at the specimen is near 10 mW (*see* **Note 8**).
7. Use "Time Series" module of LSM510 imaging software to perform a time-lapse imaging experiment. Define acquisition frequency, timing of photo-activation and length of experiments in the "Time Series." The automation first acquires a number of confocal images of Fluo-3 at the defined frequency, then two-photon uncages cm-IP_3 at 730 nm, and then continues imaging Fluo-3 to follow m-IP_3-stimulated $[Ca^{2+}]_i$ elevation.
8. To follow $[Ca^{2+}]_i$ fluctuations in three dimensions in dissected tissues or in other physiological preparations after localized two-photon uncaging, set up a z-stack by centering the uncaged area along *z*. Use "Time Series" automation to acquire 4D images (*xyz-t*) of Fluo-3 before and after photoactivation.
9. Perform additional episodes of two-photon uncaging by repeating steps 7 or 8 to uncage the same or a different area in the same cell, or other areas in different cells.

3.4. Post-acquisition Data Analysis

To analyze $[Ca^{2+}]_i$ fluctuations before and after uncaging cm-IP_3, quantify Fluo-3 intensity in cells using data generated from the digital microscopy. Changes in Fluo-3 intensity are normalized against baseline signal at resting $[Ca^{2+}]_i$. This analysis can be applied to data obtained by both wide-field and laser scanning imaging.

1. Open a time-lapse Fluo-3 image sequence. Draw regions of interest (ROI) in selected cells.

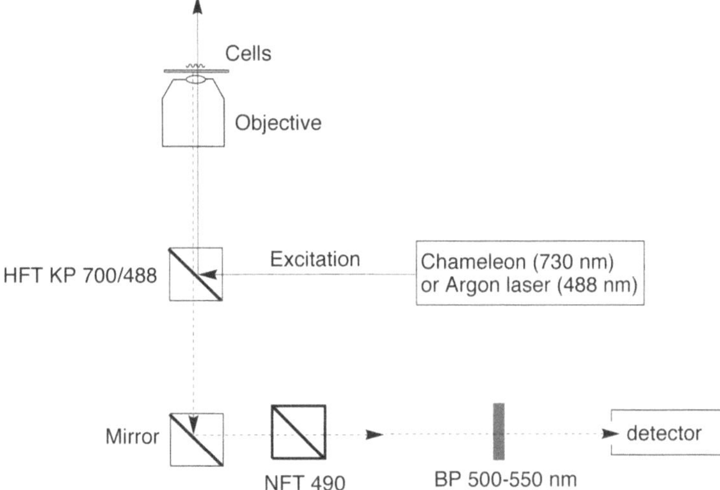

Fig. 6.2. Optical configuration of LSM 510 for confocal laser scanning imaging of Fluo-3 and two-photon uncaging. *Solid* and *dashed lines* represent excitation and emission light paths, respectively.

Table 6.1
Example settings of Zeiss LSM510 for confocal imaging of Fluo-3 and two-photon uncaging

Variable	Setting
Objective	40 × 1.3 NA (Plan-Neofluar)
Digital zoom	2.5
Laser	Argon laser emitting at 458, 477, 488, and 514 nm
Laser power	≤ 0.5% of total laser output (30 mW)
Image dimension (XY)	256 × 256
Z stepping size	1 μm
Scan speed	9 (3.2 μs pixel time)
Scan mode	Unidirectional
Amplifier offset	0.1
Amplifier gain	2.5
Averaging	1
Detector gain	≤ 650
Pinhole diameter	200

2. Measure the time course of the average fluorescence intensity (F_t) of these ROI using OpenLab, ImageJ, or other equivalent imaging software. Also measure the fluorescence intensity of an area that contains no cells as the background signal (F_b).

3. Subtract the background signal from the fluorescence signals of all ROI. Plot $(F_t-F_b)/(F_0-F_b)$ against time, where F_t is the fluorescence intensity of a ROI at time t, and F_0 is the fluorescence intensity of the same ROI at the start (time 0).

An example of such an analysis is shown in **Fig. 6.3**. The experiment involved two-photon uncaging of a small region in a HeLa cell loaded with cm-IP_3. Ca^{2+} activity rose immediately at the uncaging area, then quickly propagated throughout the cell, and gradually returned to the basal level in about a minute.

3.5. Quantification of Cellular IP_3 Mass After Loading and Uncaging cm-IP_3

To quantify the total amount of cm-IP_3 loaded into cells, or to evaluate how much m-IP_3 is released after a UV flash, we measure the amount of IP_3 in cells using the competitive IP_3 binding assay. Since the IP_3 binding assay typically detects IP_3 mass with nanomolar sensitivity, it is necessary to culture cells at high density.

1. Culture cells in 35–mm-diameter tissue culture dishes until cells become nearly confluent. Wash cells twice with HBS and incubate them in 0.4 mL of HBS. cm-IP_3/PM (2 mM × 1 μL) and pluronic (10%, 1 μL) mixed in 0.1 mL HBS is then added to the cells.
2. Incubate cells on a shaker at room temperature for 1 h. Remove the loading solution by aspiration. Wash cells once with HBS and add 0.5 mL of fresh HBS to each dish. Incubate cells on the shaker for another 40 min to allow complete digestion of PM esters.
3. To measure the total amount of cm-IP_3 loaded into cells, add ice cold trichloroacetic acid (TCA, 0.1 mL of 1 M aqueous solution) to quench cells. Place the dish on ice or in a cold room.
4. Photolyze cm-IP_3 in the cells with a hand-held UV lamp by placing the front of the light bulb approximately 5 cm above the dish. The light intensity reaching the dish surface is typically on the order of 1×10^{-8} E (cm^2 s). This can be measured by ferrioxalate actinometry (24) or using a light power meter. Under these settings, cm-IP_3 is photolyzed almost completely after 10 min of illumination.
5. Alternatively, to quantify how much IP_3 is released after a UV uncaging, cells loaded with cm-IP_3 from step 2 are illuminated with the UV lamp. We typically expose cells to UV light for 6 s, delivered either in one episode or in three episodes (2 s/episode) spaced 1 s apart. To ensure that the same amount of UV light is delivered from run to run, we use a mechanical shutter (Uniblitz) to gate UV exposure.
6. Immediately after UV uncaging, add 0.1 mL of ice-cold TCA (1 M) to cells. Shake the dish briefly to ensure

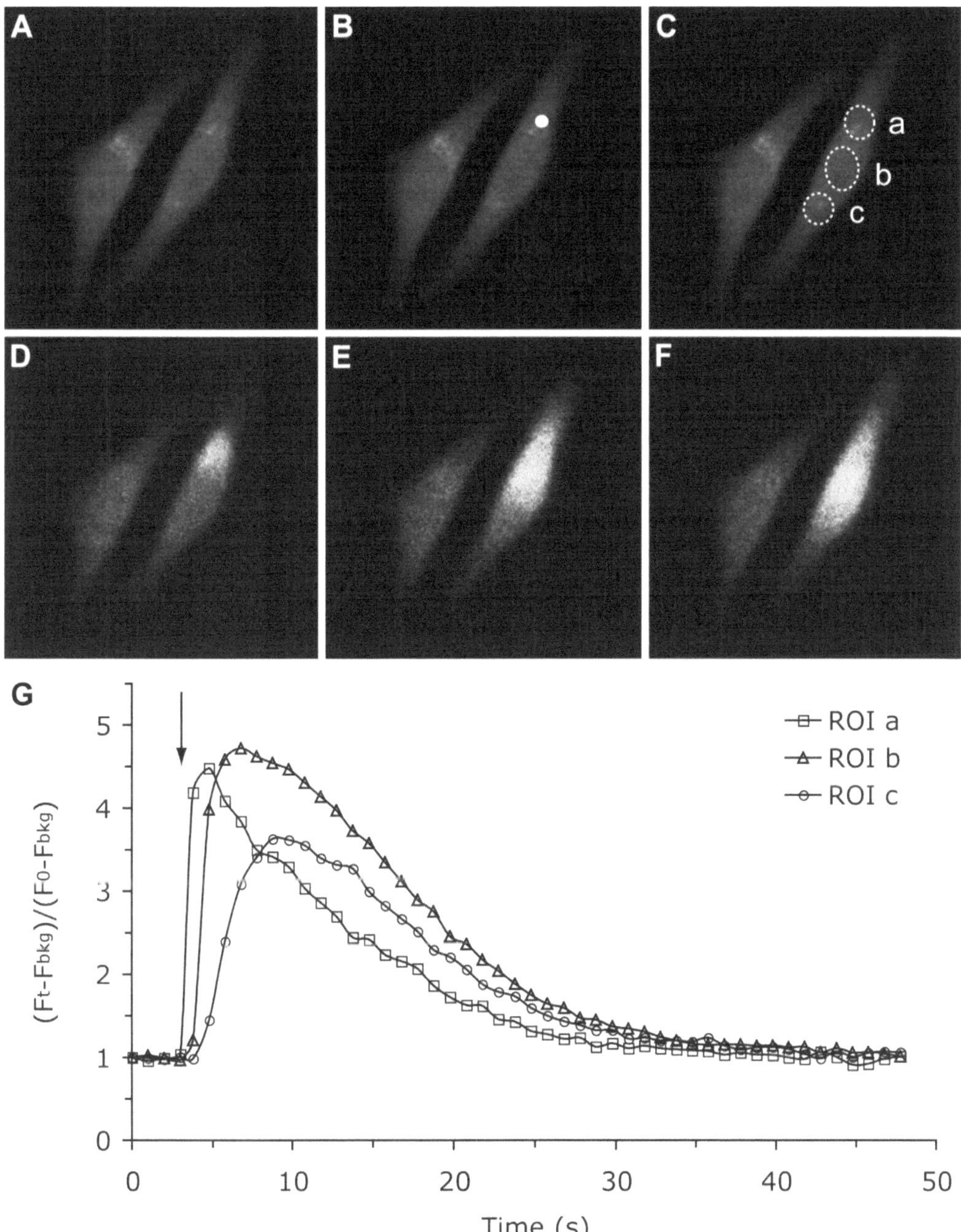

Fig. 6.3. Two-photon uncaging of cm-IP$_3$ and confocal laser scanning imaging of Fluo-3 in cultured HeLa cells. (**a**–**c**) CLSM images (Ex 488 nm, Em 500–550 nm) of HeLa cells. The uncaging area is outlined by the *white circle* in **b**. Three ROI (a, b, and c) are indicated by *dashed circles* in **c**. (**d**–**f**) Fluo-3 images approximately 0.5 s (**d**), 1.5 s (**e**), and 2.5 s (**f**) after the two-photon uncaging (730 nm) of cm-IP$_3$ in the area shown in **b**. (**g**) Time course of Fluo-3 signal normalized against its initial intensity in three ROI shown in **c**. The *arrow* indicates the time of two-photon uncaging. Modified from Figure 6 of (8) with permission from Elsevier.

thorough mixing and place it immediately on ice or in a cold room (*see* **Note 9**).

7. Leave petri dishes from step 4 or step 6 on a shaker in the cold room. After shaking the dish gently for 5 min, collect ~0.45 mL supernatant from each dish and transfer it to a 1.5-mL centrifuge tube. Spin the tubes at 4000 rpm on a bench-top centrifuge for 10 min at 4°C.
8. Transfer 0.4 mL of the supernatant to a new centrifuge tube (1.5 mL or larger). To each tube, add 0.8 mL of trichlorotrifluoroethane-trioctylamine (3 vol:1 vol) solution to neutralize and to extract TCA from the cell lysate. Vortex the tube on high speed for at least 15 s, then spin it at 4000 rpm for 1 min.
9. After centrifugation, carefully collect 0.3 mL of the top aqueous layer and transfer it to a new vial kept on ice. The pH of the solution should be around 4.5.
10. Assay the IP_3 mass in the neutralized cell lysate using the IP_3 binding assay kit, following the instruction from the manufacturer.

4. Notes

1. Pluronic is a non-ionic, mild detergent that helps to solubilize hydrophobic compounds in aqueous solutions and significantly improves the loading efficiency of cm-IP_3/PM and Fluo-3/AM.
2. If the pluronic stock solution turns cloudy during storage, warm it at ~37°C for several minutes.
3. The amount of DMSO in the loading solution should be kept below 0.5% (v/v); therefore, concentrated stock solutions are used.
4. Removal of cells from plastic should be done in less than a minute so that cells on the glass surface remain hydrated.
5. Make sure that peripheral plastic surface is dry before adding the loading solution. The surface tension keeps the solution within the glass area. Otherwise, spilling to the surrounding plastic can decrease the loading efficiency.
6. In addition to a Xenon or mercury lamp, low-cost UV light emitting diodes (25) can also be used to perform photolysis. For applications demanding rapid uncaging of less than 1 ms, strong excitation sources such as pulsed UV lasers or capacitor charged Xe flash lamps (26) should be used.
7. Two-photon laser scanning microscopy has been widely used to image $[Ca^{2+}]_i$ with Fluo-3, Fluo-4, Calcium Green-1, or Oregon Green 488 BAPTA at excitation wavelengths

of 800 nm and above. Since the efficiency of two-photon uncaging of DMNB group decreases rapidly above 750 nm (19), it should be possible to perform two-photon imaging exciting at 800 nm or above without photolyzing cm-IP_3. This would allow integration of two-photon uncaging of cm-IP_3 (near 730 nm) and two-photon imaging of $[Ca^{2+}]_i$ in the same experiment, using two Chameleon lasers set at different wavelengths. We have developed a similar approach to study cell–cell junctional coupling using a caged coumarin dye (15, 27).

8. The total amount of light energy input during two-photon uncaging can be estimated from the average laser power at the specimen and the summed laser exposure time. Using the settings in **Table 6.1** (1 pixel = 0.36 μm × 0.36 μm), the summed laser exposure time for a 3.6 μm square is about 4.8 ms (100 pixels × 3.2 μs pixel dwell time × 15 repetitions), so the total amount of the average light input is 48 μjoule. The actual uncaging duration, however, is much longer and is on the order of seconds. This is due to the mode of operation of laser line scanning in the LSM510 system. Improvements in the newer version of the hardware and software of the LSM510 imaging system makes it possible to shorten the uncaging duration.
9. M-IP_3, once generated from cm-IP_3 by photolysis, is rapidly metabolized in cells within 10 s (8). It is therefore crucial to quench cells with cold TCA solution immediately in order to measure the amount of m-IP_3.

Acknowledgments

We thank the Welch Foundation (I-1510) and the National Institute of Health for financial supports. Imaging experiments involving two-photon excitation were performed at the Live Cell Imaging Core Facility of UT Southwestern, directed by Dr. Kate Luby-Phelps.

References

1. Kaplan, J. H., Forbush, B., 3rd, and Hoffman, J. F. (1978) Rapid photolytic release of adenosine 5′-triphosphate from a protected analogue: utilization by the Na:K pump of human red blood cell ghosts. *Biochemistry* **17**, 1929–35.
2. Marriott, G. (Ed.) (1998) *Caged Compounds, Methods Enzymol*, vol 291, Academic Press, New York.
3. Goeldner, M., and Givens, R. S. (Eds.) (2005) *Dynamic Studies in Biology: Phototriggers, Photoswitches and Caged Biomolecules*, Wiley-VCH Verlag GmbH, Weinheim.
4. Papageorgiou, G., Ogden, D. C., Barth, A., and Corrie, J. E. T. (1999) Photorelease of carboxylic acids from 1-acyl-7-nitroindolines in aqueous solution: Rapid and efficient photorelease of L-glutamate. *J Am Chem Soc* **121**, 6503–4.
5. Walker, J. W., Feeney, J., and Trentham, D. R. (1989) Photolabile precursors of inositol

phosphates – preparation and properties of 1-(2-nitrophenyl)ethyl esters of myo-inositol 1,4,5-trisphosphate. *Biochemistry* **28**, 3272–80.

6. Li, W. H., Llopis, J., Whitney, M., Zlokarnik, G., and Tsien, R. Y. (1998) Cell-permeant caged $InsP_3$ ester shows that Ca^{2+} spike frequency can optimize gene expression. *Nature* **392**, 936–41.
7. Aarhus, R., Gee, K., and Lee, H. C. (1995) Caged cyclic ADP-ribose. Synthesis and use. *J Biol Chem* **270**, 7745–9.
8. Dakin, K., and Li, W. H. (2007) Cell membrane permeable esters of D-myo-inositol 1,4,5-trisphosphate. *Cell Calcium* **42**, 291–301.
9. Zipfel, W. R., Williams, R. M., and Webb, W. W. (2003) Nonlinear magic: multiphoton microscopy in the biosciences. *Nat Biotechnol* **21**, 1369–77.
10. Haugland, R. P. (2005) *The Handbook: A Guide to Fluorescent Probes and Labeling Technologies*, Invitrogen Corporation, Eugene, OR, Ch. 19, pp. 879–906.
11. Palmer, A. E., and Tsien, R. Y. (2006) Measuring calcium signaling using genetically targetable fluorescent indicators. *Nat Protoc* **1**, 1057–65.
12. West, D. J., and Williams, A. J. (2007) Pharmacological regulators of intracellular calcium release channels. *Curr Pharm Des* **13**, 2428–42.
13. Furuta, T., Wang, S. S., Dantzker, J. L., Dore, T. M., Bybee, W. J., Callaway, E. M., Denk, W., and Tsien, R. Y. (1999) Brominated 7-hydroxycoumarin-4-ylmethyls: photolabile protecting groups with biologically useful cross-sections for two photon photolysis. *Proc Natl Acad Sci U S A* **96**, 1193–200.
14. Zhao, Y., Zheng, Q., Dakin, K., Xu, K., Martinez, M. L., and Li, W. H. (2004) New caged coumarin fluorophores with extraordinary uncaging cross sections suitable for biological imaging applications. *J Am Chem Soc* **126**, 4653–63.
15. Dakin, K., and Li, W. H. (2006) Infrared-LAMP: two-photon uncaging and imaging of gap junctional communication in three dimensions. *Nat Methods* **3**, 959.
16. Zhu, Y., Pavlos, C. M., Toscano, J. P., and Dore, T. M. (2006) 8-Bromo-7-hydroxyquinoline as a photoremovable protecting group for physiological use: mechanism and scope. *J Am Chem Soc* **128**, 4267–76.
17. Momotake, A., Lindegger, N., Niggli, E., Barsotti, R. J., and Ellis-Davies, G. C. (2006) The nitrodibenzofuran chromophore: a new caging group for ultra-efficient photolysis in living cells. *Nat Methods* **3**, 35–40.
18. Gug, S., Bolze, F., Specht, A., Bourgogne, C., Goeldner, M., and Nicoud, J. F. (2008) Molecular engineering of photoremovable protecting groups for two-photon uncaging. *Angew Chem Int Ed Engl* **47**, 9525–9.
19. Brown, E. B., Shear, J. B., Adams, S. R., Tsien, R. Y., and Webb, W. W. (1999) Photolysis of caged calcium in femtoliter volumes using two-photon excitation. *Biophys J* **76**, 489–99.
20. Newton, C. L., Mignery, G. A., and Sudhof, T. C. (1994) Co-expression in vertebrate tissues and cell lines of multiple inositol 1,4,5-trisphosphate (InsP3) receptors with distinct affinities for InsP3. *J Biol Chem* **269**, 28613–9.
21. Yoneshima, H., Miyawaki, A., Michikawa, T., Furuichi, T., and Mikoshiba, K. (1997) Ca^{2+} differentially regulates the ligand-affinity states of type 1 and type 3 inositol 1,4,5-trisphosphate receptors. *Biochem J* **322 (Pt 2)**, 591–6.
22. Patterson, G. H., and Piston, D. W. (2000) Photobleaching in two-photon excitation microscopy. *Biophys J* **78**, 2159–62.
23. Kiskin, N. I., Chillingworth, R., McCray, J. A., Piston, D., and Ogden, D. (2002) The efficiency of two-photon photolysis of a "caged" fluorophore, o-1-(2-nitrophenyl)ethylpyranine, in relation to photodamage of synaptic terminals. *Eur Biophys J* **30**, 588–604.
24. Hatchard, C. G., and Parker, C. A. (1956) A new sensitive chemical actinometer. II. Potassium ferrioxalate as a standard chemical actinometer. *Proc R Acad London A* **235**, 518–36.
25. Bernardinelli, Y., Haeberli, C., and Chatton, J. Y. (2005) Flash photolysis using a light emitting diode: an efficient, compact, and affordable solution. *Cell Calcium* **37**, 565–72.
26. Rapp, G. (1998) Flash lamp-based irradiation of caged compounds. *Methods Enzymol* **291**, 202–22.
27. Yang, S., and Li, W. H. (2009) Assaying dynamic cell-cell junctional communication using noninvasive and quantitative fluorescence imaging techniques: LAMP and infrared-LAMP. *Nat Protoc* **4**, 94–101.

Chapter 7

The Application of Fluorescent Probes for the Analysis of Lipid Dynamics During Phagocytosis

Ronald S. Flannagan and Sergio Grinstein

Abstract

Phagocytosis is the process whereby specialized leukocytes ingest large particles. This is an extremely dynamic and localized process that requires the recruitment to the sites of ingestion of numerous effector proteins, together with extensive lipid remodelling. To investigate such a dynamic series of events in living cells, non-invasive methods are required. The use of fluorescent probes in conjunction with spectroscopic analysis is optimally suited for this purpose. Here we describe a method to express in RAW264.7 murine macrophages genetically encoded probes that allow for the spatio-temporal analysis of lipid distribution and metabolism during phagocytosis of immunoglobulin-opsonized beads. The fluorescence of the probes is best analysed by laser scanning or spinning disc confocal microscopy. While the focus of this chapter is on phagocytic events, this general method can be employed for the analysis of lipid distribution and dynamics during a variety of biological processes in the cell type of the investigator's choice.

Key words: Phagocytosis, macrophage, fluorescent probes, phosphatidylserine, phosphoinositide, confocal fluorescence microscopy.

1. Introduction

Biological membranes are dynamic entities that not only function as physical barriers for cells and organelles but also play important roles in metabolic and cell signalling events. The species of lipids that comprise these membranes dictate cellular processes by recruiting proteins through lipid-specific binding domains, by electrostatic interactions or by serving as substrates to generate important signalling molecules (1, 2). All of these aspects of lipid function are critically involved in phagocytosis.

D.B. Papkovsky (ed.), *Live Cell Imaging*, Methods in Molecular Biology 591,
DOI 10.1007/978-1-60761-404-3_7, © Humana Press, a part of Springer Science+Business Media, LLC 2010

Phagocytosis is a complex process by which specialized cell types, such as neutrophils, microglia and macrophages, ingest large (>0.5 μm) particles such as bacteria or apoptotic bodies. Particle engulfment by phagocytosis is essential for the clearance of infection and plays an important role in tissue remodelling (3). The ingested particles are internalized into membrane-bound vacuoles termed "phagosomes" that upon maturation become increasingly acidic and degradative in nature (reviewed in (4)).

Phagosome formation is a receptor-driven process that is triggered by recognition of intrinsic components of the target particle or of serum components – such as complement proteins or immunoglobulins – that coat the particle (5, 6). As a result of receptor engagement, tyrosine kinases are activated, effector proteins are recruited, lipids are redistributed and metabolized, and actin undergoes re-arrangement (5). The localized remodelling of the actin cytoskeleton drives the formation of pseudopods that engulf the phagocytic target. The formed vacuole undergoes extensive changes in order to acquire microbicidal properties, a process known as "maturation". The phagosomal membrane undergoes significant and rapid alterations in lipid composition throughout the stages of phagosome formation and maturation (3). Thus, phosphatidylinositol-4,5-*bis*phosphate (PI(4,5)P_2) is initially enriched at the base of the forming phagocytic cup, where it is required to initiate localized actin polymerization to drive pseudopod formation (7, 8). As actin continues to polymerize, pushing the phagocyte membrane around the target, PI(4,5)P_2 is maintained at the leading edges of the pseudopod, but is cleared from the base of the phagocytic cup. PI(4,5)P_2 clearance is caused in part by its phosphorylation by class-I phosphatidylinositol 3-kinase, generating phosphatidylinositol-3,4,5-*tris*phosphate (PI(3,4,5)P_3) (9, 10). Further clearance of PI(4,5)P_2 is caused by PI(3,4,5)P_3-dependent recruitment of PLCγ, which hydrolyses PI(4,5)P_2 to diacylglycerol (DAG) and inositol *tris*phosphate (IP_3) (7, 11). Upon fusion of the pseudopod tips and closure of the phagosome, PI(3,4,5)P_3 rapidly disappears from the phagosome membrane. At that stage phosphatidylinositol 3-phosphate (PI(3)P), another phosphoinositide, begins to accumulate (9). PI(3)P persists for several minutes on the limiting membrane of phagosomes, where it is essential for subsequent maturation of the phagosome through endosomal fusion (12). The role of phosphatidylserine (PS) during phagosome formation has not been defined, although this negatively charged lipid is abundant in the plasma membrane and the phagocytic cup (13). After sealing, PS persists on the phagosomal membrane, but whether this lipid is required for maturation is not yet known.

Understanding the cell biology of phagocytosis or any other dynamic cellular process is not a trivial task. Some of the lipids that contribute to phagosome formation exist only transiently and the entire phagocytic process occurs locally and rapidly. Conventional methods for lipid analysis, such as thin layer chromatography (TLC) or high-performance liquid chromatography (HPLC), are not optimal for investigation of such a dynamic process, as they lack the spatial and temporal resolution necessary to monitor localized and rapid events, and generally require large number of cells and use of radioisotopes. To overcome these hurdles, it is most convenient to employ fluorescent probes to detect and track the fate of specific lipids in living cells using dynamic detection methods such as confocal fluorescence microscopy. These lipid-binding probes or "biosensors" require the expression of chimeric constructs consisting of a fluorescent protein, such as green fluorescent protein (GFP), fused to a specific lipid-binding domain (**Fig. 7.1**). There exist several variants of GFP as well as variants of a red fluorescent protein (RFP) that possess unique spectral properties that can be used in combination for a variety of imaging applications (reviewed in (14)). These can be coupled to domains or motifs isolated from several proteins that bind defined species of lipids, such as PS or PI(4,5)P_2 (**Table 7.1**). There are several advantages to using fluorescent biosensors: they are relatively non-invasive and enable analysis of dynamic events in living cells. In addition, and importantly, they can be genetically encoded and therefore introduced into cells by transfection of plasmid DNA, by transduction with viruses or by microinjection of the cDNA encoding the probe.

Herein we describe a general method for the use of lipid-binding fluorescent probes in conjunction with confocal microscopy for the dynamic, non-invasive analysis of lipids during phagocytosis. By no means is this protocol limited to the

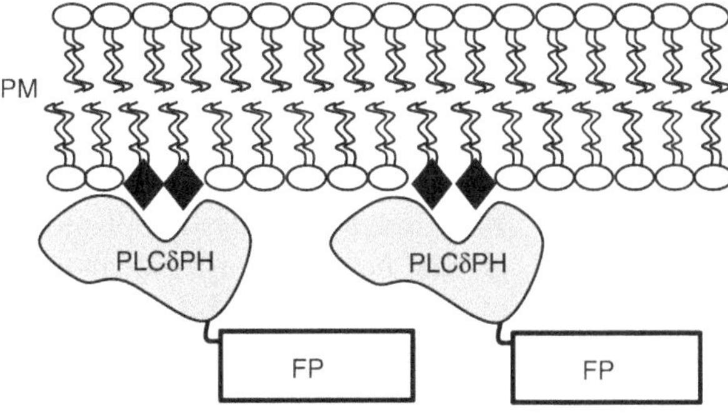

Fig. 7.1. Strategy for lipid detection using fluorescent fusion protein probes. Cells are transfected with plasmids that express a fusion protein comprised of a lipid-binding domain (e.g. PLCδPH) and a fluorescent protein (FP). The lipid-binding domain interacts reversibly with its endogenous ligand (e.g. PI(4,5)P_2 for PLCδPH), resulting in accumulation of the probe and increased fluorescence signal that can be detected spectroscopically.

Table 7.1
Lipid-binding probes that can be used in live cell imaging

Lipid-binding domain	Ligand	Protein source	References
FYVE	PI(3)P	EEA1 Hrs	(15) (15)
PX		$P40^{Phox}$	(16)
Pleckstrin homology (PH)	PI(4)P	OSH2 FAPP-1	(17) (18)
PH Tubby domain	$PI(4,5)P_2$	PLCδ Tubby	(19) (20)
PH	$PI(3,4)P_2$	TAPP1	(21)
PH	$PI(3,4)P_2/PI(3,4,5)P_3$	AKT PDK1 CRAC	(22) (23) (24)
PH	PI(3,4,5)P3	Btk GRP1 ARNO	(25) (26) (27)
C2 Annexin	PS	Lactadherin Annexin V	(28) (29)
C1	DAG	PKCγ1	(30)
BH3 CLBD	Cardiolipin	Bid MTCK1b	(31) (32)
SBD	Sphingolipids	Gp120, PrP, β-amyloid peptide	(33)
D4	Cholesterol	Perfringolysin θ toxin	(34)

study of phagocytosis and the same general procedure can be used to analyse many other processes in phagocytes and other cells. A fundamental component of the method described here is the transfection of plasmid DNA into macrophages. Primary macrophages are often refractory to transfection procedures and therefore we often use the murine macrophage cell line RAW264.7 for analysis of phagocytic events. Described below is a general protocol for the analysis of dynamic lipid behaviour during phagocytosis of IgG-coated polystyrene beads.

2. Materials

2.1. Cell Culture

1. RAW264.7 murine macrophage cell line (ATCC®, Manassas, VA).
2. 1× Dulbecco's modified Eagle's medium (DMEM) (WISENT, Mississauga, ON, Canada), supplemented with

5% (v/v) heat-inactivated fetal bovine serum (HI-FBS) (WISENT).

3. 1× solution RPMI 1640 buffered with 25 mM 4-(2-hydroxyethyl)piperazine-1-ethanesulfonic acid (HEPES) (WISENT), referred to hereafter as HPMI.
4. Sterile 1× phosphate-buffered saline (PBS).
5. 0.05% trypsin solution containing 0.53 mM ethylenediaminetetra-acetic acid (EDTA) and sodium bicarbonate (WISENT).
6. T-25 or T-75 tissue culture flasks (Sarstedt, Montreal, QC, Canada).
7. Tissue culture dishes with 6 or 12 wells (Becton Dickinson, Mississauga, ON, Canada).
8. Round glass cover-slips (18 mm diameter for 12-well dishes or 25 mm for 6-well dishes) (Fisher Scientific, Ottawa, ON, Canada).
9. Humidified tissue culture incubator at 37°C with 5% CO_2.

2.2. Transfection Reagents

1. Fugene HD transfection reagent (Roche Applied Science, Mississauga, ON, Canada).
2. Serum-free DMEM (WISENT).

2.3. Plasmid Isolation

1. Transformed bacterial stock (usually an *Escherichia coli* strain that was used during plasmid construction) in glycerol (*see* **Note 1**).
2. Sterile Luria-Bertani (LB) bacteriological medium (Tryptone, 10 g/L, NaCl 5 g/L and yeast extract 5 g/L) (BioShop Canada, Burlington, ON, Canada).
3. Antibiotic stocks for plasmid selection (ampicillin 100 mg/mL, kanamycin 40 mg/mL, chloramphenicol 30 mg/mL) (Sigma-Aldrich, Canada).
4. High Speed plasmid Maxi kit (QIAGEN Inc., Mississauga, ON, Canada).

2.4. Phagocytosis Assays

1. Inert polystyrene beads (3.87 μm diameter) with 2% (v/v) divinylbenzene (Bangs Laboratories, Fishers, IN,).
2. Lyophilized human IgG (Sigma-Aldrich), re-suspended in sterile PBS at a concentration of 50 mg/mL.

2.5. Microscopy

1. A spinning-disk confocal microscope equipped with at least a 63× magnifying objective, a light source and filter set that is appropriate for the fluorescent protein of choice and an objective heater. The system in our laboratory consists of a Zeiss Axiovert 200 M inverted microscope

(Carl Zeiss Canada, Toronto, ON, Canada) equipped with diode-pumped solid-state laser (Spectral Applied Research, Richmond Hill, ON, Canada) providing lines 440, 491, 561, 638 and 655 nm, a motorized X-Y stage (API, WA) and an piezo focus drive (Quorum Technologies, Guelph, ON, Canada). Images on this system are captured using a back-thinned EM-CCD camera (Hamamatsu) controlled by the software Volocity version 4.1.1 (Improvision Inc., Waltham, MA).

2. Live-cell imaging chamber for 18 or 25 mm cover-slips (Invitrogen Canada Inc., Burlington, ON, Canada).
3. Digital temperature regulator with a heated P insert (PeCon, Germany).
4. Images are analysed using analysis software such as Volocity or Image J (http://rsb.info.nih.gov/ij /).

3. Methods

3.1. Cell Culture

1. RAW264.7 macrophages are routinely cultured in 10 mL of DMEM supplemented with HI-FBS (5% v/v) at 37°C and 5% CO_2 in a T-25 tissue culture flask. If several transfections are to be performed it may be useful to culture the macrophages in a T-75 flask, which will yield a greater number of cells that can subsequently be seeded onto cover-slips. When the macrophages reach ~70–80% confluence they are ready to be trypsinized and can be used to seed tissue culture dishes (12- or 6-well dishes) containing sterile glass cover-slips (*see* **Note 2**).

3.2. Transfection of Cells

3.2.1. Day 1

1. Prior to trypsinizing the macrophages, add to a 12-well tissue culture dish sterile 18-mm glass cover-slips (one per well) using sterile tweezers. To each well containing a cover-slip add 1 mL of pre-warmed (37°C) DMEM plus HI-FBS. If using a 6-well dish, 2 mL of medium should be added to each well containing a 25-mm cover-slip.
2. The RAW264.7 macrophages in the T-25 flask that have reached 70–80% confluence are washed once with 8 mL of sterile PBS pre-warmed to 37°C. The PBS is then carefully aspirated.
3. Add 1 mL of pre-warmed (37°C) 0.05% trypsin/0.53 mM EDTA to the T-25 flask (for T-75 flasks add up to 2.5 mL trypsin/EDTA). Gently rotate the flask to ensure that the trypsin solution is dispersed evenly over the bottom of the flask and incubate for 1–5 min while observing for detachment of the cells (*see* **Note 3**).

4. When the majority of the RAW cells appear to be detached, gently add 8 mL of pre-warmed DMEM plus HI-FBS to wash the cells off the bottom of the flask. The addition of medium containing FBS inactivates the trypsin so that further proteolysis does not occur. Carefully pipette up and down to disperse any clumped cells. This will also ensure that any weakly adherent macrophages are detached from the bottom of the flask (*see* **Note 4**).
5. To the 12-well dish containing pre-warmed DMEM with HI-FBS add 1–4 drops of the macrophage suspension from step 4. For a 6-well dish, up to six drops of the macrophage suspension may be necessary to achieve the desired confluence. The exact number of macrophages per drop is influenced by the efficiency of trypsinization and the initial confluence of the cell culture. While it is not necessary to precisely enumerate the RAW cells in the suspension, after an overnight incubation (18 h) the adherent cells should be approximately 50% confluent (*see* **Note 5**).
6. Incubate the dish containing cover-slips seeded with RAW cells overnight in a tissue culture incubator at 37°C with 5% CO_2.
7. For plasmid isolation, 500 mL of sterile LB containing the appropriate antibiotic is inoculated with bacteria from a glycerol stock carrying the desired plasmid. The bacterial culture is then incubated overnight at 37°C, unless specific growth conditions are indicated. Note that this culture should be started 1 day prior to splitting the RAW cells. The overnight bacterial culture is lysed and the plasmid DNA isolated by directly following the QIAGEN instruction manual for the High Speed Maxi Prep kit. Typically 0.75–1 μg/μL of plasmid DNA is obtained. Once the plasmid is isolated it can be stored and used repeatedly for transfections.

3.2.2. Day 2

8. Remove Fugene HD and serum-free DMEM from the 4°C refrigerator and warm up to room temperature.
9. Transfections are performed as described in Roche's protocol that is supplied with the reagent. Typically, 2 μg of plasmid DNA and 3 μL of Fugene HD are used per well for transfections. Importantly, the ratio of DNA to transfection reagent should be determined for each construct and batch of RAW cells, as some variability can occur. When two plasmids are being co-transfected, the combined amount of DNA added to each well should not exceed what is used for only one plasmid (i.e. 1 μg of each plasmid).
10. Add 100 μL of the transfection solution prepared in the previous step drop-wise to each tissue culture well and

incubate the plate (12- or 6-well) for 16–24 h at 37°C with 5% CO_2.

3.3. Bead Opsonization

1. This is to be performed the day that phagocytosis will be examined. Dilute 50 μL of the suspension of 3.87 μm polystyrene beads into 200 μL of sterile PBS (*see* **Note 6**).
2. To the bead suspension from step 1 add 16 μL of human IgG reconstituted in sterile PBS. Mix the bead suspension continuously for 1–2 h at room temperature.
3. Wash the IgG-opsonized beads three times with 1 mL of sterile PBS and re-suspend in a final volume of 800 μL.

3.4. Synchronized Phagocytosis Assay and Image Acquisition

1. Turn on the digital heater to warm objective (63× or 100×) and the P insert. These must be at 37°C prior to acquiring images of cells undergoing phagocytosis. At this time ensure that the spinning disk confocal microscope and the light source are turned on. It is also useful at this time to set up the experimental acquisition parameters for the experiment using the Volocity software.
2. Prior to initiating the phagocytosis assay, warm an aliquot of HPMI to 37°C.
3. Remove a cover-slip from one well containing transfected cells and add to a new 12- or 6-well tissue culture dish after having been incubated for at least 16 h at 37°C with 5% CO_2. Place the plate with the remaining transfectants back into the incubator.
4. Wash the cover-slip twice with cold (4°C) 1× PBS (*see* **Note 7**).
5. To this well add 1 mL of cold (4°C) HPMI and keep plate on ice (*see* **Note 8**).
6. Next, add 15 μL of the IgG-opsonized bead suspension to this well and transfer the chilled tissue culture dish with the beads to a centrifuge with a plate rotor that has been cooled to 4°C. Centrifuge the plate for 1 min at 250×*g* to sediment the beads onto the transfected RAW cells, to synchronize phagocytic events.
7. Transfer cover-slip from step 6 to a pre-chilled live-cell imaging chamber and keep on ice. Add 500 μl of frcsh 4°C HPMI to the chamber.
8. Rapidly transfer the live-cell imaging chamber to the spinning-disk confocal microscope and focus on a single cell. The focal plane should be such that you acquire images through the middle portion of the cell. It is also possible to acquire Z-stacks for the cell of interest; however, this may lead to photobleaching over time. A fluorescence image and

a differential interference contrast (DIC) image should be captured for each acquisition (*see* **Note 9**).

9. Once the cell of interest is in focus, the cold HPMI should be aspirated and fresh, warm (37°C) HPMI should be carefully added to the chamber. With the addition of warmed medium phagocytosis will proceed rapidly and imaging should commence immediately. Setting up the acquisition protocol as a time-lapse experiment will enable multiple images of the same cell to be acquired at set intervals during the phagocytic process (*see* **Note 10**). Alternatively, images can be acquired manually at pre-determined time points, though this is slightly more laborious.

3.5. Fluorescence Measurements

The detection of plasmalemma and internal membrane components can be made using a variety of plasmid-encoded fluorescent probes. Furthermore, analysis of the dynamic behaviour of many different lipid components can be monitored separately or simultaneously by microscopy. Proper analysis of the acquired images is of utmost importance so that misinterpretation of data is avoided. Qualitative conclusions can be drawn from careful visual analysis of acquired images; however, only gross changes in the distribution of a cellular probe can be described with any certainty. To strengthen the data it is best to perform quantitative analysis of all fluorescence images (*see* **Note 11**).

Quantitative analysis of fluorescence images is usually carried out by measuring the mean fluorescence intensity per pixel of a particular region of the cell called the region of interest (ROI). Fluorescence intensity can be affected by a number or variables, including probe expression levels, the brightness of the light source used to excite the probe, the exposure time during the acquisition of the image, and photobleaching of the fluorophore over time. For these reasons it is necessary to standardize fluorescence intensity measurements for each cell being analysed. To achieve this, a region of the cell that is predicted not to undergo changes during the experiment should be selected as an internal reference for fluorescence intensity. In our case, during phagocytosis it is useful to compare fluorescence of the ROI to a remote region of the plasma membrane where phagocytic events have not occurred and no particles are bound (i.e. the bulk plasma membrane). The manner in which fluorescence intensities are normalized depends on the properties of the probe being used. In instances where there is readily detectable cytosolic fluorescence, as can be seen in **Fig. 7.2** using the PH-PLCδ/eGFP probe, it is best to correct the measured fluorescent intensities by subtracting the cytosolic fluorescence intensity. To allow for comparison between cells and experiments, fluorescent intensities can

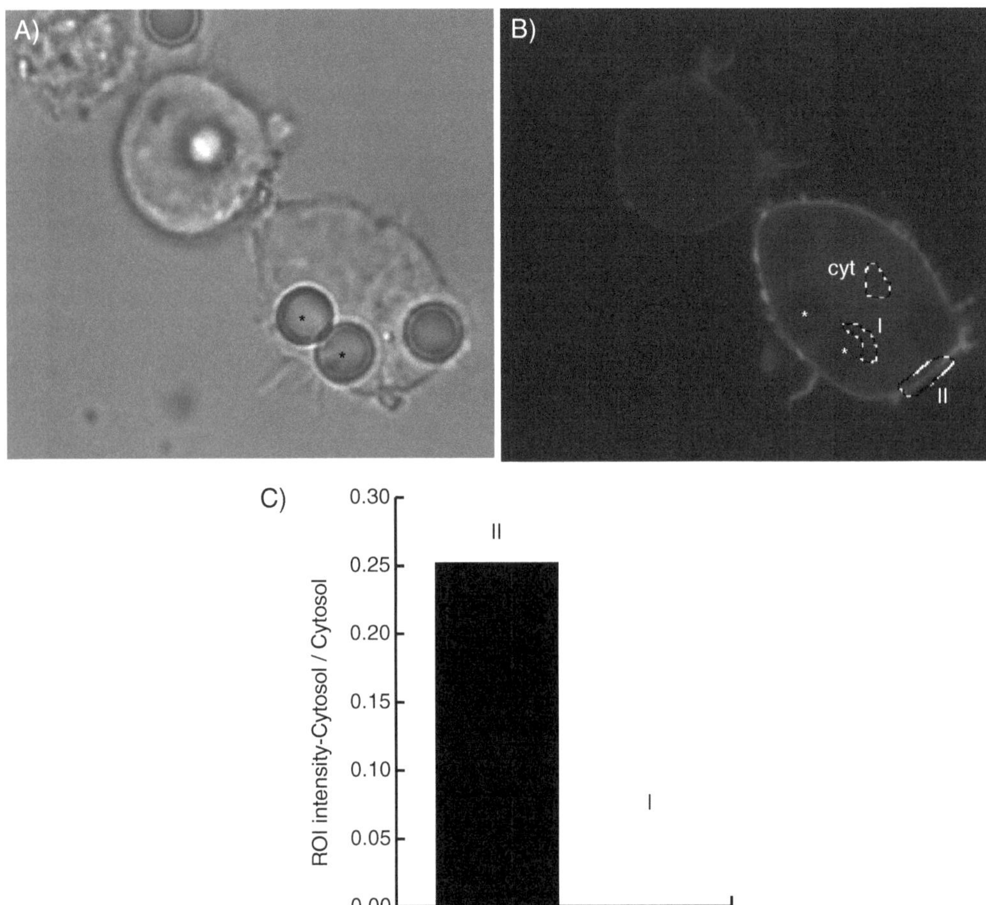

Fig. 7.2. Distribution of $PI(4,5)P_2$ in RAW264.7 cells having internalized IgG-opsonized beads. (**a**) Differential interference contrast (DIC) image of RAW macrophages that have internalized 3.87-μm polystyrene beads. (**b**) Confocal image of the fluorescence of the RAW cells from panel A, which had been transfected with a plasmid expressing the PH domain of PLCδ fused to eGFP. Accumulation of the probe at the plasma membrane indicates that $PI(4,5)P_2$ is abundant in this compartment of the cell. The regions of interest (ROI) used in the analysis of this image are shown and labelled as follows: I, phagosome membrane region; II, plasma membrane region; cyt, cytosolic region. (**c**) The normalized mean fluorescence intensity per pixel of the plasma (II) and phagosomal (I) membranes.

be normalized by dividing the corrected ROI intensities by the measured cytosolic emission.

1. Using the image analysis software such as Volocity select an experimental ROI that encompasses the membrane that has engaged an IgG-opsonized bead. For the earliest time points imaged it is necessary to compare the region of fluorescence selected with the DIC overlaid in order to ensure that a region with a bead engaged is being analysed.

2. Measure the fluorescence intensity (mean fluorescence intensity per pixel) for the experimental ROI.
3. Define a control ROI on the region of the cell that is not expected to undergo changes and measure its fluorescence intensity.
4. Define a cytosolic ROI and measure the emission intensity.
5. Subtract from the experimental ROI and the control ROI the cytosolic fluorescence, which will also account for background fluorescence.
6. Using the corrected fluorescence intensities for the experimental and control regions of interest determine the ratio of fluorescence intensity of ROI/cytosolic ROI (i.e. phagosomal membrane-cytosol/cytosol). The ratios obtained for several different cells between experimental conditions can now be compared.

Alternatively, when binding of the fluorescent probe within the cell is almost exclusively to membranes, measured fluorescence intensities can be normalized by dividing background-corrected ROI emission intensities by the emission intensity of the bulk plasma membrane.

4. Notes

1. Freeze pure cultures of each *E. coli* strain transformed with a plasmid that is transfected into mammalian cells. This should be done using sterile LB medium containing 25% (v/v) glycerol and then stored at –80°C. These glycerol stocks serve as a continuous source of plasmid DNA that can be easily isolated by inoculating bacterial culture medium. Some laboratories transform competent *E. coli* each time a plasmid has to be isolated, but this increases the time and effort required for plasmid preparation.
2. RAW cells that have been passaged excessively (> than ≈35 passages) should not be used, as the cells may become senescent and behave differently from early passage cells.
3. Trypsinization of RAW cells is not always efficient, as these cells can adhere strongly to the tissue culture flask. It is important not to overtrypsinize the cells by incubating for excessive amounts of time because the cells may become activated. If trypsinizing for 5 min at 37°C does not cause detachment of the vast majority of cells, the adherent cells can be mechanically dislodged by gently scraping the bottom of the dish with a sterile tissue culture scraper.

4. When scraping or pipetting to disperse any clumped cells after trypsinization it is important not to handle them too roughly. RAW cells can become activated by mechanical stimulation. It is almost impossible to avoid some activation, but this can be kept to a minimum.
5. While the Roche protocol for use of Fugene HD recommends transfecting cells that have grown to 80% confluence, cells grown to 50% confluence are better suited for phagocytic assays. Overcrowding of RAW cells makes imaging single cells difficult, can cause rounding of the macrophages and can alter their ability to perform phagocytosis.
6. When using polystyrene beads as an opsonin, it is important to use beads containing DVB. In our hands opsonization of polystyrene beads without the DVB additive is not as efficient. As an alternative to inert beads, red blood cells (i.e. sheep red blood cells) can be used and opsonized with anti-red cell specific antibodies.
7. To prevent untimely phagocytic events the cells are chilled to 4°C. This temperature halts phagocytosis until the cells are warmed again to temperatures above ~16°C.
8. HEPES-buffered medium is suitable for incubations where the RAW cells are not kept under CO_2. In the absence of HEPES buffering and CO_2, the pH of the tissue culture medium containing bicarbonate will change progressively, affecting cellular processes that are pH-dependent.
9. There are several advantages to using a spinning-disk confocal microscope over a laser scanning confocal microscope (LSCM) for live cell imaging. First, because the laser beam passes through pinholes in the spinning Nipkow disk the entire specimen is excited simultaneously, making extremely rapid image acquisition possible. Additionally, the low-intensity beams used by the Nipkow disk reduce phototoxicity and photobleaching, favouring time-lapse imaging of live cells. One disadvantage of the spinning disk confocal system is that it is not appropriate for experiments where photoactivation or photobleaching are required.
10. For time-lapse imaging of synchronized phagocytosis we typically capture one frame every 5–10 s for 5–8 min.
11. The quantitative analysis described above can be performed using several different imaging software programs such as Volocity (Improvision) or Image J (which is available free of charge from the National Institutes of Health at http://rsbweb.nih.gov/ij/).

Acknowledgements

Research conducted in the laboratory of S.G. is supported by the Heart and Stroke Foundation of Canada (HSF), the Cystic Fibrosis Foundation of Canada and the Canadian Institutes of Health Research (CIHR). R.S.F holds a CIHR Fellowship. S.G. is the current holder of the Pitblado Chair in Cell Biology.

References

1. Vereb, G., Szöllősi, J., Matkó, J., Nagy, P., Farkas, T., Vígh, L., Mátyus, L., Waldmann, T. A., and Damjanovich, S. (2003) Dynamic, yet structured: The cell membrane three decades after the Singer–Nicolson model, *Proc Natl Acad Sci USA* **100**, 8053–8.
2. Yeung, T., and Grinstein, S. (2007) Lipid signaling and the modulation of surface charge during phagocytosis, *Immunol Rev* **219**, 17–36.
3. Yeung, T., Ozdamar, B., Paroutis, P., and Grinstein, S. (2006) Lipid metabolism and dynamics during phagocytosis, *Curr Opin Cell Biol* **18**, 429–37.
4. Kinchen, J. M., and Ravichandran, K. S. (2008) Phagosome maturation: going through the acid test, *Nat Rev Mol Cell Biol* **9**, 781–95.
5. Swanson, J. A., and Hoppe, A. D. (2004) The coordination of signaling during Fc receptor-mediated phagocytosis, *J Leukoc Biol* **76**, 1093–103.
6. van Lookeren Campagne, M., Wiesmann, C., and Brown, E. J. (2007) Macrophage complement receptors and pathogen clearance, *Cell Microbiol* **9**, 2095–102.
7. Botelho, R. J., Teruel, M., Dierckman, R., Anderson, R., Wells, A., York, J. D., Meyer, T., and Grinstein, S. (2000) Localized biphasic changes in phosphatidylinositol-4,5-bisphosphate at sites of phagocytosis, *J Cell Biol* **151**, 1353–68.
8. Coppolino, M. G., Dierckman, R., Loijens, J., Collins, R. F., Pouladi, M., Jongstra-Bilen, J., Schreiber, A. D., Trimble, W. S., Anderson, R., and Grinstein, S. (2002) Inhibition of phosphatidylinositol-4-phosphate 5-kinase Ialpha impairs localized actin remodeling and suppresses phagocytosis, *J Biol Chem* **277**, 43849–57.
9. Vieira, O. V., Botelho, R. J., Rameh, L., Brachmann, S. M., Matsuo, T., Davidson, H. W., Schreiber, A., Backer, J. M., Cantley, L. C., and Grinstein, S. (2001) Distinct roles of class I and class III phosphatidylinositol 3-kinases in phagosome formation and maturation, *J Cell Biol* **155**, 19–25.
10. Marshall, J. G., Booth, J. W., Stambolic, V., Mak, T., Balla, T., Schreiber, A. D., Meyer, T., and Grinstein, S. (2001) Restricted accumulation of phosphatidylinositol 3-kinase products in a plasmalemmal subdomain during Fc gamma receptor-mediated phagocytosis, *J Cell Biol* **153**, 1369–80.
11. Lennartz, M. R. (1999) Phospholipases and phagocytosis: the role of phospholipid-derived second messengers in phagocytosis, *Int J Biochem Cell Biol* **31**, 415–30.
12. Lindmo, K., and Stenmark, H. (2006) Regulation of membrane traffic by phosphoinositide 3-kinases, *J Cell Sci* **119**, 605–14.
13. Yeung, T., Terebiznik, M., Yu, L., Silvius, J., Abidi, W. M., Philips, M., Levine, T., Kapus, A., and Grinstein, S. (2006) Receptor activation alters inner surface potential during phagocytosis, *Science* **313**, 347–51.
14. Wang, Y., Shyy, J. Y., and Chien, S. (2008) Fluorescence proteins, live-cell imaging, and mechanobiology: seeing is believing, *Annu Rev Biomed Eng* **10**, 1–38.
15. Raiborg, C., Bremnes, B., Mehlum, A., Gillooly, D. J., D'Arrigo, A., Stang, E., and Stenmark, H. (2001) FYVE and coiled-coil domains determine the specific localisation of hrs to early endosomes, *J Cell Sci* **114**, 2255–63.
16. Ellson, C. D., Gobert-Gosse, S., Anderson, K. E., Davidson, K., Erdjument-Bromage, H., Tempst, P., Thuring, J. W., Cooper, M. A., Lim, Z. Y., Holmes, A. B., Gaffney, P. R., Coadwell, J., Chilvers, E. R., Hawkins, P. T., and Stephens, L. R. (2001) PtdIns(3)P regulates the neutrophil oxidase complex by binding to the PX domain of p40(phox), *Nat Cell Biol* **3**, 679–82.
17. Roy, A., and Levine, T. P. (2004) Multiple pools of phosphatidylinositol 4-phosphate detected using the pleckstrin homology domain of Osh2p, *J Biol Chem* **279**, 44683–9.

18. Godi, A., Di Campli, A., Konstantakopoulos, A., Di Tullio, G., Alessi, D. R., Kular, G. S., Daniele, T., Marra, P., Lucocq, J. M., and De Matteis, M. A. (2004) FAPPs control Golgi-to-cell-surface membrane traffic by binding to ARF and PtdIns(4)P, *Nat Cell Biol* **6**, 393–404.
19. Stauffer, T. P., Ahn, S., and Meyer, T. (1998) Receptor-induced transient reduction in plasma membrane PtdIns(4,5)P2 concentration monitored in living cells, *Curr Biol* **8**, 343–6.
20. Santagata, S., Boggon, T. J., Baird, C. L., Gomez, C. A., Zhao, J., Shan, W. S., Myszka, D. G., and Shapiro, L. (2001) G-protein signaling through tubby proteins, *Science* **292**, 2041–50.
21. Kimber, W. A., Trinkle-Mulcahy, L., Cheung, P. C. F., Deak, M., Marsden, L. J., Kieloch, A., Watt, S., Javier, R. T., Gray, A., Downes, C. P., Lucocq, J. M., and Alessi, D. R. (2002) Evidence that the tandem-pleckstrin-homology-domain-containing protein TAPP1 interacts with Ptd(3,4)P2 and the multi-PDZ-domain-containing protein MUPP1 in vivo, *Biochem J* **361**, 525–36.
22. Servant, G., Weiner, O. D., Herzmark, P., Balla, T., Sedat, J. W., and Bourne, H. R. (2000) Polarization of chemoattractant receptor signaling during neutrophil chemotaxis, *Science* **287**, 1037–40.
23. Komander, D., Fairservice, A., Deak, M., Kular, G. S., Prescott, A. R., Peter Downes, C., Safrany, S. T., Alessi, D. R., and van Aalten, D. M. F. (2004) Structural insights into the regulation of PDK1 by phosphoinositides and inositol phosphates, *EMBO J* **23**, 3918–28.
24. Dormann, D., Weijer, G., Parent, C. A., Devreotes, P. N., and Weijer, C. J. (2002) Visualizing PI3 kinase-mediated cell-cell signaling during Dictyostelium development, *Curr Biol* **12**, 1178–88.
25. Várnai, P., Rother, K. I., and Balla, T. (1999) Phosphatidylinositol 3-kinase-dependent membrane association of the Bruton's tyrosine kinase pleckstrin homology domain visualized in single living cells, *J Biol Chem* **274**, 10983–9.
26. Klarlund, J. K., Tsiaras, W., Holik, J. J., Chawla, A., and Czech, M. P. (2000) Distinct polyphosphoinositide binding selectivities for pleckstrin homology domains of GRP1-like proteins based on diglycine versus triglycine motifs, *J Biol Chem* **275**, 32816–21.
27. Venkateswarlu, K., Oatey, P. B., Tavaré, J. M., and Cullen, P. J. (1998) Insulin-dependent translocation of ARNO to the plasma membrane of adipocytes requires phosphatidylinositol 3-kinase, *Curr Biol* **8**, 463–6.
28. Yeung, T., Gilbert, G. E., Shi, J., Silvius, J., Kapus, A., and Grinstein, S. (2008) Membrane phosphatidylserine regulates surface charge and protein localization, *Science* **319**, 210–3.
29. Andree, H. A., Reutelingsperger, C. P., Hauptmann, R., Hemker, H. C., Hermens, W. T., and Willems, G. M. (1990) Binding of vascular anticoagulant alpha (VAC alpha) to planar phospholipid bilayers, *J Biol Chem* **265**, 4923–8.
30. Colón-González, F., and Kazanietz, M. G. (2006) C1 domains exposed: from diacylglycerol binding to protein-protein interactions, *Biochim Biophys Acta* **1761**, 827–37.
31. Liu, J., Weiss, A., Durrant, D., Chi, N., and Lee, R. M. (2004) The cardiolipin-binding domain of bid affects mitochondrial respiration and enhances cytochrome c release, *Apoptosis* **9**, 533–41.
32. Schlattner, U., Gehring, F., Vernoux, N., Tokarska-Schlattner, M., Neumann, D., Marcillat, O., Vial, C., and Wallimann, T. (2004) C-terminal lysines determine phospholipid interaction of sarcomeric mitochondrial creatine kinase, *J Biol Chem* **279**, 24334–42.
33. Mahfoud, R., Garmy, N., Maresca, M., Yahi, N., Puigserver, A., and Fantini, J. (2002) Identification of a common sphingolipid-binding domain in Alzheimer, prion, and HIV-1 proteins, *J Biol Chem* **277**, 11292–6.
34. Ohno-Iwashita, Y., Shimada, Y., Waheed, A. A., Hayashi, M., Inomata, M., Nakamura, M., Maruya, M., and Iwashita, S. (2004) Perfringolysin O, a cholesterol-binding cytolysin, as a probe for lipid rafts, *Anaerobe* **10**, 125–134.

Chapter 8

Imaging of Mitotic Cell Division and Apoptotic Intra-Nuclear Processes in Multicolor

Kenji Sugimoto and Shigenobu Tone

Abstract

To follow the cell division cycle in the living state, certain biological activity or morphological changes must be monitored keeping the cells intact. Mitotic events from prophase to telophase are well defined by morphology or movement of chromatin, nuclear envelope, centrosomes, and/or spindles. To paint or simultaneously visualize these mitotic subcellular structures, we have been using ECFP-histone H3 for chromatin and chromosomes, EGFP-Aurora-A for centrosomes and kinetochore spindles and DsRed-fused truncated peptide of importin alpha for the outer surface of nuclear envelope as living cell markers. Time-lapse images from prophase through to early G1 phase can be obtained by constructing a triple-fluorescent cell line (Sugimoto et al., *Cell Struct. Funct.* **27**, 457–467, 2002). Here, we describe the multicolor imaging of mitosis of a human breast cancer cell line, MDA435, and a further application to characterizing the apoptotic chromatin condensation process in isolated nuclei by simultaneously visualizing kinetochores with EGFP and chromatin with a fluorescent dye, SYTO 59.

Key words: Apoptosis, mitosis, kinetochore, time-lapse imaging, chromatin condensation.

1. Introduction

The mitotic cycle of mammalian cells consists of interphase and five mitotic stages, as illustrated in **Fig. 8.1a** (1). Centrosomes duplicate in G1/S phase and "prekinetochores" doubles in G2 phase (2). Chromatin condensation occurs and mitotic chromosomes are observed first in prophase, while the duplicated centrosomes move to the opposite poles and maturate into asters. After nuclear envelope breakdown, the asters capture the kinetochores of mitotic chromosomes in prometaphase and then form a typical mitotic spindle (3). Chromosomes are congressed to

D.B. Papkovsky (ed.), *Live Cell Imaging*, Methods in Molecular Biology 591,
DOI 10.1007/978-1-60761-404-3_8,

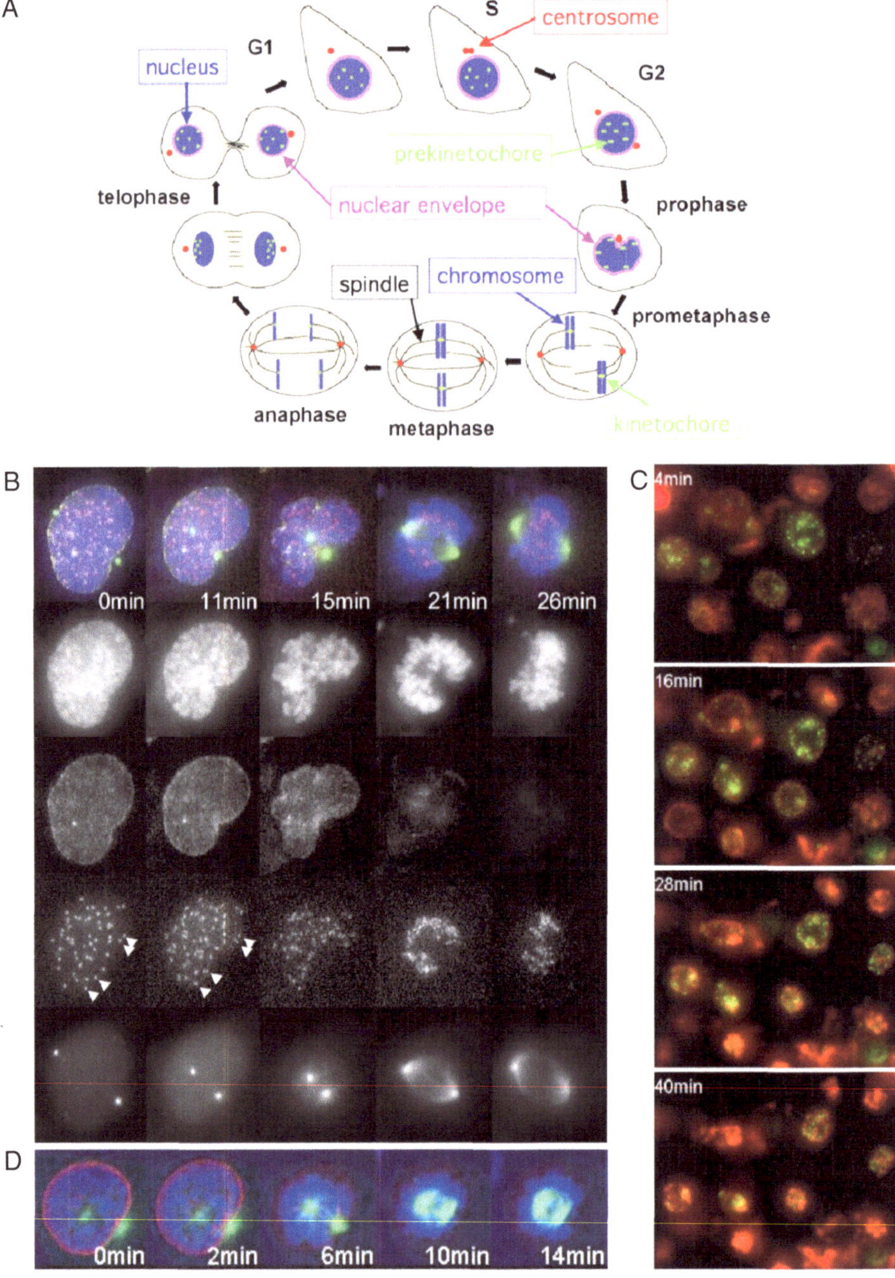

Fig. 8.1. Imaging of mitotic cell division and apoptotic intra-nuclear processes. (**a**) A schematic illustration of mitotic cycle of mammalian cells. (**b**) Live cell image of prophase-to-metaphase of a quadruple-fluorescent MDA435 cell line. ECFP-histone H3, EGFP-Aurora-A, mKO-CENP-A, and DsRed-importin-alpha were stably expressed to visualize chromatin (*blue*), centrosomes/asters/kinetochore spindles (*green*), kinetochores (*red*), and nuclear envelope (*yellow*). Time-lapse images were captured by a conventional wide-field microscope system (1). The separate images of ECFP, EGFP, mKO, and DsRed were also shown below. (**c**) Time-lapse imaging of cell-free apoptosis. MDA-AF8 nuclei were incubated with S/M extract and imaged by fluorescent microscopy (10). DNA was visualized using SYTO 59 (*red*) and kinetochores with EGFP-CENP-A (*green*). Representative images are shown. (**d**) Time-lapse images of triple-fluorescent MDA435 cell line captured by a laser confocal system (*see* **Fig. 8.3**). Fluorescent images are ECFP-histone H3 (*blue*), EGFP-alpha-tubulin (*green*), and DsRed-importin-alpha (*red*) (1). (Reproduced from (1) and (10) with permission from Elsevier.)

and aligned at the spindle equator in metaphase. Sister chromatid separation is a visual sign for anaphase onset and the following process is straightforward. The daughter chromosomes are segregated to the poles in anaphase and the nuclear envelope reforms around the sister chromatids in telophase. These events after anaphase onset occur within 10 min or so, followed by spindle deformation and chromosome decondensation in early G1 phase.

To accurately identify each mitotic stage, it would be very convenient if the mitotic apparatus has been differentially visualized or painted by a set of multiple fluorescent proteins with various fluorescent wavelengths from cyan to far-red (4). Recently, we constructed three quadruple-fluorescent MDA435 cell lines in which chromatin, kinetochores, nuclear envelope, and either the inner centromere, or microtubules, or centrosomes/spindles are simultaneously visualized with ECFP, EFGP, mKO, and DsRed (1). Each mitotic stage of the individual cells could be identified simply by capturing live cell images under a microscope without any requirement of cell-fixing or staining steps (**Fig. 8.1b**).

Dynamic changes in the compaction of nuclear chromatin are one of the characteristic phenomena of apoptotic execution (5). During apoptosis, however, the level of chromatin condensation is even greater than that observed in mitosis. The chromatin drastically undergoes a phase change from a heterogeneous, genetically active network to an inert, highly condensed form that is fragmented and packaged into "apoptotic bodies." Cell-free systems with isolated nuclei have been developed to study the molecular mechanisms of apoptotic execution (6–9). The protocol we described here was originally developed in Dr. Earnshaw's laboratory in Edinburgh (6, 7). We further applied the live cell imaging to a cell-free system to characterize the biochemical mechanism underlying apoptotic chromatin condensation (10, **Fig. 8.1c**). Three distinct stages of apoptotic nuclear condensation were revealed by the time-lapse images of isolated nuclei prepared from a kinetochore-fluorescent cell line, MDA-AF8 (11).

2. Materials

2.1. Cell Culture and Electroporation

1. Dulbecco's modified Eagle's medium (DMEM) (Nissui Pharmaceutical, Tokyo, Japan).
2. Trypsin (0.25%) (Gibco/BRL, Bethessa, MD).

3. Plasmid DNAs: pEGFP-AF8 for a kientochore marker, prepared with spin column such as a "Quantum Prep Plasmid Miniprep Kit" (Bio-Rad, Hercules, CA). Use pECFP-histone H3 for chromatin/chromosome, pEGFP-Aurora-A for centrosomes/kinetochore spindle, and pDsRed-importin-alpha for nuclear envelope (12). pTK-Hyg and pPur (Clontech, Palo Alto, CA) are for hygromycin B and puromycin, respectively.
4. K-PBS: 30 mM NaCl, 120 mM KCl, 8 mM Na_2HPO_4, 1.5 mM KH_2PO_4, 5 mM $MgCl_2$
5. Electroporation cuvette (4 mm width) (Bio-Rad).
6. Electroparation apparatus: Gene Pulser and capacitance extender (Bio-Rad) (*see* **Note 1**).

2.2. Screening of Fluorescent Cells

1. Geneticine (G418 sulfate) (Nakalai Tesque, Kyoto, Japan): dissolved with 0.5 M HEPES buffer, pH 7.2, to final concentration of 0.8–1.6 mg/mL and filtrated through 0.22 mm filter.
2. Hygromycin B (50 mg/mL, Roche Diagnostics GmbH, Nonnenwald, Penzberg, Germany).
3. Puromycin (Sigma): dissolved in PBS(–) to make a stock solution (0.5 mg/mL).
4. Cloning ring (7 mm in diameter) (Iwaki Glass Co. Ltd., Chiba, Japan).
5. Coverslip (12×12 mm) (Matsunami Glass Ind. Ltd., Osaka, Japan).
6. 4% paraformaldehyde (Nakalai Tesque): dissolved in PBS(–) at 60°C and neutralized to pH 7.4–7.6 with 2 N NaOH.
7. DAPI-containing glycerol: dissolve 0.1 g DABCO (Sigma) with 1 mL 0.2 M Tris–HCl (pH 7.5) and mix well with 9 mL glycerol (nonfluorescent grade, Nakalai Tesque). Add DAPI stock solution (100 μg/mL) to a final concentration 1 μg/mL.

2.3. Preparation of Nuclei

1. 1× nuclei solution: 10 mM KCl, 10 mM PIPES buffer (pH 7.4), 1.5 mM $MgCl_2$, 1 mM DTT, 10 μM cytochalasin B, 1× CLAP, 100 μM PMSF.
2. 1000× CLAP: 5 mg each of chymotrypsin, leupeptin, aprotinin, and pepstatin A were dissolved in 5 mL of DMSO.
3. Syringe attached by 21G needle (needle tips were bent by 10–20°).
4. 30 % sucrose solution: 3 g sucrose dissolved with 10 mL 1× nuclei solution.

5. Buffer A: 250 mM sucrose, 80 mM KCl, 20 mM NaCl, 5 mM EGTA, 15 mM PIPES (pH 7.4), 1 mM DTT, 0.5 mM spermidin, 0.2 mM spermin, 100 μM PMSF, 1× CLAP, 50% Glycerol.
6. Incomplete KPM: 50 mM KCl, 50 mM PIPES (pH 7.0), 10 mM EGTA, 1.92 mM $MgCl_2$.
7. Complete KPM: 1 mM DTT, 20 μM cytochalasin B, 1× CLAP, 100 μM PMSF were added to incomplete KPM.
8. Aphidicolin (Sigma).
9. Nocodazole (Sigma).
10. 10× MDB buffer: 0.5 *M* NaCl, 20 μM $MgCl_2$, 50 mM EGTA, 100 mM PIPES, 100 mM DTT.
11. SYTO 59 (Molecular Probes, Invitrogen Corp., Carlsbad, CA).
12. Human stable cell line MDA-AF8 expressing EGFP-CENP-A (11).
13. ATP plus a regeneration system: mix equal amount of ATP/creatine just before use.
14. Phosphate solution (40 mM ATP, 200 mM phosphocreatine, 10 mM PIPES, pH 7.0) and creatine kinase solution (1 mg/mL creatine kinase, 10 mM PIPES, pH 7.0, 1 mM DTT, 50 mM NaCl).
15. 35-mm glass base dish (Iwaki Glass Co. Ltd.).

2.4. Time-Lapse Imaging System

1. Inverted fluorescent microscope: Eclipse TE300 or TE2000-U (Nikon, Tokyo, Japan).
2. Objective lens: PlanApo VC 60×, NA1.40 (Nikon) (*see* **Note 2**).
3. Stage-top incubator: INU-NI-F1 (Tokai Hit, Fujinomiya, Sizuoka, Japan).
4. CCD camera: ORCA-ER (Hamamatsu Photonics, Hamamatsu, Sizuoka, Japan).
5. Filter wheels and *z*-axis motor: BioPoint MAC5000 (Ludl Electronic Products, Hawthorne, NY).
6. Excitation and emission filters: 436/20 and 472/30 nm for ECFP, 484/15 and 520/35 nm for EGFP, 532/10 and 556/20 nm for mKO, and 580/13 and 630/60 nm for DsRed (Chroma Technology, Rockingham, VT; Semrock, Rochester, NY) (*see* **Note 3**).
7. Dichroic: 86006bs for CFP/YFP/DsRed (Chroma Technology) (*see* **Note 4**).
8. Image capturing and analyzing software: LuminaVision for MacOSX (Mitani Corporation, Fukui, Japan).

3. Methods

3.1. Cell Culture and Electroporation

1. Culture human MDA435 cells in DMEM with 10% fetal bovine serum (FBS, Biowest, Nuaille, France) in 90-mm dishes (usually four dishes for each preparation).
2. Rinse the cells with 4 mL PBS(−).
3. Trypsinize the cells with 1 mL 0.25% trypsin.
4. Suspend the cells with 4 mL of DMEM with 10% FBS.
5. Transfer into 15 mL tube.
6. Count the cell numbers with a hematometer.
7. Spin the cells 1000 rpm 5 min at room temperature (RT) with table-top centrifuge.
8. Resuspend the pellet with 12 mL PBS(−).
9. Recentrifuge 1000 rpm 5 min.
10. Repeat the resuspension and centrifugation step.
11. Resuspend the cells with ice-cold K-PBS to 1.2×10^7 cells/mL.
12. Transfer 0.45 mL aliquots of cell suspension into each eppendorf tube.
13. Add 16 μg of plasmid DNA and keep on ice for 10 min. For selection with drugs other than G418, add 7 μg of the appropriate selection plasmid as well.
14. Transfer into an ice-cold 0.4 cm electroporation cuvette.
15. Pulse with 960 μF at 0.22 kV with Gene Pulser and capacitance extender.
16. Keep the cuvette on ice for 10 min.
17. Add 0.5 mL of DMEM
18. Put a lid on the cuvette and turn upside down for mixing.
19. Keep at RT for 10 min.
20. Dilute the cell suspension into 4 mL of DMEM with 10%FBS in 50-mm dish.
21. Add 1–2 mL aliquots of the dilution into 8 mL of DMEM with 10%FBS in 90-mm dish.
22. Drop 0.5 mL aliquot of the cell suspension onto 18× 18 mm coverslip in 35-mm dish.
23. Incubate 35- and 90-mm dishes in CO_2 incubator for 1–2 days.
24. Add G418 stock solution to 0.8–1.2 mg/mL and culture for 2–3 weeks to obtain single colonies. For the

construction of multi-color cell lines, stable transformants are sequentially selected by hygromycin B (200 μg/mL) and puromycin (20–50 μg/mL).

3.2. Screening of Fluorescent Transfomants

1. Rinse the cells with 4 mL PBS(−).
2. Put cloning rings (7 mm in diameter) on each single colony.
3. Trypsinize the cells with a few drops of 0.25% trypsin.
4. Suspend the cells with a portion of 2 mL of DMEM with 10% FBS in each well of 12-well plates.
5. Grow the cells on coverslip (12 × 12 mm) for several days.
6. Transfer each coverslip into another 12-well plate.
7. Fix the cells with ice-cold 4% paraformaldehyde for 20 min.
8. Rinse with PBS(−) for 5 min twice.
9. Pick up and put each coverslip upside down on 5 μL drop of DAPI-containing glycerol on slide glasses.
10. Observe the cells under a fluorescent microscope.

3.3. Preparation of Isolated Nuclei

1. Culture human cells in DMEM with 10% FBS in three T75 cultures for each preparation.
2. Spin the cells at 1000 rpm for 5 min at RT with table-top centrifuge.
3. Resuspend the pellet in 30 mL PBS(−) into a 50-mL tube.
4. Recentrifuge at 1000 rpm for 5 min at 4°C.
5. Resuspend in 2 mL 1× nuclei solution.
6. Transfer cell suspension (1 mL) into two eppendorf tubes
7. Recentrifuge at 3000 rpm for 5 min at 4°C.
8. Resuspend in 1.5 mL each with 1× nuclei solution.
9. On ice for 20 min (with occasional shaking).
10. Up and down 30 times with shringe attached by 21G needle (*see* **Notes 1**).
11. Layer 750 μL of suspension over 500 μL 30% sucrose solution (3 g with 10 mL 1× nuclei solution).
12. Take care not to disturb sucrose solution.
13. Centrifuge with swing rotor at 3000 rpm 10 min at 4°C.
14. Discard supernatant.
15. Resuspend the pellet in 1 mL 1× nuclei solution.
16. 5000 rpm 5 min at 4°C.
17. Discard supernatant.

18. Resuspend the pellet in 120 μL buffer A.
19. Aliquot 20 μL × 5 tubes; store at –20°C.

3.4. Preparation of S/M Extracts

1. Culture chicken DU249 cells (12×T150 flasks) and presynchronize in S phase with aphidicolin at 2 μg/mL for 12 h.
2. Release from the block for 6 h and synchronize in mitosis with nocodazole at 100 ng/mL for 3 h.
3. Harvest mitotic cells with 2000 rpm for 5 min at 4°C to make S/M extracts (extracts for apoptosis induction) from floating cells obtained by selective detachment after the nocodazole treatment.
4. Wash cells with 40 mL incomplete KPM.
5. Centrifuge and resuspend with 40 mL of complete KPM.
6. Centrifuge and resuspend with 1 mL of complete KPM.
7. Transfer to a grinder (KONTES 20 or 21).
8. Centrifuge and aspirate supernatant.
9. Freeze in liquid nitrogen.
10. Store at –80°C.
11. Thaw in cold water and grind with pestle on ice just during thawing (*see* **Note 6**).
12. Freeze again in liquid nitrogen, thaw in cold water, and grind again.
13. Transfer to ultracentrifuge tubes (7×20 mm polycarbonate tube).
14. Ultracentrifuge at 55,000 rpm for 1 h at 4°C using Beckman TL-100, TLA100 rotor.
15. Recover clear extract.
16. Take 1 μL to measure protein concentration.
17. Make 20 μL aliquots and freeze in liquid nitrogen.
18. Store at –80°C.

3.5. How to Stain the Nucleus by Adding a Fluorescent Dye

1. Suspend MDA-AF8 nuclei with 200 μL 1× MDB buffer.
2. Centrifuge at 5000 rpm 5 min at 4°C.
3. Discard supernatant.
4. Stain nuclei with SYTO 59 for 20 min at 0.5 μM at RT.
5. Centrifuge at 5000 rpm 5 min at 4°C.
6. Wash twice with 1× MDB buffer.
7. Centrifuge at 5000 rpm 5 min at 4°C.
8. Discard supernatant.

3.6. Induction of Apoptsis

1. Suspend MDA-AF8 nuclei stained with SYTO 59 into a 10 μL reaction (S/M extracts) supplemented with 1 μL ATP plus a regeneration system.

3.7. Time-lapse Imaging

There are two different systems to obtain the time-lapse images, wide-field and confocal microscopies. Here, we mainly describe the basic components for the former system, such as a high sensitive cooled CCD camera, excitation and emission filter wheels, *z*-axis motor, and a stage-top incubator (**Fig. 8.2**). For the

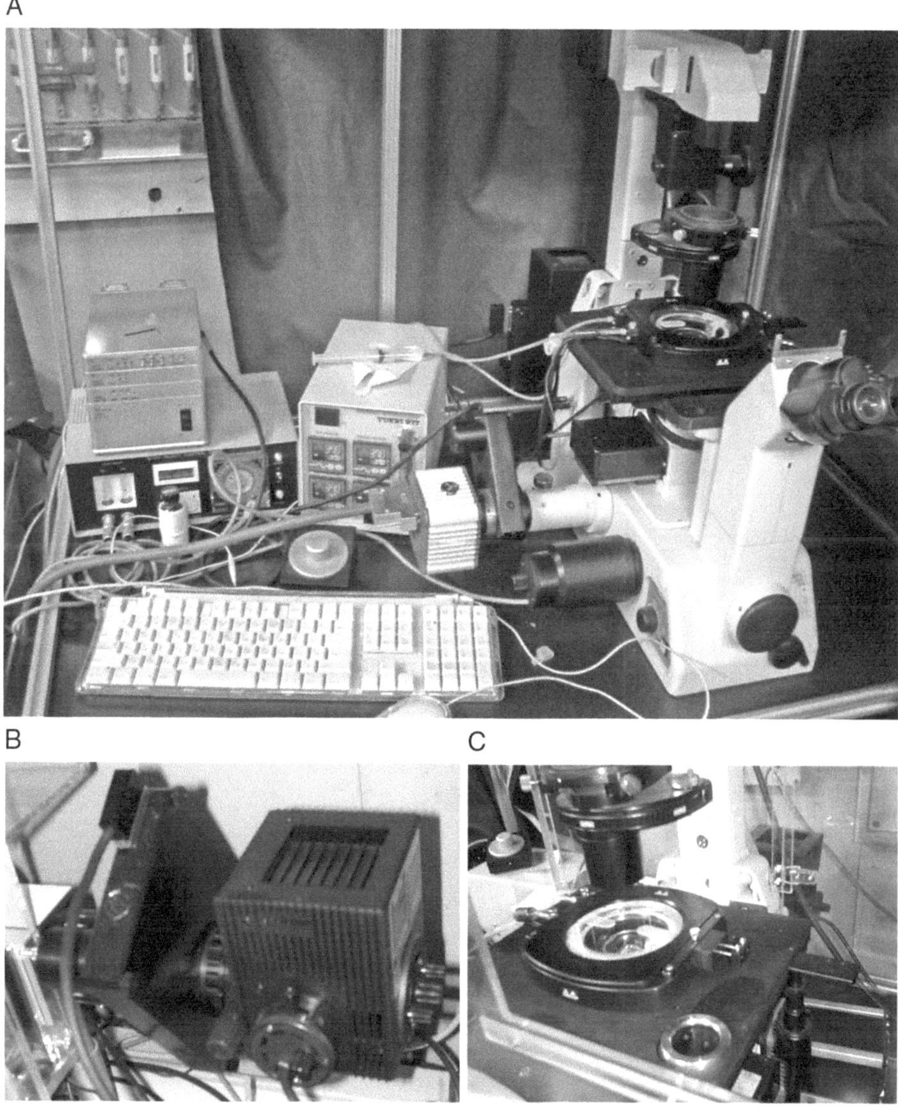

Fig. 8.2. A typical conventional wide-field microscope system. (**a**) The basic components contain a stage-top incubator INU-NI-F1 (Tokai Hit), a high sensitive cooled CCD camera ORCA-ER (Hamamatsu Photonics), a BioPoint MAC5000 excitation and emission filter wheels, and *z*-axis motor (Ludl Electric Products). (**b**) A 100 W halogen lamp is used a light source.

illumination, we have used a halogen lamp (1 and 13, *see* **Fig. 8.2b**), instead of Hg or Xe lamp. To obtain the confocal images, this system can be easily system-upped by applying combined laser beams (440, 488, and 561 nm) and setting a confocal scan unit (CSU10, Yokogawa Electric Corp., Tokyo, Japan) between the emission filter wheel and the CCD camera (**Fig. 8.3**). To operate this type of confocal microscopy system, we can use the same image capturing software (LuminaVision for MacOSX) as used for a wide-field microscopy (*see* **Note 7**).

1. Place the mixture on the bottom of a 35-mm glass base dish on the stage of a fluorescent microscope.

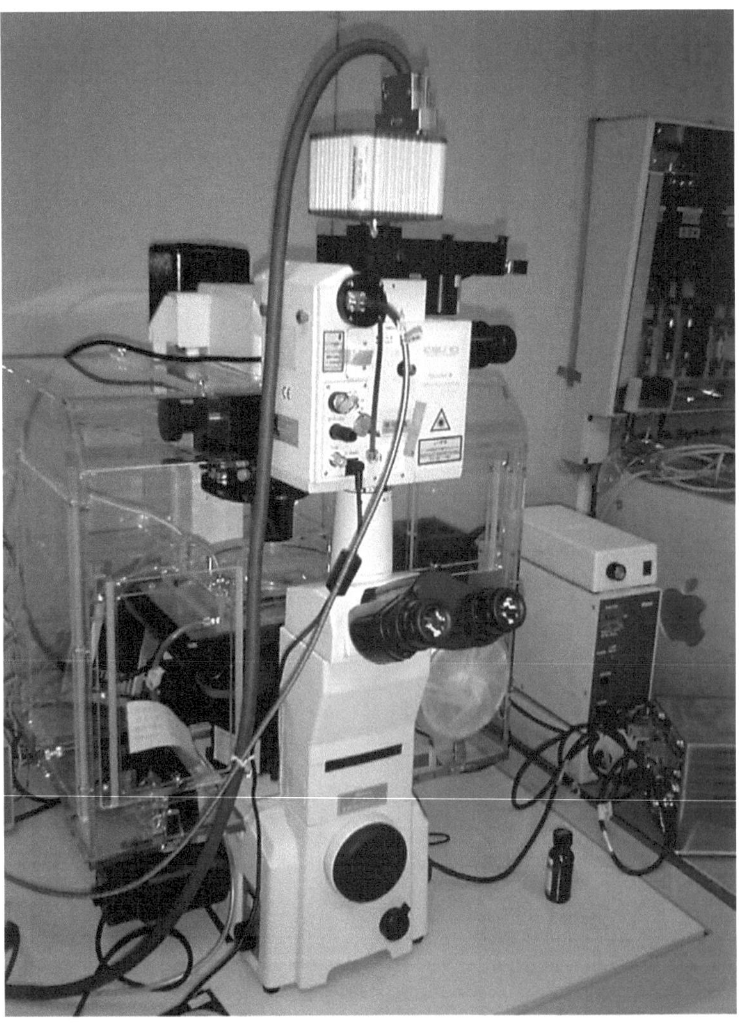

Fig. 8.3. System up to a laser confocal system. A laser scan unit CSU10 (Yokogawa Electric Corp.) is set between the microscope and the emission filter wheel. The combined laser beams of 440, 488, and 561 nm were used as a light source to illuminate ECFP, EGFP, and DsRed (*see* **Fig. 8.1d**).

2. Capture the images with a software (LuminaVision for MacOSX) controlling a cooled CCD digital camera (200–500 ms exposure for each filter) and rotating the excitation and emission filter wheels and *z*-axis motor with MAC5000, for collecting z-series optical sections (0.2-μm intervals).

4. Notes

1. Recently, a new 24- to 96-well type version, MXCell, is released. It is much convenient for determining the optimum condition to introduce plasmid DNA with this system.
2. The "VC" type of PlanApo 60× is recommended, rather than normal one. It is worth trying it if you want to obtain clearer images.
3. The filters with higher transmission (>95%) such as ion beam sputtering (IBS) thin-film coating filters (BrighLine, Semrock) or modified magnetron sputter-coated ones (ET Series, Chroma) are recommended. Sometimes, it is the easiest and most effective way to improve the signal of images, rather than obtaining an highly expensive EM-CCD camera.
4. The Dichroic 86006bs can be used for CFP/GFP/DsRed as well (14). Recently, higher transmitting multiband filter sets are available for CFP/YFP/HcRed (Semrock) and CFP/YPF/mCherry (89006, Chroma), but not yet for four living colors such as CFP/GFP/mKO/DsRed.
5. Check lysis with a phase-contrast microscope. If lysis is not complete, add 10 or 20 more strokes.
6. We prefer manual grinding to machine-driven grinding. Excess grinding will destroy the apoptotic activity.
7. It may be better to use the other excellent software (15) such as MetaMorph (Molecular Device, Sunnyvale, CA) or Volocity (Improvision a PerkinElmer Company, Coventry, UK) to operate the recent devices such as a precisExcite LED excitation (CoolLED Ltd., Andover, UK) or AURA or Spectra Light Engine (Lumencor, Inc., Beaverton, OR) and a Laser Merge Module, LMM5 (Spectral Applied Research, Ontario, Canada) or extremely sensitive EM-CCD cameras such as ImagEM (Hamamatsu Photonics) or Cascade II (Photometrics a division of Roper Scientific, Tucson, AZ). Recently, ORCA-R2 (Hamamatsu Photonics) is available for the new version of ORCA series, instead of ORCA-ER or AG.

Acknowledgments

The author would like to thank Mrs. Kouichi Amami (Tsukasa Giken, Osaka, Japan), Hideo Hirukawa (Yokogawa Electric Corp., Tokyo, Japan) and Ken Salisbury (formerly, Improvision, Coventry, UK) for the technical assistance for these microscopic analyses. This work is dedicated to the deceased Katsuji Rikukawa (Nikon Instech Co. Ltd., Japan).

References

1. Sugimoto, K., Senda-Murata, K., and Oka, S. (2008) Construction of three quadruple-fluorescent MDA435 cell lines that enable monitoring of the whole chromosome segregation process in the living state. *Mut Res* **657**, 56–62.
2. Brenner S., Pepper, D., Berns, M.W., Tan, E., and Brinkley, B. (1981) Kineochore structure, duplication, and distribution in mammalian cells: analysis by human autoantibodies from scleroderma patients. *J Cell Biol.* **91**, 95–102.
3. Rieder, C.L. (1982) The formation, structure, and composition of the mammalian kinetochore and kinetochore fiber. *Int Rev Cytol* 79, 1–58.
4. Shaner N.C., Steinbach, P.A. and Tsien R.Y. (2004) A guide to choosing fluorescent proteins. *Nat Methods* **2**, 905–909.
5. Wyllie, A.H., Morris, R.G., Smith, A.L., and Dunlop, D. (1984) Chromatin cleavage in apoptosis: association with condensed chromatin morphology and dependence on macromolecular systheswis. *J Pathol* **142**, 67–77.
6. Lazebnik, Y.A., Cole, S., Cooke, C.A., Nelson, W.G., and Earnshaw, W.C. (1993) Nuclear events of apoptosis in vitro in cell-free mitotic extracts: a model system for analysis of the active phase of apoptosis, *J Cell Biol* **123**, 7–22.
7. Samejima, K., Tone, S., Kottke, T.J., Enari, M., Sakahira, H., Cooke, C.A., Durrieu, F., Martins, L.M., Nagata, S., Kaufmann, S.H. and Earnshaw, W.C. (1998) Transition from caspase-dependent to caspase-independent mechanisms at the onset of apoptotic execution, *J Cell Biol* **143**, 225–239.
8. Liu, X., Kim, C.N., Yang, J., Jemmerson, R., and Wang, X. (1996) Induction of apoptotic program in cell-free extracts: requirement for dATP and cytochrome c, *Cell* **86**, 147–157.
9. Widlak, P., Palyvoda, O., Kumala, S., and Garrard, W.T. (2002) Modeling apoptotic chromatin condensation in normal cell nuclei. Requirement for intranuclear mobility and actin involvement, *J Biol Chem* **277**, 21683–21690.
10. Tone, S., Sugimoto, K., Tanda, K., Suda, T., Uehira, K., Kanouchi, H., Samejima, K., Minatogawa, Y. and Earnshaw, W.C. (2007) Three distinct stages of apoptotic nuclear condensation revealed by time-lapse imaging, biochemical and electron microscopy analysis of cell-free apoptosis, *Exp Cell Res* **313**, 3635–3644.
11. Sugimoto, K., Fukuda, R. and Himeno, M. (2000) Centromere/kinetochore localization of human centromere protein A (CENP-A) exogeneously expressed as a fusion to green fluorescent protein. *Cell Struct Funct* **25**, 253–261.
12. Sugimoto, K., Urano, T., Zushi, H., Inoue, K., Tasaka, H., Tachibana, M. and Dotsu, M. (2002) Molecular dynamics of Aurora-A kinase in living mitotic cells simultaneously visualized with histone H3 and nuclear membrane protein importin-alpha. *Cell Struct Funct* **27**, 457–467.
13. Sugimoto, K. (2006) Setting up a fluorescent microscope for live cell imaging in three colors. *Jikken Igaku* (*Expeimental Medicine*, in Japanese), **24**, 87–92.
14. Fukada, T., Inoue, K., Urano, T., and Sugimoto, K. (2004) Visualization of chromosomes and nuclear envelope in living cells for molecular dynamics studies. *BioTechniques* **37**, 552–556.
15. Sugimoto, K. (2006) Live cell imaging analysis of three-color cell-lines. *Jikken Igaku* (*Experimental Medicine*, in Japanese), **24**, 397–401.

Chapter 9

Manipulation of Neutrophil-Like HL-60 Cells for the Study of Directed Cell Migration

Arthur Millius and Orion D. Weiner

Abstract

Many cells undergo directed cell migration in response to external cues in a process known as chemotaxis. This ability is essential for many single-celled organisms to hunt and mate, the development of multicellular organisms, and the functioning of the immune system. Because of their relative ease of manipulation and their robust chemotactic abilities, the neutrophil-like cell line (HL-60) has been a powerful system to analyze directed cell migration. In this chapter, we describe the maintenance and transient transfection of HL-60 cells and explain how to analyze their behavior with two standard chemotactic assays (micropipette and EZ-TAXIS). Finally, we demonstrate how to fix and stain the actin cytoskeleton of polarized cells for fluorescent microscopy imaging.

Key words: Migration, chemotaxis, neutrophil, HL-60, actin cytoskeleton, amaxa transfection, micropipette, EZ-TAXIS assay.

1. Introduction

Directed cell migration toward chemical cues, or chemotaxis, is essential in eukaryotic cells for development, immune response, and wound healing (1). The human neutrophil is a particularly powerful model for eukaryotic chemotaxis. Neutrophils seek infectious bacteria to engulf at wound sites as part of the innate immune system. They follow gradients of formylated peptides released by the bacteria (1). Many open questions remain in neutrophil and eukaryotic cell migration. How do cells interpret shallow gradients or initially establish polarity? How is their cytoskeleton rearranged during polarization and movement, and what

D.B. Papkovsky (ed.), *Live Cell Imaging*, Methods in Molecular Biology 591,
DOI 10.1007/978-1-60761-404-3_9, © Humana Press, a part of Springer Science+Business Media, LLC 2010

limits this rearrangement to one part of the cell and not another? What signaling components and circuitry are required to accomplish theses processes?

The human promyelocytic leukemia (HL-60) cell line was developed as a simple model system to study neutrophil cell migration without the need to derive cells from primary tissue (2). The immortal cell line can be propagated for extended periods of time in culture and may be frozen for longer term storage. This is impossible with bone marrow or peripheral blood derived neutrophils. When needed, neutrophil-like cells can be derived from HL-60 cells through differentiation with DMSO or retinoic acid (3). Differentiated HL-60 cells (dHL-60) can then be used in chemotaxis assays and to visualize the cytoskeleton of a polarized cell (4). Amaxa nucleofection may be used in dHL60 cells to knock genes down (5–8) or transfect cells with a fluorescent tag to analyze protein localization (our unpublished results). Cell behavior can be analyzed either in response to a point source of chemoattractant (9–11), or using the EZ-TAXIS system, which allows simultaneous measurement of directionality and speed for six different assay conditions (12, 13).

2. Materials

2.1. HL-60 Cell Culture and Differentiation

1. Culture medium: Roswell Park Memorial Institute (RPMI) 1640 plus L-glutamine and 25 mM HEPES (Fisher Scientific) supplemented with antibiotic/antimycotic (Invitrogen) and 15% heat-inactivated fetal bovine serum (FBS) (Invitrogen); store at 4°C (*see* **Note 1**).
2. Dimethyl sulphoxide (DMSO), endotoxin, and hybridoma tested (Sigma).

2.2. Amaxa Nucleofection of HL-60 Cells

1. Recovery medium: VZB-1003 monocyte medium (Amaxa Inc.) supplemented with 2 mM glutamine (Invitrogen) and 20% FBS.
2. 100 μL of transfection solution containing VCA-1003 supplement (Amaxa) mixed with 2 μg DNA per reaction (*see* **Note 2**).

2.3. Plating Cells for Microscopy

1. Fibronectin from bovine plasma (Sigma); store lyophilized protein at –20°C.
2. Ca^{2+}/Mg^{2+} free phosphate-buffered saline (PBS): 137 mM NaCl, 2.7 mM KCl, 10 mM Na_2HPO_4, 1.8 mM KH_2PO_4 (Invitrogen).

3. Gold Seal cover glass 20 × 40 mm No. 1.5 (Fisher Scientific).
4. Lab-Tek 8-well Permanox™ chamber slide (Nunc).
5. Chemoattractant: 10 mM formylated Methione-Leucine-Phenylalanine (fMLP),(Sigma-Aldrich) in DMSO; store at –20°C. Prepare 100 μM working stocks in RMPI and store at 4°C up to 1 month (*see* **Note 3**).

2.4. Micropipette Assay

1. Glass capillary with filament (World Precision Instruments) (*see* **Note 4**).
2. Alexa594 working stock: 10 mM Alexa Fluor 594 hydrazide sodium salt (Invitrogen) in DMSO; store at 4°C and protect from light.
3. Chemoattractant solution: 200 nM fMLP, 10 μM Alexa594 in RPMI culture medium; protect from light.

2.5. EZ-TAXIScan Assay

1. RPMI culture medium.
2. Chemoattractant solution: 200 nM fMLP in RMPI culture medium.

2.6. Staining the Actin Cytoskeleton

1. Intracellular buffer (2×): 280 mM KCl, 2 mM $MgCl_2$, 4 mM EGTA, 40 mM HEPES, pH 7.5, 0.4% low endotoxin albumin from human serum (Sigma) (*see* **Note 5**).
2. Fixation buffer (2×): 640 mM sucrose, 7.4% formaldehyde (Sigma) in 2× intracellular buffer; store at 4°C (*see* **Note 6**).
3. Stain buffer: 0.2% Triton X-100, 2 μL/mL rhodamine phalloidin (Invitrogen) in intracellular buffer (*see* **Note 7**).

3. Methods

3.1. Maintenance of HL-60 Cell Culture Line

1. Unless imaging, all cell work is performed under a biological safety cabinet.
2. HL-60 cells are passaged when the cells reach a density between 1 and 2 million cells/mL in 25 cm^2 cell culture flasks with 0.2 μm vent cap. Split cells to 0.15 million cells/mL in a total volume of 10 mL prewarmed culture medium. Cells will need to be passaged again after 2–3 days (**Fig. 9.1**). Maintain cells at 37°C and 5% CO_2 in standard tissue culture incubator (*see* **Note 8**).
3. Differentiate cells in culture medium containing 1.3% DMSO. Because DMSO is more viscous and denser than culture medium, premix medium with DMSO before adding

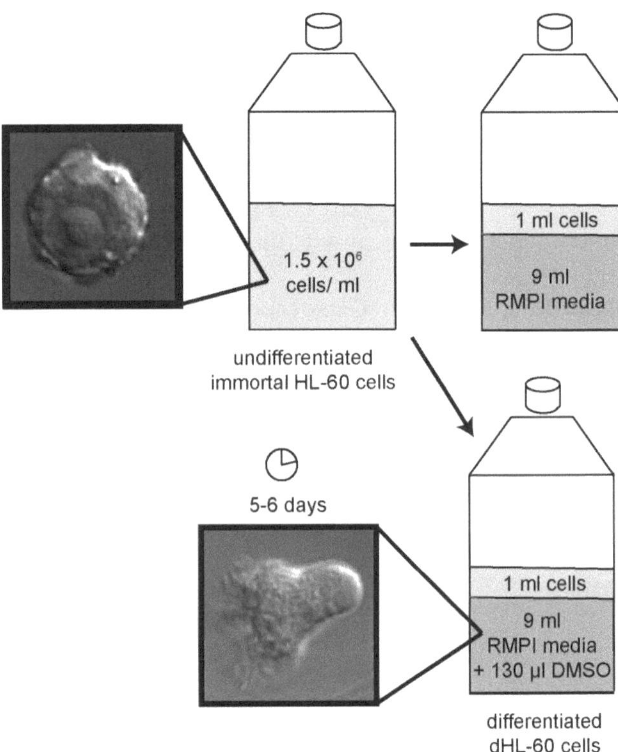

Fig. 9.1. Passaging and differentiating HL-60 cells. When cells reach a density between 1 and 2 million cells/mL, split to 0.15 million cells/mL in a total volume of 10 mL prewarmed culture medium. Differentiate cells in culture medium plus 1.3% DMSO; cells take ~5 days to become migratory.

cells. Cells stop proliferating upon differentiation and typically achieve a density of 1–2 million cells/mL at 7 days post-differentiation (**Fig. 9.1**). Cells are most active 5–6 days post-differentiation, but can still respond even after 8 days (*see* **Note 9**).

4. To freeze cells, pellet cells by spinning at 100×*g* for 10 min. Aspirate medium and resuspend in chilled culture medium plus 10% DMSO at 10 million cells/mL. Aliquot 1.8 mL each into cryovials, place in Nalgene cryofreezing container with isopropanol at –80°C for 2 days, and then transfer to liquid nitrogen storage (*see* **Note 10**).
5. Thaw cells by quickly warming a cryovial at 37°C just until last bit of ice has melted. Dilute thawed cells in 10 mL of prewarmed culture medium and spin at 100×*g* for 10 min. Remove supernatant, resuspend pellet in 20 mL culture medium, and recover in a 75-cm^2 culture flask.

3.2. Transient Transfection of DNA into HL-60 Cells

1. Prepare ~2 mL of recovery medium per transfection in a 6-well plate and let equilibrate at 5% CO_2 and 37°C for 15 min or more. Add 500 μL of equilibrated recovery medium to an eppendorf tube per transfection (*see* **Note 11**).

2. Spin 5 million cells in a 10-mL Falcon tube at 100×*g* for 10 min. Use a separate Falcon tube for each transfection.
3. After the spin, remove all medium with aspirator and gently resuspend cells in 100 μL transfection solution with a L-1000 pipette (*see* **Note 12**).
4. As quickly as possible, transfer transfection solution to nucleofection cuvette. Electroporate in single chamber nucleofector on program Y-001.
5. Immediately remove cuvette, use a plastic pipette to obtain 500 μL recovery medium from eppendorf tube, flush chamber, and replace medium containing cells in eppendorf tube.
6. Incubate for 30 min in eppendorf tube at 37°C (*see* **Note 13**).
7. Transfer 500 μL recovery medium and cells with L-1000 pipette to 1.5 mL recovery medium in a 6-well plate. Expression in viable cells occurs in ~2 h with best behavior <8 h after transfection (**Fig. 9.2**).

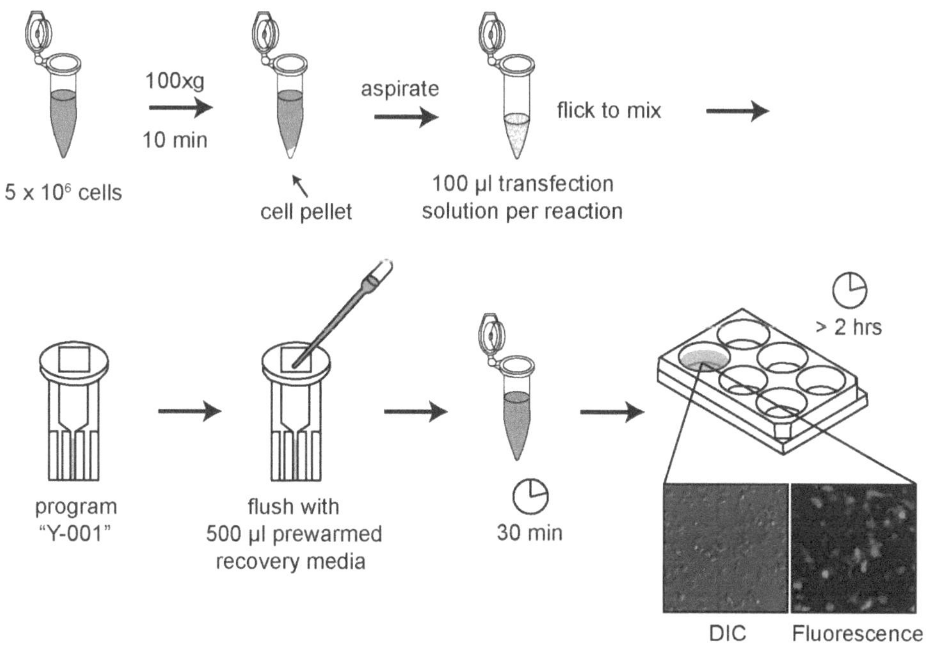

Fig. 9.2. Transient transfection of HL-60 cells with Amaxa nucleofection. Spin ~5 million cells at 100×*g*. Aspirate supernatant and resuspend pellet in 100 μL transfection solution per reaction and nucleofect with Amaxa program "Y-001." Flush with prewarmed recovery medium and incubate in an eppendorf tube for 30 min. Transfer to a 6-well dish with 1.5 mL of recovery medium; expression occurs after 2 h. Shown is an example of HL-60 cells 5 h after transfection with GFP visualized with DIC and fluorescence microscopy.

3.3. Preparing for Live Cell Microscopy

1. Dissolve 1 mL of sterile water in 1 mg fibronectin and let sit at room temperature for 1 h. Avoid shaking or physically mixing because this may denature the fibronectin.

2. Add 4 mL of Ca^{2+}/Mg^{2+} free PBS to fibronectin solution to obtain 200 μg/mL solution (**Fig. 9.3a**): store fibronectin solution at 4°C for up to 2 weeks.

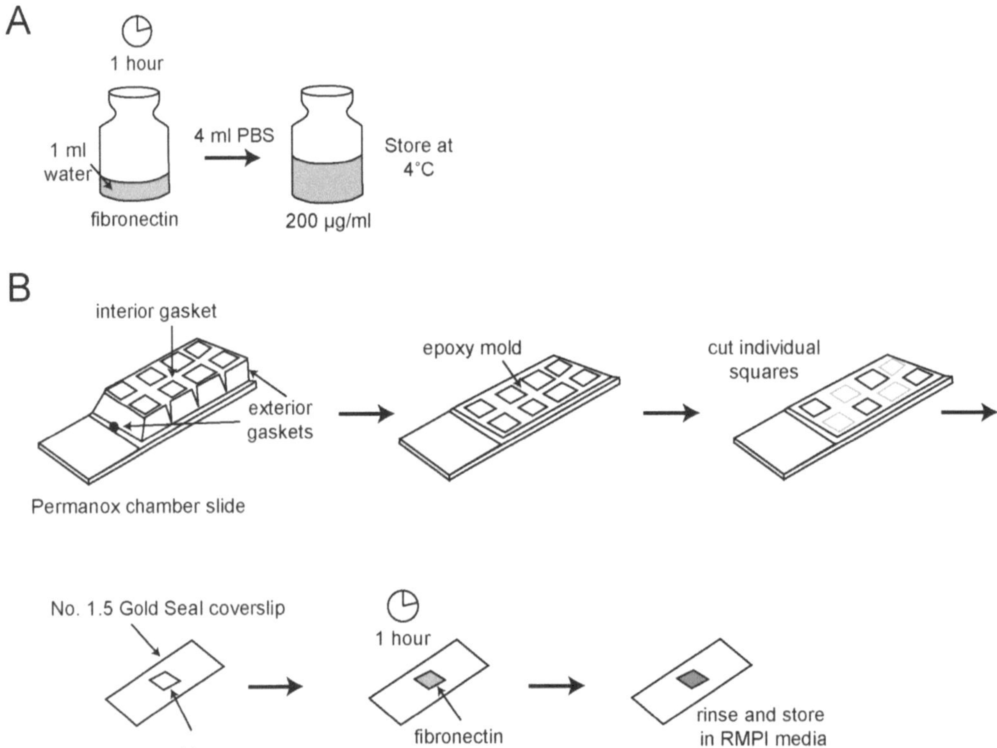

Fig. 9.3. Preparing a coverslip for live cell microscopy. (**a**) Dissolve 1 mg of bovine fibronectin in sterile water. After 1 h, add 4 mL of PBS and store 200 μg/mL fibronectin solution at 4°C. (**b**) Remove gaskets from plastic permanox 8-well chamber. Cut epoxy mold squares and stick to No. 1.5 gold seal cover glass. Add 125 μL of fibronectin, let sit for 1 h, rinse once with RPMI culture medium, and store in RPMI medium until ready to image.

3. Remove exterior gaskets from Permanox chamber slide and the interior gasket with a razor blade. Discard 8-well plastic walls to expose underlying epoxy mold. With razor, cut single-well epoxy chambers and adhere to glass coverslips (*see* **Note 14**).
4. Add 125 μL fibronectin solution to epoxy chamber on coverslip. Incubate at room temperature for 1 h.
5. Aspirate fibronectin solution and rinse once in RPMI culture medium. Add 250 μL RPMI culture medium until ready to plate cells. Fibronectin coated slides may be stored up to 2 days at 4°C (**Fig. 9.3b**).
6. For cell plating, remove culture medium on slide and replace with 250 μL differentiated cells (in RPMI culture medium) or transfected cells (in monocyte medium). Incubate at 37°C for 5–15 min (*see* **Note 15**).

7. Aspirate cells at the corner of the epoxy chamber (avoid touching the coverslip), rinse in RPMI, and replace with 125 μL RPMI. Cells are most migratory at 37°C, but also function at room temperature.
8. During image acquisition, 125 μL of 200 nM fMLP may be added to coverslip (final concentration is 100 nM) to show random migration in response to uniform chemoattractant.

3.4. Micropipette-Stimulated Cell Migration

1. Pull glass filaments on Model P-87 micropipette puller (Sutter) (Program: heat = 750, pull = 0, velocity = 20, time = 250, pressure = 100, loops 2 or 3 times) to achieve ~2- to 3-μm needle diameter.
2. Backfill needles with chemoattractant solution and connect to a needle holder. Flick to remove air bubbles at the tip. The needle holder is held by a micromanipulator and agonist flow controlled by adjusting balance pressure between 0 and 3 psi on an IM-300 injection system (Narishige) (*see* **Note 16**).
3. Place cells seeded on a coverslip on microscope and find focus of cells in brightfield.
4. Orient needle in light path and switch to the back focal plane of the objective (usually labeled "B" for Bertrand lens). Find the needle and lower it until just above the cells, then switch back to the normal focal plane for viewing.
5. Image dye in fluorescence or total internal reflection fluorescence microscopy using appropriate filter sets with ~20 ms exposures and near maximal multiplication on an EM-CCD camera (*see* **Note 17**). An example of a cell migrating toward a micropipette filled with chemoattractant is shown in **Fig. 9.4**.

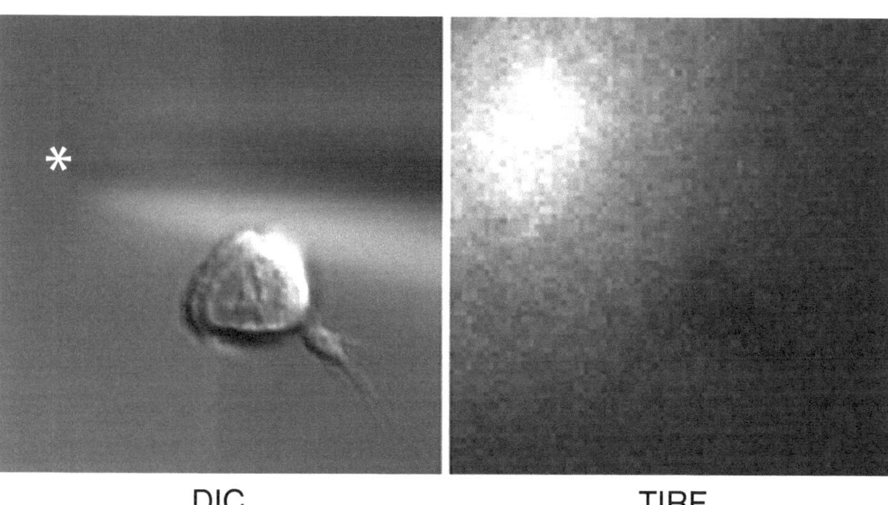

Fig. 9.4. An example of an HL-60 cell crawling toward a micropipette visualized with DIC and TIRF microscopy. Asterisk indicates micropipette tip.

3.5. Ascertaining Directionality and Velocity of Migrating Cells with the "EZ-TAXIS" Assay System

1. Place "40 Glass" (41 Glass is used in conjunction with a #1.5 coverslip) in holder base. Fill with 3 mL of culture medium, place small "o" ring beneath wafer housing, and add wafer housing into holder base. Place large "o" ring on top of wafer housing. Gently close inner lever.
2. Place EZ-TAXIScan chip gently into wafer housing with forceps. Chip should fit snuggly on top of glass and between wafer holder (*see* **Note 18**). Place rubber gasket (to protect chip) beneath the wafer clamp and place wafer clamp on wafer housing. Gently close outer lever.
3. Place assembled device on top of preheated EZ-TAXIS microscope (**Fig. 9.5a, b**).

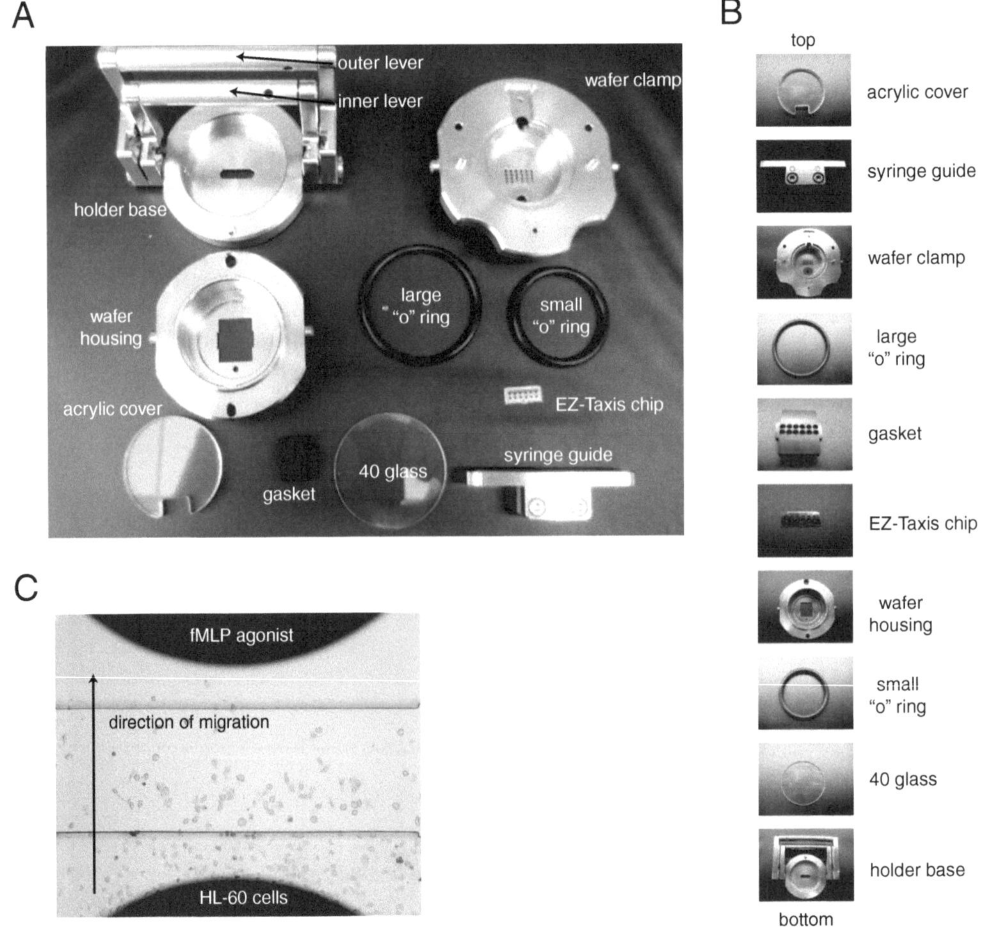

Fig. 9.5. The components of the EZ-TAXIS system are shown in (**a**), with the individual components in their order of assembly from top to bottom shown in (**b**). (**c**) An example of HL-60 cells migrating toward chemoattractant in the EZ-TAXIS assay visualized with brightfield microscopy.

4. Using syringe guide, add 1 μL of cells to lower chamber using microsyringe. Use a plastic pipette to draw cells into imaging surface.
5. Add 1 μL of 100 nM – 1 μM chemoattractant to upper chamber. Begin image acquisition immediately (**Fig. 9.5c**).

3.6. Staining the Actin Cytoskeleton

1. Stimulate adherent cells with micropipette or uniform chemoattractant. Alternatively, you can stimulate cells in suspension.
2. Add one volume of 2× fixation buffer to cells to give a final concentration of 320 mM sucrose and 3.7% formaldehyde. Fix for 20 min at 4°C (*see* **Note 19**).
3. Aspirate supernatant (adherent cells) or spin down cells at 400×*g* for 1 min and then aspirate supernatant (suspended cells).
4. Permeabilize cells with 125 μL stain buffer and protect from light for 20 min (**Fig. 9.6**).

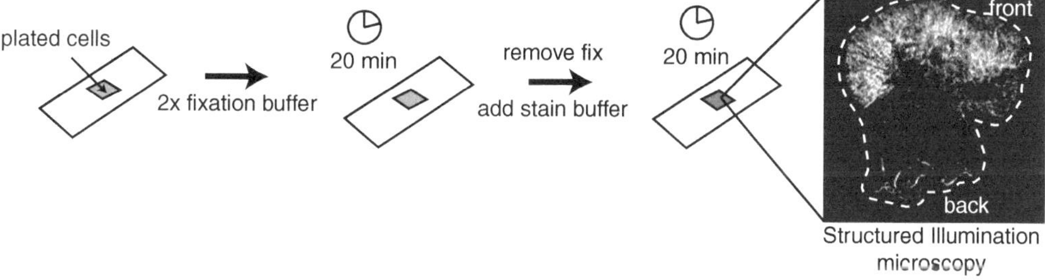

Fig. 9.6. Staining the actin cytoskeleton. Add 2× fixation buffer to plated cells and fix for 20 min at 4°C. Remove fixation buffer and replace with stain buffer for 20 min; protect from light. Shown is an example of an HL-60 cell stained with rhodamine phalloidin visualized with structured illumination microscopy.

5. Remove supernatant, replace with intracellular buffer, and image (*see* **Note 20**).

4. Notes

1. HL-60 cells acidify their medium quite rapidly, so HEPES is essential to counterbalance the pH.
2. Poor expression occurs in retroviral LTR-containing vectors.
3. If an exact concentration of fMLP is required, make stocks fresh weekly because formyl peptide will slowly oxidize in aqueous solutions over time.

4. Capillaries with filaments are easier to load than capillaries without filaments.
5. HL-60 cells may be spuriously stimulated with albumin containing endotoxins. It is important to add the albumin fresh to the intracellular buffer and let sit at room temperature for 30 min before mixing because the albumin will oxidize in aqueous solution to products that may also activate HL-60 cells.
6. Sucrose is important to maintain osmolarity.
7. Unlike Alexa Fluor 594, rhodamine phalloidin increases its fluorescence upon binding to actin filaments. This makes it preferable to Alexa Fluor 594. However, for imaging in the green channel, we would use Alexa Fluor 488 or similar fluorophore over the rapidly bleaching FITC.
8. Keep a basin of water on the bottom shelf of the incubator to maintain humidity.
9. Cells may also be differentiated with retinoic acid (14).
10. As an alternative to Nalgene freezing chambers, two styrofoam 15 mL Falcon tube holders may be sandwiched around tubes to provide insulation during slow freezing process.
11. Our protocol differs significantly from the standard Kit V protocol from Amaxa because we have optimized for cells to retain chemotactic function (not necessarily highest transfection efficiency).
12. It is important to remove every drop of medium from the Falcon tube because FBS will interfere with transfection. Large-tip pipettes minimize shearing of cells and increases the number of healthy cells post-transfection.
13. Cells will adhere to each other and form a thin, white film during the 30-min incubation in eppendorf tube. This step increases recovery efficiency. Additionally, cells are fragile after transfection and they behave poorly if centrifuged.
14. An 8-well chamber will yield four molds. Only bottom surface (side touching permanox chamber slide) of epoxy mold is sticky. Additionally, Lab-Tek 8-well cover glass #1.5 chamber slide (Nunc) may be used for epifluorescence and brightfield microscopy. However, the epoxy that hold the chamber walls to the coverslip is autofluorescent and thus less suitable for TIRF.
15. Cells may be plated in culture medium containing chemoattractant for a higher proportion of adherent cells.
16. We use a universal needle holder, MINJ4 (Tritech) connected to a MM-89 micromanipulator (Narishige). Addi-

tionally, it is difficult to handle capillaries with gloves on. Once the solution is loaded into the needle, flick the needle vigorously to remove microscopic bubbles. Do not try to force air bubbles through the tip by increasing air pressure. Invariably, the bubbles will become stuck and the needle rendered useless.

17. For quantification of gradient, TIRF imaging is superior to epifluorescence microscopy because TIRF reveals the agonist concentrations in the same plane as the cell. The signal obtained from epifluorescence is a combination of the agonist the cell experiences plus out of focus blur from fluorescent dye above the cell. Confocal imaging can also be used to quantitate the gradient.
18. Chips come in four different sizes (4, 5, 6, and 8 μm) depending on the gap space between the chip and glass. We have used the 4–6 μm chips with success.
19. Formaldehyde can partially permeabilize cells, allowing water influx due to high internal protein concentration.
20. Can repeat wash if background too high. If pressed for time, you can immediately image after adding stain buffer. However, this will give high background staining. You can also mount cells in Vectashield (Vector Laboratories) for decreased spherical aberrations and photobleaching.

Acknowledgements

This work was supported by the NIH (R01GM084040), a UCSF/UC Berkeley Nanomedicine Development Center Award, a Searle Scholars Award to O. Weiner, and a National Defense Science and Engineering Graduate Fellowship to A. Millius.

References

1. Ridley, A. J., Schwartz, M. A., Burridge, K., Firtel, R. A., Ginsberg, M. H., Borisy, G., et al. (2003) Cell migration: integrating signals from front to back. *Science* **302**, 1704–1709.
2. Collins, S. J., Gallo, R. C., and Gallagher, R. E. (1977) Continuous growth and differentiation of human myeloid leukaemic cells in suspension culture. *Nature* **270**, 347–349.
3. Collins, S. J., Ruscetti, F. W., Gallagher, R. E., and Gallo, R. C. (1978) Terminal differentiation of human promyelocytic leukemia cells induced by dimethyl sulfoxide and other polar compounds. *Proc Natl Acad Sci USA* **75**, 2458–2462.
4. Rao, K. M., Currie, M. S., Ruff, J. C., and Cohen, H. J. (1988) Lack of correlation between induction of chemotactic peptide receptors and stimulus-induced actin polymerization in HL-60 cells treated with dibutyryl cyclic adenosine monophosphate or retinoic acid. *Cancer Res* **48**, 6721–6726.
5. Carter, B. Z., Milella, M., Tsao, T., McQueen, T., Schober, W. D., Hu, W., et al. (2003) Regulation and targeting of

antiapoptotic XIAP in acute myeloid leukemia. *Leukemia* **17**, 2081–2089.
6. Nakata, Y., Tomkowicz, B., Gewirtz, A. M., and Ptasznik, A. (2006) Integrin inhibition through Lyn-dependent cross talk from CXCR4 chemokine receptors in normal human $CD34^+$ marrow cells. *Blood* **107**, 4234–4239.
7. de la Fuente, J., Manzano-Roman, R., Blouin, E. F., Naranjo, V., and Kocan, K. M. (2007) Sp110 transcription is induced and required by *Anaplasma phagocytophilum* for infection of human promyelocytic cells. *BMC Infect Dis* 7, 110.
8. Gattenloehner, S., Chuvpilo, S., Langebrake, C., Reinhardt, D., Muller-Hermelink, H. K., Serfling, E., et al. (2007) Novel RUNX1 isoforms determine the fate of acute myeloid leukemia cells by controlling CD56 expression. *Blood* **110**, 2027–2033.
9. Weiner, O. D., Servant, G., Welch, M. D., Mitchison, T. J., Sedat, J. W., and Bourne, H. R. (1999) Spatial control of actin polymerization during neutrophil chemotaxis. *Nat Cell Biol* **1**, 75–81.
10. Servant, G., Weiner, O. D., Herzmark, P., Balla, T., Sedat, J. W., and Bourne, H. R. (2000) Polarization of chemoattractant receptor signaling during neutrophil chemotaxis. *Science* **287**, 1037–1040.
11. Wang, F., Herzmark, P., Weiner, O. D., Srinivasan, S., Servant, G., and Bourne, H. R. (2002) Lipid products of PI(3)Ks maintain persistent cell polarity and directed motility in neutrophils. *Nat Cell Biol* **4**, 513–518.
12. Ferguson, G. J., Milne, L., Kulkarni, S., Sasaki, T., Walker, S., Andrews, S., et al. (2007) PI(3)Kγ has an important context-dependent role in neutrophil chemokinesis. *Nat Cell Biol* **9**, 86–91.
13. Nishio, M., Watanabe, K., Sasaki, J., Taya, C., Takasuga, S., Iizuka, R., et al. (2007) Control of cell polarity and motility by the $PtdIns(3,4,5)P_3$ phosphatase SHIP1. *Nat Cell Biol* **9**, 36–44.
14. Breitman, T. R., Selonick, S. E., and Collins, S. J. (1980) Induction of differentiation of the human promyelocytic leukemia cell line (HL-60) by retinoic acid. *Proc Natl Acad Sci USA* 77, 2936–2940.

Chapter 10

A Method for Analyzing Protein–Protein Interactions in the Plasma Membrane of Live B Cells by Fluorescence Resonance Energy Transfer Imaging as Acquired by Total Internal Reflection Fluorescence Microscopy

Hae Won Sohn, Pavel Tolar, Joseph Brzostowski, and Susan K. Pierce

Abstract

For more than a decade, fluorescence resonance energy transfer (FRET) imaging methods have been developed to study dynamic interactions between molecules at the nanometer scale in live cells. Here, we describe a protocol to measure FRET by the acceptor-sensitized emission method as detected by total internal reflection fluorescence (TIRF) imaging to study the interaction of appropriately labeled plasma membrane-associated molecules that regulate the earliest stages of antigen-mediated signaling in live B lymphocytes. This protocol can be adapted and applied to many cell types where there is an interest in understanding signal transduction mechanisms in live cells.

Key words: B lymphocyte, fluorescence resonance energy transfer (FRET), total internal reflection fluorescence (TIRF) imaging, B cell receptor (BCR) signaling, planar lipid bilayer, immune synapse.

1. Introduction

B lymphocytes play a pivotal role in the adaptive immune system by producing antibodies against pathogens (1). Binding of antigens to B cell receptors (BCRs) leads to BCR clustering that triggers the recruitment of Lyn, the first kinase in the BCR signaling cascade (2). Current evidence indicates that B cells contact antigen on the surface of antigen presenting cells (APCs). At the interface of the contact between B cell and APC, an immune synapse (IS), a structure associated with B cell activation, is

D.B. Papkovsky (ed.), *Live Cell Imaging*, Methods in Molecular Biology 591,
DOI 10.1007/978-1-60761-404-3_10, © Humana Press, a part of Springer Science+Business Media, LLC 2010

organized. Within the IS, various receptor-ligand interactions take place, including the association of the BCR with the antigen and the lymphocyte function-associated antigen-1 (LFA-1) with the intercellular cell adhesion molecule-1 (ICAM-1) (3, 4). Live cell imaging techniques provide both the temporal and spatial information concerning the dynamic association of the BCR with signaling molecules over the time and length scale that are critical to decipher the earliest events that occur during the contact of live B cells with APCs bearing antigen. The advent of fluorescent protein technologies has made such observations possible by standard microscopy, and the development of sensitive imaging techniques, such as TIRF imaging, has enabled the study of signaling proteins at a single molecule level (5, 6).

TIRF imaging is a spatially limited technique that is particularly well suited to the study of molecules near the plasma membrane. In TIRF imaging, fluorophores in specimens are excited with collimated light, typically from a laser, that is introduced at a specific angle. The incident beam of excitation light, brought either through a prism or a high numerical aperture, oil-immersion objective lens, is totally internally reflected to form an evanescent field at the interface of the coverslip and aqueous media bathing the cells. In the case of through-lens systems, an evanescent field is produced along the optical axis by virtue of the difference in refractive index between oil/glass and aqueous medium of the cell. The evanescent field exponentially decays as it enters the cell, penetrating to a depth of ~100 nm, with depth being dependent on the angle of incidence and the wavelength of excitation light. As a result, TIRF imaging greatly reduces background fluorescence outside the plane of focus to the point where fluorescent emission from single fluorophores can be detected.

To study the interaction of membrane-associated proteins between B cells and APCs by TIRF imaging, planar lipid bilayers bearing mobile ligands have been developed to mimic APC (7). This system provides a means to the study the spatial and temporal dynamics of receptor-signaling molecule interaction in live B cells introduced to the lipid bilayer-coated chamber coverslip. In contrast to standard confocal microscopy, TIRF imaging enables the entire surface area of the cell in contact with the planar lipid bilayer to be imaged as interfering fluorescent signals from the interior of the cell are eliminated. In addition, because highly sensitive CCD cameras are used to capture images, frame rates can greatly exceed (>30 fps) those obtained on a typical scanning confocal microscope.

We have taken advantage of TIRF microscope technology and combined it with FRET imaging to study dynamic signal transduction events in live B cells (8, 9). FRET is the non-radioactive transfer of energy from a donor fluorophore to an acceptor fluorophore by dipole–dipole coupling. The result is a quenching of donor fluorescence and a concomitant increase (sensitization) in acceptor emission. Energy transfer can take place only over a limited distance (within 10 nm); thus, the closer the donor and acceptor fluorophores, the more efficient is the energy transfer. FRET efficiency (E) is exquisitely sensitive to the distance separating donor and acceptor fluorophores, being inversely dependent by the 6th power (10). This dependency provides a nanometer-scale resolution that can be measured via standard light microscopes. Several techniques have been developed to detect FRET, which include acceptor photobleaching, fluorescence lifetime imaging (FLIM) of the donor fluorophore, and acceptor-sensitized emission (also called three cube FRET) (11–13). While the sensitized emission approach requires the most careful controls and corrections of all the methods, it allows FRET measurements to be performed on relatively inexpensive equipment (compared to FLIM) and allows for dynamic measurements in live cells (compared to acceptor photobleaching, which is an end point method that destroys the fluorescence of the cell). To measure FRET by the sensitized emission, images from three channels are acquired: donor channel (D), acquired with the donor laser excitation wavelength and donor emission filter; FRET channel (F), acquired with donor laser excitation wavelength and acceptor emission filter; and acceptor channel (A), acquired with acceptor laser excitation wavelength and acceptor emission filter. Acquired images are processed to calculate FRET efficiency and FRET efficiency can be expressed as either FRET efficiency for the acceptor (Ea) or FRET efficiency for the donor (Ed) according to the following equations (14):

$$\mathrm{Ea} = \frac{\mathrm{FRa}}{Ka} \qquad [1]$$

$$\mathrm{Ed} = \frac{\mathrm{FRd}}{Kd + \mathrm{FRd}} \qquad [2]$$

where FRa and FRd are determined by

$$\mathrm{FRa} = \frac{F - \beta * D - \gamma * (1 - \beta * \delta) * A}{A * \gamma * (1 - \beta * \delta)} \qquad [3]$$

$$\mathrm{FRd} = \frac{F - \beta * D - \gamma * (1 - \beta * \delta) * A}{D - \delta * F} \qquad [4]$$

In these equations, D, F and A are background-subtracted fluorescence intensities in the donor, FRET, and acceptor channels, respectively. β is correction factor for donor spectral bleedthrough into FRET channel (F/D) when excited by 442 nm laser and is determined from cells expressing only the donor fluorophore. γ and δ are correction factors obtained from cells expressing only the acceptor fluorophore for acceptor crosstalk captured in FRET channel (F/A) and spectral bleed-through of acceptor emission back into donor channel (D/F), respectively. *Ka* and *Kd* are conversion constants between FRET ratios and FRET efficiencies determined from cells expressing a FRET positive control. The positive control is a fusion of the donor and the acceptor fluorophore at distance close enough to produce non-zero FRET at a 1:1 molar ratio. In the protocol below, we monitor the association of the BCR with Lyn in B cells contacting antigen-bound membranes by FRET/TIRF imaging, using CH27 cells which express Lyn-CFP (donor) and Igα-YFP (acceptor), and biotinylated antigens which are bound on the biotin-lipid membrane through a streptavidin bridge, to reveal the initiation of BCR signaling in response to the membrane-bound antigens in a spatial and temporal way.

2. Materials

2.1. Cell Culture

1. CH27, a mouse B lymphoma cell line (*see* **Note 1**).
2. Cell culture medium: Iscove's modified Dulbecco medium (IMDM) supplemented with 15% fetal bovine serum (FBS) (Gibco/BRL, Bethesda, MD), 10 mM of MEM non-essential amino acids solution (GIBCO), 50 mM β-mercaptoethanol, 10,000 units/mL of penicillin and 10,000 μg/mL of streptomycin solution (Pen Strep, 100×) (GIBCO).
3. Plasmids: Lyn-CFP (the FRET donor), Igα-YFP (the FRET acceptor), and Lyn16-CFP-YFP (the FRET positive control). These plasmids can be obtained from authors' laboratory (*see* **Note 2**). It is recommended to purify plasmid DNAs with an endotoxin-free column prior to transfections (QIAGEN, Valencia, CA).
4. Cell line Nucleofector kit V and NuclefectorII machine (Amaxa Inc., USA). Keep the reagent at 4°C and use within 90 days after mixing the two components.
5. 12-well culture plates (Corning, Corning, NY).

2.2. Preparation of Planar Lipid Bilayers Bearing Antigen

2.2.1. Cleaning of All Glass Items

1. Fresh KOH/EtOH cleaning solution: 2.1 M KOH, 85% EtOH (*see* **Note 3**). Caution: this solution is corrosive and KOH liberates toxic gas when it contacts water; avoid exposure.
2. Rinsing solution: 95% EtOH, 100% EtOH.
3. NextGen^{TM} V 5 mL clear glass vials with a V-shaped bottom (Wheaton Science Products, Millville, NJ).

4. Coplin jars with lids.
5. Roll & Grow™ Plating Glass Beads (MP Biomedicals, Solon, OH).
6. Deionized ultrapure water.

2.2.2. Preparation of Small Unilamellar Vesicles (SUVs)

1. 25 mM 1,2-Dioleoyl-*sn*-Glycero-3-phosphocholine (DOPC) (Avanti Polar Lipids, Alabaster, AL). This is provided in chloroform as a 20 mg/mL solution.
2. 10 mM 1,2-Dioleoyl-*sn*-Glycero-3-phosphoethanolamine-cap-biotin (DOPE-cap-biotin) (Avanti Polar Lipids). This is provided in chloroform as a 10 mg/mL solution. Keep both DOPC and DOPE-cap-biotin at –20°C (*see* **Note 4**).
3. 200 μL and 1 mL size Hamilton syringes.
4. High purity chloroform. Caution: this is toxic, flammable, and can easily evaporate. Store at room temperature in a safety storage cabinet.
5. Compressed argon gas with gas regulator (Model: 10-575-110, Fisher Scientific, Pittsburg, PA). Insert a Pasteur pipette into the 0.45-μm syringe-type filter (Corning) connected to the gas regulator via tubing for drying lipid mixtures.
6. Water-bath-type sonicator (Model: G112SP1G, Laboratory Supplies, Hicksville, NY).
7. Phosphate-buffered saline (PBS), pH7.4: Prepare 10×stock with 1.55 M NaCl, 16.9 mM KH_2PO_4, and 26.5 mM Na_2HPO_4, pH7.4 (adjusted with HCl), and dilute with fresh ultrapure deionized water and filter with a disposable bottle-top vacuum filter.
8. Beckman rotor, SW55Ti and Ultra-Clear 13×51 mm ultracentrifuge tubes (Beckman Instruments Inc., Palo Alto, CA).
9. Individually wrapped serological glass pipettes and Pasteur pipettes (Kimble Glass, Owens, IL).

2.2.3. Preparation of Antigen-Tethered Planar Lipid Bilayer

1. SUV working solution (0.1 mM): Dilute 20 μL of 5 mM SUV stock solution with 1 mL of freshly prepared PBS. The SUV stock solution is stable for several months in a refrigerator, but planar lipid bilayers must be prepared freshly on the same day of image acquisition from diluted SUV working solution.
2. 24×50 mm #1.5 coverglasses.
3. Nanostrip solution (OM Group Ultra Pure Chemicals, Derbyshire, UK). This is a strong acid and will react with carbohydrate. Protect skin and eyes when using. Store at room temperature in the safety storage cabinet.

4. Lab-Tek chamber #1.0 Borosilicate coverglasses (Nalge Nunc International, Rochester, NY) (*see* **Note 5**).
5. Sylgard 164 Silicone Elastomer adhesive (Dow Corning, Midland, MI).
6. Biotinylated goat Fab anti-mouse IgM and streptavidin (Jackson Immuno Research Lab, West Grove, PA): Prepare for 20 μM and 5 mg/mL stock solution, respectively, in 50% glycerol and keep at –20°C. Both are stable for at least 1 year in this condition.

2.3. FRET/TIRF Imaging

2.3.1. Imaging Reagents

1. Imaging buffer: 1×PBS or 1×Hank's balanced salt solution (HBSS) containing 0.1% FBS (Gibco/BRL). Freshly prepare with ultrapure deionized water and filter it with disposable bottle-top filter system.
2. FluoSpheres polystyrene microspheres, 1.0 μm (430/465 nm) (Molecular Probes, Eugene, OR). Store at 2–6°C and protect from light.

2.3.2. Imaging Equipment

1. Olympus IX-81 microscope equipped with a TIR illumination port and an IX2 ZDC autofocus laser system (Olympus USA, Center Valley, PA).
2. Olympus TIRF microscope objective lenses: 60×, 1.45 NA PlanApo and 100×, 1.45 NA PlanApo.
3. 442/514 nm polychroic mirror for Olympus filter cube (Chroma Technology, Rockingham, VT).
4. 442 and 514 nm laser-line (notch) blocking filters (Semrock, Rochester, NY). Required for additional laser blocking in the emission path.
5. FC/APC single-mode fiber optic cable (OZ Optics, Ottawa, ON, Canada).
6. FW1000 excitation and emission filter wheels (ASI, Eugene, OR).
7. MS2000 automated X-Y, piezo Z stage (ASI, Eugene, OR).
8. Two channel optical splitter, DualView (MAG Biosystems, Tucson, AZ) equipped with a 505 nm dichroic filter and HQ485/30 nm and HQ560/50 nm emission filters (Chroma Technology, Rockingham, VT) to collect CFP and YFP signals, respectively.
9. 40 mW, 442 nm diode laser (BlueSky Research, Milpitas, CA).
10. 300 mW argon gas 488/514 nm laser (Dynamic Laser, Salt Lake City, UT).

11. Cascade 512II EM-CCD, –70°C cooled camera (Photometrics, Tucson, AZ).
12. Acousto-optical tunable filter and multichannel controller (AOTF) (NEOS, Melbourne, FL).
13. Computer equipped with: Intel Core Duo, 2.66 GHz processor, 4 GB RAM, Windows XP, 500 GB hard drive, four or more 9 pin com ports, three or more open PCI slots.
14. MetaMorph system control software (Molecular Devices, Downingtown, PA).
15. Image-Pro Plus Software 6.3 (MediaCybernetics, Bethesda, MD) and Microsoft Office Excel for image analysis.

3. Methods

To obtain robust and reproducible FRET data, accurate determination of the correction factors, β, γ, and δ is necessary. The FRET channel contains not only the sensitized acceptor emission by FRET but also spectral contaminants from the direct excitation and emission of CFP and YFP that the correction factors eliminate. Moreover, an accurate determination of the conversion constants, *Ka* or *Kd*, are required to allow FRET data to be compared between different experiments. Therefore, image acquisitions of control and experimental samples must be performed for each imaging session and the constants calculated from control cells for each experiment.

3.1. Cell Culture and Preparation of Cells Expressing Lyn-CFP and/or Igα-YFP, and Lyn16-CFP-YFP

1. Maintain CH27 cells in 15% IMDM culture media in T-25 culture flasks at 37°C under 7% CO_2 in a humidified incubator. Passage cells by 1:15 dilution of cells into fresh media when confluent (*see* **Note 6**).
2. Prepare purified plasmids, Lyn-CFP, Igα-YFP, Lyn16-CFP-YFP, so that they are ready for transient transfection using Amaxa Nucleofection 1 day before imaging (*see* **Note 7** for other transfection options).
3. Passage cells 3 days before imaging by diluting cells 1:15 in fresh media in T-75 flask.
4. One day before imaging, when cells reach a density of $\sim 1.5–2.0 \times 10^6$ cells/mL, transfer 3.0×10^6 cells to a 15-mL centrifuge tube for each transfection. Pellet cells at $90 \times g$ at room temperature for 10 min (*see* **Note 8**).
5. While centrifuging, set Amaxa Nucleofector II device to program A-23. For CH27 cells, best transfection

efficiencies are obtained with Nucleofector Kit V reagent and program A-23, which is recommended for the control GFP plasmid in Amaxa protocol.

6. Aspirate medium completely from the cell pellet with a 2-mL pipette connected to a vacuum waste tank (*see* **Note 9**).
7. Add 100 μL of reagent V pre-mixed with supplement and resuspend the cell pellet completely by gently shaking and pipetting once or twice using micropipette. Perform steps 7–10 for each sample separately.
8. Use 4 μg of plasmid DNA for each transfection. Use 2 μg of DNA per each plasmid if transfecting two plasmids. Mix DNA with the resuspended cells by brief, gentle shaking.
9. Carefully transfer the sample to an Amaxa-certified cuvette. Avoid air bubbles while pipetting, and close with blue cap.
10. Insert the cuvette into the holder of Nucleofector II device, and press >>X<< button to start program A-23.
11. Immediately after completion, remove the cuvette from holder and gently add about 1 mL of warm culture media to the bottom of the cuvette using plastic pipette provided in the kit, and transfer to a 12-well culture plate containing 1 mL of warm media per well.
12. Incubate overnight at 37°C under 7% CO_2 in a humidified incubator.
13. Check cells the next day for viability and transfection efficiency using inverted epi-fluorescence microscope. Confirm expression levels and distribution of Lyn-CFP and Igα-YFP (*see* **Note 10**).

3.2. Preparation of Planar Lipid Bilayers Bearing Antigen

3.2.1. Cleaning of All Glass Items in KOH/EtOH Cleaning Solution

1. Separate glass vials from caps and immerse ten vials in freshly prepared KOH/EtOH cleaning solution for 10 min. Make sure that the vials are completely filled with the cleaning solution; remove air bubbles if required.
2. Remove the cleaning solution and transfer the vials to a 2-L beaker containing 1 L of 95% EtOH. Rinse briefly by moderately swirling the beaker several times.
3. With vigorous shaking, rinse each vial thoroughly under a flow of ultrapure deionized water.
4. Bake vials at 160°C for 1 h.
5. Place all caps in 100% EtOH in a beaker and vigorously shake for a few minutes. Then discard the solution.

6. Rinse three times with a copious amount of ultrapure deionized water to remove EtOH.
7. Dry caps at 60°C.
8. Likewise, clean glass beads and three Coplin jars with KOH/EtOH solution, rinse, and oven-dry.

3.2.2. Preparation of SUVs

1. Fill three cleaned vials with 4 mL of 100% EtOH and another three vials with 4 mL of chloroform in a fume hood.
2. Place the tip of a Hamilton syringe in the first EtOH vial and move the piston up and down several times to clean. Repeat the process sequentially in the next two EtOH vials followed by three chloroform vials. Air-dry the syringe on Kimwipes.
3. Using two syringes, transfer to a cleaned vial 1 mL of 25 mM DOPC and 25 μL of 10 mM DOPE-cap-biotin in chloroform, resulting in a 100:1 ratio of DOPC:DOPE-cap-biotin.
4. Tighten the cap and mix briefly by tapping. For vials containing stock lipid solutions, immediately fill them with argon gas. Store lipid stocks at –20°C.
5. Dry the lipid mixture in a glass vial under a stream of argon gas. Slowly rotate the vial until the solution is dried up and a thin film of lipid is formed on the walls (*see* **Note 11**).
6. Completely evaporate residual chloroform by putting the vial under high vacuum at room temperature overnight (*see* **Note 12**).
7. For lipid hydration, degas ~20 mL of fresh 1×PBS in a nanostrip-cleaned side-arm flask (*see* **Note 13**). Degassing is complete when solution stops releasing bubbles when shaken.
8. Hydrate the lipid film with 5 mL of degassed PBS at room temperature. Fill the vial with argon gas and cap tightly.
9. Vortex for about 30 s until the lipid film is completely dissociated from the wall and makes an opaque solution. Keep on ice.
10. Fill the sonicator with a mixture of ice and water to achieve bath temperature of <4°C. Continuously sonicate the lipid mixture in 10-min rounds until it becomes clear (*see* **Note 14**). This indicates the formation of SUVs.
11. Transfer the SUVs into a 5-mL ultracentrifuge tube to fit Beckman SW55Ti rotor and clarify the SUV solution by ultracentrifugation at 4°C for 1 h at 46,800 *g* (*see* **Note 15**).

12. Transfer supernatant containing the SUVs into a new 5-mL tube using a 5-mL sterile serological pipette and continue clarifying with ultracentrifugation at 4°C for 8 h at 54,700 *g* using the same rotor. Carefully transfer supernatant into a new 15-mL plastic tube.
13. Fill the tube with argon gas; cap and seal it with parafilm. The SUVs are ready for use (*see* **Note 16**).

3.2.3. Preparation of Antigen-Tethered Planar Lipid Bilayers

1. Fill each of three Coplin jars with Nanostrip, deionized ultrapure water, and 100% EtOH, respectively.
2. Place 24×50 #1.5 coverslips into the jar containing Nanostrip for at least 1 h.
3. Transfer the coverslips with forceps into the jar containing deionized ultrapure water and agitate several times. Discard, fill with fresh ultrapure water, and repeat the process ten times.
4. Transfer the coverslips into the jar containing 100% EtOH for several minutes.
5. Take out a coverslip with forceps and blow-dry it completely with a stream of argon gas. Place the coverslip on top of a monolayer of cleaned glass beads in a sterile 100-mm round culture dish.
6. Tear off the bottom coverslip of an 8-well Lab-Tek chamber and attach the Nanostrip-cleaned coverslip to chamber with Sylgard 164 (*see* **Note 17**).
7. Dilute 20 μL of 5 mM SUVs into 1 mL of PBS for a 0.1 mM final concentration of SUVs. Dispense 200 μL of diluted SUVs into each of the wells. Wait for 10 min.
8. Rinse planar lipid bilayer in each well with about 20 mL of 1×PBS, keeping it in solution at all times. After the last rinse, fill the chamber to the top with PBS (*see* **Note 18**).
9. Remove 400 μL of PBS from the filled well and add 250 μL of 5 μg/mL solution of streptavidin in PBS. Mix twice by gently pipetting up and down with a micropipette (*see* **Note 19**).
10. Incubate for 10 min and wash excess streptavidin by repeating step 8.
11. Again, remove 400 μL and add 250 μL of 10 nM biotinylated goat Fab anti-mouse IgM (antigen) and incubate 20 min (*see* **Note 20**). Save some chambers for no antigen controls.
12. Wash unbound excess antigen by repeating step 8.

13. Immediately before imaging, exchange PBS with imaging buffer. Follow step 8 only with 10 mL of imaging buffer (*see* **Note 21**).

3.3. Image Acquisition Using Total Internal Reflection Fluorescence (TIRF) Microscope

For FRET imaging in our experiments, images in the CFP, FRET, and YFP channels are captured as a set from control cells expressing Lyn-CFP-only, Igα-YFP-only, Lyn16-CFP-YFP, a positive FRET control, and from experimental cells expressing both fluorophores. To study the dynamic association of two molecules using FRET imaging, time-lapse imaging is used and three images are obtained at each time interval. The procedure for acquiring images for FRET analysis is adapted to the equipment in our facility: an inverted Olympus IX-81, through-lens TIRF microscope that is equipped with Olympus 60×1.45 N.A. and 100×1.45 N.A. objective lenses, an optical splitter (DualView) for simultaneous imaging of the CFP and FRET channels, and a Cascade 512 II EM-CCD camera. Three-channel image set for each time interval is acquired under the following excitation (laser) and emission wavelengths:

CFP = excitation 442 nm, emission 470–500 nm

FRET = excitation 442 nm, emission 535–585 nm

YFP = excitation 514 nm, emission 535–585 nm.

Under identical conditions, all control cells are imaged each day. Lyn16-CFP-YFP expressing cells are also subjected to complete YFP photobleaching. CFP intensity is recorded before and after photobleaching to determine FRET efficiency (explained in detail below).

1. At least 1 h prior to the preparation of lipid bilayers and biological specimens, turn on power on all equipment including lasers. It is important to equilibrate the stage and lens heaters to 37°C and to stabilize gas laser.
2. Choose 442 or 514 nm laser lines (*see* **Note 22**).
3. Set 442/514 nm excitation filter (clean-up) in front of the AOTF (*see* **Note 23**).
4. Choose 442/514 nm polychroic mirror to reflect excitation light to the specimen.
5. Select the emission path to the camera port and ensure that appropriate emission filters are selected in the DualView image splitter (*see* **Note 24**).
6. Choose Olympus 100×1.45 N.A. oil-immersion objective lens.
7. Select the following camera settings as a starting point: 50 ms exposure time; full chip (512 ×512 pixel image size); 16bit gray scale; no pixel binning; no frame averaging; EM-

mode; 5 MHz transfer speed; gain 2 (standard conversion gain); EM-gain 3000; overlap mode (clear pre-sequence: clears CCD charge buildup twice before image acquisition) (*see* **Note 25**).

8. Focus at specimen on the coverslip in a live acquisition mode. Remove the specimen and wipe away the oil (*see* **Note 26**).
9. Focus 442 and 514 laser lines to the back focal plane of selected objective lens. First, ensure that field diaphragm is open on TIR illumination port and tilt back bright field illumination swing arm to free a path for laser beam. Micrometer on TIR illumination port controls the beam angle. Return the micrometer (rightward) to its neutral setting so the beam is projected on the ceiling. Loosen the locking bolt and slide the fiber optic coupler until a small, focused spot followed by a series of diffraction rings is observed on the ceiling (*see* **Note 27**).
10. Align dual image splitter (DualView; MAGS Biosystems, Tuscon, AZ) onto CCD chip. A step-by-step alignment protocol is provided by the manufacturer. In steps 10a–c below, we provide an overview and commentary. It is important to align the apparatus before every imaging session. While alignment is usually stable, it is best to leave filter cassette in place once the alignment is complete. For FRET imaging, DualView is equipped with a filter cassette containing a 505 dichroic beamsplitter, and an HQ485/30 (470–500 nm), and HQ560/50 (535–585 nm) emission filters for simultaneous acquisition of CFP (left) and FRET (right) image, respectively.
    a. Choose open position in the emission path filter wheel. Emission filters required for the experiment are contained in the DualView (*see* **Note 28**).
    b. Place the alignment grid provided by the manufacturer in specimen holder and focus grid lines in bright field in a live acquisition mode. Follow the procedure outlined by the manufacturer to align the X, Y position. The grid is used for major adjustments and is typically needed when filter cassette is changed; otherwise, fine adjustments between experiments are made with alignment beads (*see* step 10c). For final adjustments with the grid, display live, false-colored, overlapped image of left and right sides of the CCD. We use red and green color combination to display yellow overlap (alignment). If further alignment is required, split the distance of adjustment by moving left/right or up/down knobs equally (*see* **Note 29**).

c. Perform fine adjustments with FluoSpheres (430/465 nm); these beads fluoresce in both CFP and FRET channels. The beads can be mounted or used in solution after settling to the bottom of a Lab-Tek chamber. Image beads in a live acquisition mode. Overlap left and right image in the software and adjust the DualView so that beads exactly overlap near the center of the image. Digitally magnifying the image so that individual pixels are viewed is helpful. Due to optical aberration, the beads at the periphery will not overlap (*see* **Note 30**).

11. Return to the standard acquisition mode and capture images of several fields of beads. These images will be used for alignment during image processing. Adjust laser power appropriately, so as not to saturate the image (*see* **Note 31**).
12. Allow time for stage heater to re-equilibrate (*see* **Note 32**).
13. Place the chamber containing the antigen-tethered planar lipid bilayer freshly prepared in imaging buffer onto the oil-mounted lens and allow for 10 min to reach 37°C.
14. Set imaging software to acquire two images for each time interval. For the first image, specimen is excited with the 442 nm laser and emission in the CFP and FRET channels is captured simultaneously. For the second image, specimen is excited with the 514 nm laser to capture YFP channel. Set interval time for 2 s and capture 500 frames.
15. Prepare CFP-only and YFP-only control cells at 1×10^4 cells/mL in the imaging buffer, mix at 1:1 ratio, and drop 20 μL of the mixture into a control chamber with no antigen positioned over the 100×objective lens. These control cells will be imaged with Lyn16-CFP-YFP FRET positive control cells (*see* steps 20–22). Find cells under transmitted light using a live acquisition mode (*see* **Note 33**).
16. Turn off transmitted light illumination source, excite the specimen with both 442 and 514 nm laser lines, and find a field that includes both CFP and YFP single-positive cells (*see* **Note 34**).
17. Illuminate the specimen by TIR. Turn the micrometer to the left on the TIR illumination port while using fine focus adjustment to image the cells proximal to the coverslip. As TIR angle is increased, a point will be reached where image intensity increases relative to the initial 0° TIR angle setting. Move past this point, again while adjusting focus, until image intensity is lost. Then slowly turn back the micrometer to the right to enter TIR-mode. Note micrometer setting for TIR angle (*see* **Note 35**).

18. Adjust laser power so that image is not saturated (*see* **Note 36**).
19. Perform test time-lapse acquisition for the calculation of correction factors and bleaching rate.
20. Prepare Lyn16-CFP-YFP FRET positive control cells at 2.0×10^6 cells/mL in imaging buffer and slowly drop 50 μL of cells into the same chamber containing single positive control cells.
21. View cells introduced to planar lipid bilayer in a live acquisition mode. Adjust focus and wait until they firmly settle down on the bilayer.
22. Acquire a set of CFP, FRET, and YFP channel images (pre-bleach) and then completely bleach YFP with 514 nm line at full power with TIR angle 0°. Return the micrometer to TIR angle noted in step 17 and acquire another three sequential image sets (post-bleach) under the same acquisition conditions to obtain Ka and $E_{\text{bleaching}}$.
23. Move chamber coverglass so that the center of the chamber containing antigens-tethered PLB can be placed over the lens and repeat steps 15–17 for image acquisition of experimental cells expressing both Lyn-CFP and Igα-YFP.
24. Prepare experimental cells expressing both Lyn-CFP and Igα-YFP cells at 2×10^6 cells/mL in imaging buffer and slowly drop 50 μL of cells into the chamber containing single positive control cells on antigen-tethered lipid bilayer.
25. View progress of introduced cells to the planar lipid bilayer in a live acquisition mode. Adjust the focus. Begin time-lapse image acquisition when blurred fluorescence images start to appear on the screen under a live image mode (*see* **Note 37**).

3.4. Image Processing for FRET Imaging

The principles of FRET calculations have been described elsewhere (12, 14, 15). For our protocol using B cells expressing Igα-YFP and Lyn-CFP, the use of FRET efficiency for the acceptor (Ea), that is FRET efficiency normalized for fluorescence intensity of YFP, is advantageous because YFP rather than CFP is limiting in the FRET pair, Lyn-CFP and Igα-YFP. For our microscope system, the δ factor is typically negligible ($\delta\approx0$), so the equation for Ea can be simplified to the following:

$$\text{Ea} = \frac{\text{FRET} - \beta * \text{CFP} - \gamma * \text{YFP}}{\gamma * \text{YFP} * Ka} \qquad [5].$$

To calculate FRET, raw images in each channel must be processed by background subtraction, image filtering, and masking of background for each channel, followed by image arithmetic on a pixel-by-pixel basis according to the equation (10.5). These processing steps are necessary to obtain the final FRET image and avoid false-positive or false-negative results. Processing of images can be performed using image processing software that is capable of image alignment and arithmetic, such as the Image Pro Plus (Media Cybernetics).

1. Open image that contains the CFP-FRET dual view of fluorescence beads.
2. Separate it into CFP and FRET channel images by creating new area of interests (AOIs) for the left and right half of the image.
3. Align separated CFP and FRET images using alignment function in image processing software. Alternatively, find the best alignment manually by overlaying the two images and moving one with respect to the other. Record the amount of horizontal and vertical shift, rotation, and scaling that was applied to obtain complete alignment of beads in the two images.
4. If necessary, merge individual image files of a sequence of CFP-FRET and YFP images at each time frame by stacking them in series.
5. Separate all DualView images based on the alignment determined in step 3. This will produce three sequences of images: CFP, FRET, and YFP.
6. For each image, set AOI to background (an area without cells) and determine average background value. Subtract that value from respective image using arithmetic operations.
7. Smoothen images by applying a low-pass (averaging) filter to reduce the noise of individual pixels. The size, passes, and strength of the filter are determined empirically to smoothen the image, but not to blur cellular details completely (*see* **Note 38**).
8. Determine the threshold value in each channel image to distinguish the foreground of the cell from background in that particular channel, using the mean plus three standard deviations of pixel intensity of the background. Mask the background by thresholding the image at that intensity. As a result, background pixels in all three (CFP, FRET, and YFP) channels should have zero intensity.
9. Determine β from cells expressing CFP only. $\beta = \frac{\text{FRET}}{\text{CFP}}$, where FRET and CFP are background-subtracted intensity

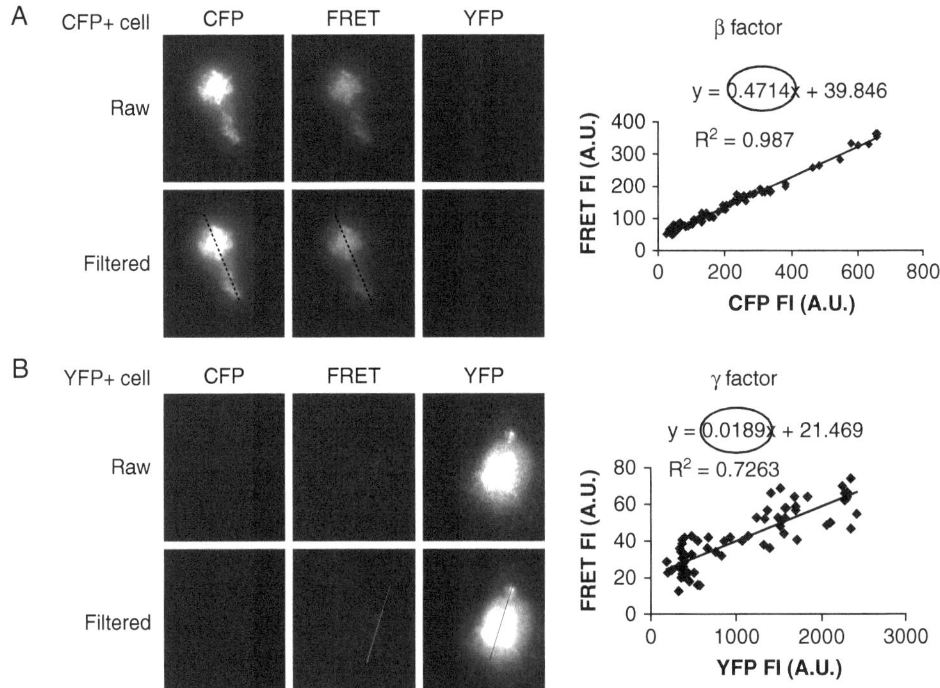

Fig. 10.1. Determination of correction factors: **(a)** for CFP bleed-through (β) and **(b)** for YFP crosstalk (γ) from direct excitation of CFP and YFP by 442 nm laser, respectively. Cells expressing either Lyn-CFP alone (CFP+ cell) or Igα-YFP alone (YFP+ cell) were settled on planar lipid bilayer. Cells were excited sequentially with 442 and 514 nm laser light by TIRF microscope. Fluorescent emission was split through DualView to acquire CFP and FRET images (442 nm excitation) and the YFP image (514 nm excitation). Images were aligned, split to obtain raw CFP, FRET, and YFP images from the control cells, and then background subtracted and a Lo-pass filter applied (Filtered) as described in **Section 3**. To calculate β, intensity data was obtained from a line drawn over the same relative position in the cell in CFP and FRET filtered image and was plotted as FRET vs CFP (**a**) for each pixel in the line. β was determined from the slope of the line in the plot. γ (**B**) was calculated similarly using YFP and FRET filtered images. Similar results are obtained by averaging fluorescence intensities from several control cells.

values in AOIs from the cells in FRET and CFP channels as excited by the 442 nm laser (**Fig. 10.1a**).

10. Likewise, determine γ from cells expressing YFP only. $\gamma = \frac{\text{FRET}}{\text{YFP}}$, where FRET and YFP are background-subtracted intensity values in AOIs from the cells in FRET and YFP channels as excited by the 442 and 514 nm laser, respectively (**Fig. 10.1b**).
11. Determine FRa, FRET efficiency ($E_{\text{bleaching}}$), and *Ka* from images before and after YFP photobleaching of cells expressing Lyn16-CFP-YFP using the following series of equations:

$$\text{FRa} = \frac{N_{\text{sen}}}{\text{YFP} * \gamma}, \quad [6]$$

where N_{sen} is the sensitized acceptor emission by FRET:

$$N_{sen} = \text{FRET} - \beta * \text{CFP} - \gamma * \text{YFP} \quad [7]$$

$$E_{\text{bleaching}} = \frac{\text{Dequenching}}{\text{CFP}_{\text{after}}} \quad [8]$$

where dequenching is difference of fluorescence intensities of CFP before ($\text{CFP}_{\text{before}}$) and after acceptor photobleaching ($\text{CFP}_{\text{after}}$) (*see* **Note 39**) (*see* **Fig. 10.2**):

$$\text{Dequenching} = \text{CFP}_{\text{after}} - \text{CFP}_{\text{before}} \quad [9]$$

$$Ka = \frac{\text{FRa}}{E_{\text{bleaching}}} \quad [10]$$

a. First, record in a MS Excel sheet the mean fluorescence intensities of CFP, FRET, and YFP from the same AOIs drawn over the cell contact areas from each channel image before YFP photo-bleaching (before-bleach), and subtract background value from each. An example table of raw and background subtracted intensity values, which was obtained from images is provided in **Fig. 10.2b**.

b. Calculate N_{sen} using equation [7] from background-subtracted FRET, CFP, and YFP mean intensities.

c. Calculate dequenching by equation [9].

d. Make intensity plots for N_{sen} vs YFP and dequenching vs $\text{CFP}_{\text{after}}$ from the values calculated in step 12c. Determine FRa and $\text{E}_{\text{bleaching}}$ from the slopes of the relationships of N_{sen} vs YFP and dequenching vs $\text{CFP}_{\text{after}}$, respectively (**Fig. 10.2b**).

e. Obtain *Ka* using equation [10] (*see* **Note 40**).

12. Finally, produce FRET efficiency (Ea) images from sample cells by pixel-by-pixel arithmetic image calculation using equation [5] (*see* **Fig. 10.3a**).

13. Because the resulting Ea image from step 12 shows artificial FRET in the region outside the cell contact zone (TIRF imaging plane) due to background noise, mask the Ea image by following steps:

    a. Apply a flatten enhancing filter to both CFP and YFP images to augment the borders along contact zones.

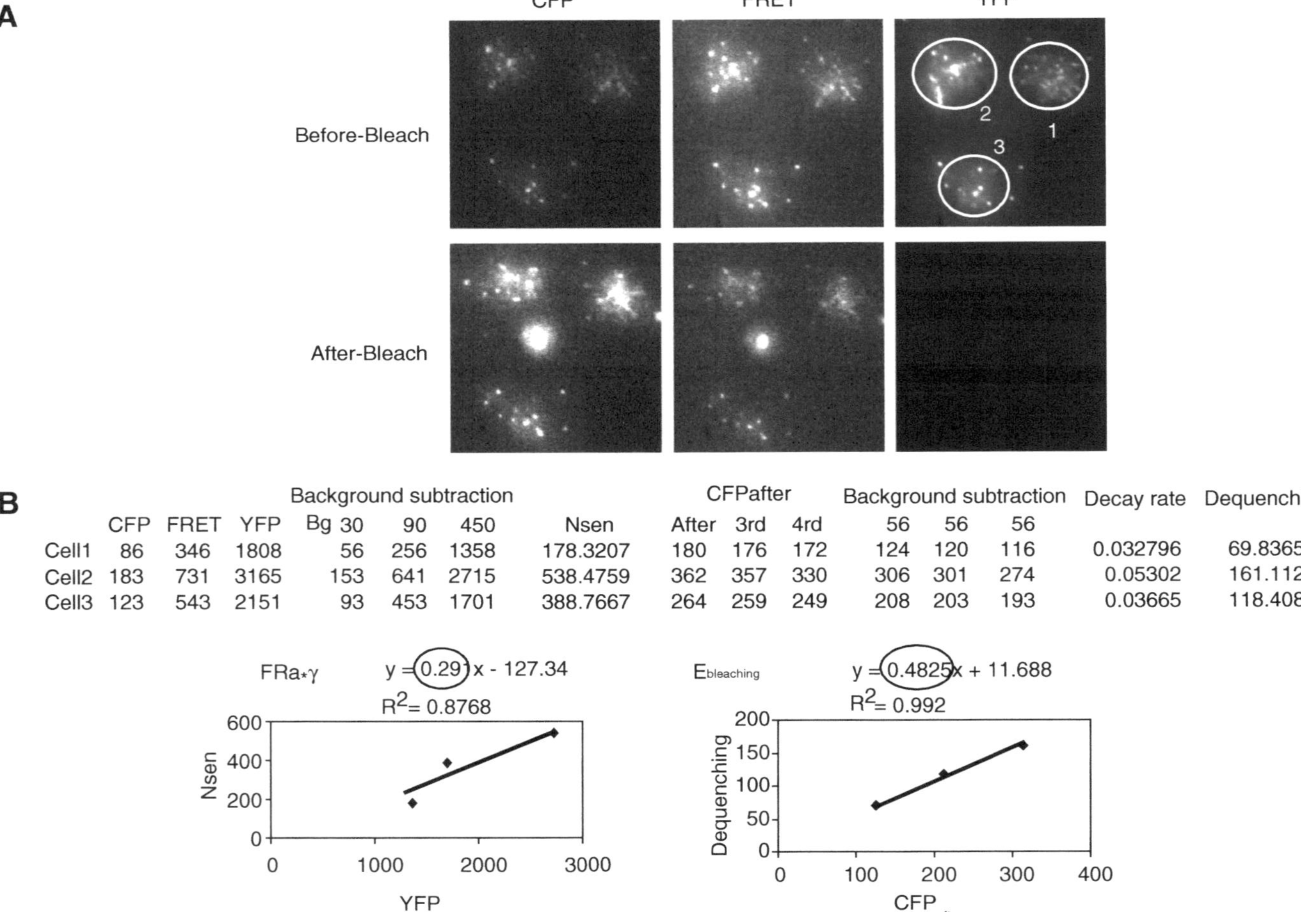

	CFP	FRET	YFP	Background subtraction Bg 30	90	450	Nsen	CFPafter After	3rd	4rd	Background subtraction 56	56	56	Decay rate	Dequenching
Cell1	86	346	1808	56	256	1358	178.3207	180	176	172	124	120	116	0.032796	69.83656
Cell2	183	731	3165	153	641	2715	538.4759	362	357	330	306	301	274	0.05302	161.1121
Cell3	123	543	2151	93	453	1701	388.7667	264	259	249	208	203	193	0.03665	118.4084

Fig. 10.2. Determination of $E_{bleaching}$ and *Ka* from Lyn16-CFP-YFP FRET positive control. (**a**) CFP, FRET, and YFP images before and after YFP photobleaching were acquired from cells expressing Lyn16-CFP-YFP on planar lipid bilayer and processed as described in **Section 3**. AOIs from three cells were drawn over cell contact areas from each image before (Pre-Bleach) and after bleaching (Post-Bleach). Note that images are not saturated on the 16 bit gray scale. (**b**) Procedure to obtain *Ka* from the CFP, FRET, and YFP images in a Microsoft Excel sheet is shown. Shown are a table containing the values that include background-subtracted YFP, Nsen, CFP_{after} and Dequenching, calculated from the mean fluorescence intensities of AOIs drawn in (**a**), and intensity plots for N_{sen} vs YFP and Dequenching vs CFP_{after} from the values of the table. FRa and $E_{bleaching}$ were determined from the slope of the line of N_{sen} vs YFP and Dequenching vs CFP_{after}, respectively. Finally, *Ka* was obtained from the equation (10).

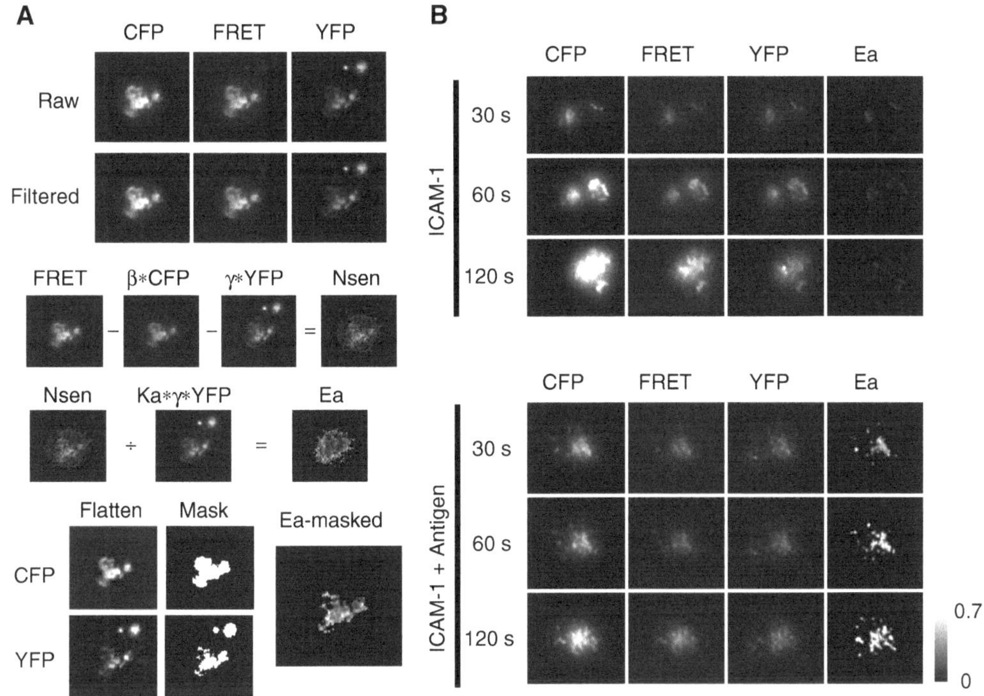

Fig. 10.3. Membrane-bound-antigen-induced association of Lyn with BCR. Cells were dropped onto planar lipid bilayer with or without antigen and time-lapse images were acquired at 37°C for 15 min. **(a)** Procedure for image processing to obtain the FRET efficiency (Ea) image from experimental cells expressing both Lyn-CFP and Igα-YFP is shown. Images taken 1 min after cell contact to the antigen-bound membrane were processed and shown are the representative images in each step described in **Section 3**. β^*CFP, γ^*YFP, *Ka*$^*\gamma^*$YFP, $N4_{sen}$ and Ea images were obtained from pixel-by-pixel image calculation according to the appropriate equations. Raw; background-subtracted images, Filtered: Lopass-filtered images, Flatten: flatten-filtered images, Mask: masked images from flattened images, Ea-masked: masked image of Ea with both CFP and YFP masks. **(b)** CFP, FRET, YFP and Ea images at each indicated time after cell contact in the absence (ICAM-1) or presence of antigen (ICAM-1 + Antigen) on a planar lipid bilayer are shown. Images were processed as in **(a)** and shown are the filtered CFP, FRET, and YFP images and the masked Ea image. Ea is shown as *gray* scale is from 0 to 0.7. The fluorescence intensities of CFP, FRET, and YFP images are not comparable.

b. Set the threshold for the enhanced Flatten images as done in step 8 and create mask image for both.

c. Then, apply the combined CFP and YFP masks over the Ea image resulting in the masked Ea image. An sample result is shown in **Fig. 10.3b**.

4. Notes

1. CH27 cells, which express on their surface a phosphorylcholine (PC) specific BCR, can be obtained from authors' lab.

2. Although specific design of chimeric constructs needs to be considered for each gene, Lyn and Igα were fused to the N-terminus of CFP or YFP using pECFP-N1 or pEYFP-N1 vector (BD Biosciences Clontech) by inserting these genes in-frame at the Bam HI site upstream of the CFP or YFP vector. In both cases, there is a 6 amino acid spacer between the gene of interest and the tag which gave us the best expression of these fusion proteins at plasma membrane. Full length Igα and mouse Lyn cDNA clones are available commercially (Open Biosystems, Huntsville, AL) for PCR templates. The membrane targeted FRET positive control plasmid, Lyn16-CFP-YFP, has been previously described (15).
3. To make cleaning solution, dissolve 120 g KOH into 120 mL of deionized water in a beaker. Add about 1 L of 95% EtOH to the KOH solution and mix thoroughly using a stirrer. Wear suitable protective clothing, gloves, and eye/face protection when making the cleaning solution. Do not keep it in a glass bottle with glue-jointed bottom for an extended period of time.
4. After each use, fill the bottle with argon and wrap the cap with parafilm to avoid evaporation of chloroform. The lipids are easily oxidized when exposed to the air.
5. This particular Lab-Tek chamber is used because the coverslip can be easily removed by hand.
6. CH27 cells grow better in 7% CO_2 than 5%. When first thawed from liquid nitrogen, incubate them at 7%. Once cells are growing, they can be cultured at either condition. When confluent after about 2–3 days, the media turns from pink to orange-yellow color. Otherwise, keep track of cell density by counting cells. When cells reach about 1.5×10^6 cells/mL, dilute cells into fresh media. Cells are usually passaged every 3 days.
7. Plasmid transfection conditions must be determined empirically for each particular B cell line. We found that creating stable transfectants with a ratio of ~1:1 of CFP to YFP worked best in FRET experiments. Retroviral infection system using MSCV vector and PT67 packaging cell line (Clontech) followed by FACS sorting for positives worked best for CH27 cells. Stable cell lines expressing both CFP and YFP can be established by sequential or simultaneous infection followed by repetitive sorting for CFP and/or YFP positive cells. Usually, we use "spin infection" by centrifugation at 1844 *g* for 90 min followed by concentration of the viral supernatant by centrifugation at 2300 *g* for 20 min using a centrifugal filter device (Amicon Ultra-15

with 10 k MW cut-off, Millipore). Stable cell lines for positive clones can be established by drug selection or sorting by FACS.

8. To achieve maximal transfection efficiency, it is important to maintain cells with good viability (over 90%) during passage. CH27 cells show maximum transfection efficiency when cell density reaches $\sim 1.5–2.0 \times 10^6$ cells/mL on the day of transfection.
9. It is very important to remove any residual liquid from the pellet to obtain good transfection efficiency. A sterile 200 μL micropipette tip at the end of aspiration pipette works well to remove any remaining medium. Cell pellets become loose with time, which can lead to accidental aspiration of cells. We recommend a maximum of four samples to be spun at a time.
10. If transfection efficiencies are less than 10%, chances are low to capture cells expressing both CFP and YFP in the same imaging field. Sort for positive cells by FACS if available or use stable cell transfectants.
11. It is important to keep the pressure low enough as to not to splash the mixture on the walls of the vial.
12. Cover the vial with a Kimwipe to inhibit dust from getting inside.
13. To clean side arm flask, swirl about 50 mL of Nanostrip in the flask for a few minutes, discard, and then rinse with a copious amount of deionized ultrapure water. Dry the flask in an oven.
14. To avoid overheating the sonicator bath, do not sonicate for more than 10 min. Pause sonication, place the vial on ice, and add more ice to the bath to adjust water temperature.
15. An opaque pellet will be seen at the bottom of the tube after the first round of centrifugation. Pellet size depends on the extent of sonication.
16. SUVs are stable for several months, when stored at 4°C under argon. To minimize repeated exposure of SUV stock to air and to prolong stability, store in 1 mL aliquots in 15-mL sterile plastic tubes under argon sealed with parafilm. The SUV mixture can be filtered through a 0.2-μm syringe-type Whatman polysulfone filter for further purification; resulting concentration of lipid is about half of the original determined by measurement of OD at 234 nm using UV–vis spectrophotometer.
17. Mix Sylgard components ~1:1 on a clean surface with a 200-μL micropipette tip and use quickly before harden-

ing. With the chamber upside down, fill the channels at the base of each well. Syringe with a short, thick (18 G) needle can be used. Avoid applying too much silicone. Using forceps or with gloved hands holding only the edges, place the coverslip over the bottom of the chamber and apply gentle pressure along the channels. Place chambers upright on a Kimwipe, cover with about 0.5 kg weight, and leave for 30 min before use. Prepare chambers on the day of use as dust will re-accumulate and prevent the formation of a lipid bilayer.

18. To avoid exposing the lipid bilayer to air during washing, remove solution with a 200-μL micropipette tip attached to aspiration pipette while simultaneously adding 1×PBS with a slow, consistent flow rate using a 10-mL serological pipette attached to auto-pipetter.
19. To monitor mobility/integrity of the lipid bilayer, mix 1:9 Alexa488-labeled streptavidin with unlabeled streptavidin and add 250 μL to one well.
20. Dilute antigen with 1×PBS to 10 nM and microcentrifuge at maximum speed for about 10 min to eliminate aggregates.
21. Planar lipid bilayers are less stable in the presence of serum. It lasts at most for 1–2 h. Its mobility can be tested by a simple FRAP experiment with fluorescently labeled planar lipid bilayer in a chamber.
22. We recommend MetaMorph software to control imaging and laser (via AOTF) systems.
23. Despite the 10^4 blocking power of AOTF, we found it necessary to clean-up (block) non-selected laser lines emanating from the gas laser as this light was detected by our EM-CCD. Filters work best when located on the laser table. Ours are held in a filter wheel placed after the AOTF; filter position on the table is flexible and is only defined by what laser lines require for blocking. Interestingly, unwanted reflections detected by the EM-CCD were seen when excitation filters were placed in the TIR illumination port or filter cube holder of the microscope.
24. We place 442 nm notch filter in the microscope filter cube to further block escaping excitation light from the camera to decrease background signal and spurious reflection patterns.
25. Once optimized for cells, it is important to keep all settings the same. Maximal useful EM-gain is idiosyncratic for each specific camera type and needs to be determined empirically. Best EM-gain is reached when the ratio of signal to

background no longer improves with further increase of gain.

26. The goal is to place objective lens at correct working distance before focusing laser lines. Changing the height of objective lens will change the distance of objective's back focal plane relative to laser focusing lens in TIR illumination port. Laser focusing can be achieved directly through the specimen as long as the cell number is low, which minimizes light scatter.
27. Laser light will exit the objective lens as collimated beam, which can be observed through a concentrated solution of fluorophore if so desired. We replaced the original Olympus TIR illumination port and tube lens with a 100 and 200 mm focal length achromatic lens, respectively, from JML Optical. These second-party lenses offer superior color correction, allowing both beams to co-focus. Be aware that this only improves laser co-focusing and does not affect the differences in penetration depth of specific excitation wavelengths.
28. Laser-line blocking filters (notch filters) can be placed in this filter wheel, polychroic filter cube of the microscope, or in DualView filter cassette.
29. For a CCD camera with a 512×512 pixel dimension, most image acquisition software, including MetaMorph, designates the first pixel position in the upper left corner as 0,0 and the bottom right as 511,511; therefore, we divide horizontal axis from 0 to 255 for the left side and 256 to 511 for the right.
30. To our knowledge, diffraction-limited beads (<300 nm) that fluoresce well from CFP channel into the FRET channel when mounted on coverslips are not available commercially. We therefore use 1-μm-sized beads.
31. Most imaging software can be set to display saturated pixels in red.
32. We occasionally monitor well temperature with YSI 4600 thermometer equipped with a 400 series probe.
33. Adjust tungsten lamp intensity appropriately.
34. For finding cells, TIR illumination angle can be kept at 0°.
35. Ideally, it is better to work with subregions because even illumination of the entire specimen plane (512×512 region) at an optimal TIR angle is not possible.
36. For our experiments, we use ~2 and ~1.0 mW for 442 and 514 nm laser lines, respectively (measured with a power meter through 100×lens). It is important to minimize

photobleaching during time lapse by keeping laser power to a minimum while still achieving adequate signal.

37. To compensate for focus drift, our system is equipped with the Olympus laser autofocusing device, which works well for long time-lapse experiments.
38. Typically within small regions of interest where FRET is measured, we found that flat field correction was unnecessary. To correct for uneven illumination of the field of view, each pixel value of the image is divided by the mean intensity of normalized flat field image, which is pixel-by-pixel ratio image between CFP and FRET channels obtained using a fluorophore solution in a chamber that emits in both CFP and FRET channels.
39. With some exception such as degradation of fusion protein in certain cells, FRET positive control cells show empirically about 55±10% $E_{\text{bleaching}}$ regardless of imaging conditions due to the fact that the fusion protein with 2 amino acid linker between CFP and YFP maintains the fixed 1:1 molar ratio between donor and acceptor and constant FRET.
40. If YFP back bleed-through (δ correction factor) and bleaching rate of CFP between scans is considerable for your system, the following adjustments to the equations are required.

$$E_{\text{bleaching}} = \frac{\text{CFP}_{\text{after}} - \frac{\text{CFP}_{\text{before}} - \delta * \text{FRET}}{1 - \beta * \delta} * (1 - \text{Br})}{\text{CFP}_{\text{after}} + \frac{\text{CFP}_{\text{before}} - \delta * \text{FRET}}{1 - \beta * \delta} * \text{Br}} \quad [11]$$

where Br is average bleaching rate of pure CFP between two subsequent images (i and i+1) in time lapse: $\text{Br} = 1 - (\text{CFP}_{i+1}/\text{CFP}_i)$. This is estimated from time course of residual CFP imaging after YFP has been bleached.

FRa is also obtained from the following equation rewritten from equation [3]:

$$\text{FRa} = \frac{N_{\text{sen}}}{\text{YFP} * \gamma * (1 - \beta * \delta)}, \quad [12]$$

where $N_{\text{sen}} = \text{FRET} - \beta * \text{CFP} - \gamma * (1 - \beta * \delta) * \text{YFP}$
Finally, Ka is obtained from equation [10].

Acknowledgments

This work was supported by the Intramural Research Program of the National Institute of Allergy and Infectious Diseases, National Institutes of Health.

References

1. Fearon, D. T., and Carroll, M. C. (2000) Regulation of B lymphocyte responses to foreign and self-antigens by the CD19/CD21 complex, *Annu Rev Immunol* **18**, 393–422.
2. Dal Porto, J. M., Gauld, S. B., Merrell, K. T., Mills, D., Pugh-Bernard, A. E., and Cambier, J. (2004) B cell antigen receptor signaling 101, *Mol Immunol* **41**, 599–613.
3. Pierce, S. K. (2002) Lipid rafts and B-cell activation, *Nat Rev Immunol* **2**, 96–105.
4. Tolar, P., Sohn, H. W., and Pierce, S. K. (2008) Viewing the antigen-induced initiation of B-cell activation in living cells, *Immunol Rev* **221**, 64–76.
5. Axelrod, D. (1981) Cell-substrate contacts illuminated by total internal reflection fluorescence, *J Cell Biol* **89**, 141–145.
6. Kandere-Grzybowska, K., Campbell, C., Komarova, Y., Grzybowski, B. A., and Borisy, G. G. (2005) Molecular dynamics imaging in micropatterned living cells, *Nat Methods* **2**, 739–741.
7. Brian, A. A., and McConnell, H. M. (1984) Allogeneic stimulation of cytotoxic T cells by supported planar membranes, *Proc Natl Acad Sci U S A* **81**, 6159–6163.
8. Evans, J., and Yue, D. T. (2003) New turf for CFP/YFP FRET imaging of membrane signaling molecules, *Neuron* **38**, 145–147.
9. Sohn, H. W., Tolar, P., and Pierce, S. K. (2008) Membrane heterogeneities in the formation of B cell receptor-Lyn kinase microclusters and the immune synapse, *J Cell Biol* **182**, 367–379.
10. Truong, K., and Ikura, M. (2001) The use of FRET imaging microscopy to detect protein-protein interactions and protein conformational changes in vivo, *Curr Opin Struct Biol* **11**, 573–578.
11. Ciruela, F. (2008) Fluorescence-based methods in the study of protein-protein interactions in living cells, *Curr Opin Biotechnol* **19**, 338–343.
12. van Rheenen, J., Langeslag, M., and Jalink, K. (2004) Correcting confocal acquisition to optimize imaging of fluorescence resonance energy transfer by sensitized emission, *Biophys J* **86**, 2517–2529.
13. Zal, T., and Gascoigne, N. R. (2004) Photobleaching-corrected FRET efficiency imaging of live cells, *Biophys J* **86**, 3923–3939.
14. Tolar, P., Sohn, H. W., and Pierce, S. K. (2005) The initiation of antigen-induced B cell antigen receptor signaling viewed in living cells by fluorescence resonance energy transfer, *Nat Immunol* **6**, 1168–1176.
15. Sohn, H. W., Tolar, P., Jin, T., and Pierce, S. K. (2006) Fluorescence resonance energy transfer in living cells reveals dynamic membrane changes in the initiation of B cell signaling, *Proc Natl Acad Sci USA* **103**, 8143–8148.

Chapter 11

Sample Preparation for STED Microscopy

Christian A. Wurm, Daniel Neumann, Roman Schmidt, Alexander Egner, and Stefan Jakobs

Abstract

Since the discovery of the diffraction barrier in the late nineteenth century, it has been commonly accepted that with far-field optical microscopy it is not possible to resolve structural details considerably finer than half the wavelength of light. The emergence of STED microscopy showed that, at least for fluorescence imaging, these limits can be overcome. Since STED microscopy is a far-field technique, in principle, the same sample preparation as for conventional confocal microscopy may be utilized. The increased resolution, however, requires additional precautions to ensure the structural preservation of the specimen. We present robust protocols to generate test samples for STED microscopy. These protocols for bead samples and immunolabeled mammalian cells may be used as starting points to adapt existing labeling strategies for the requirements of sub-diffraction resolution microscopy.

Key words: Fluorescence microscopy, Stimulated emission depletion microscopy, Nanoscopy, Superresolution, Immunofluorescence, Sample preparation.

1. Introduction

Stimulated emission depletion (STED) microscopy (1) and the related approaches to fundamentally break the diffraction barrier in far-field optical microscopy have been covered by several comprehensive reviews (2–5) and will not be discussed here in detail. In brief, STED microscopy typically uses a focused laser beam for excitation, which is overlapped with a second laser beam, the so-called "STED beam" that exhibits no light intensity at its focal center but strong intensities at the periphery. When excited, fluorophores exposed to the "STED beam" are almost instantly transferred back to their ground state by means of stimulated

D.B. Papkovsky (ed.), *Live Cell Imaging*, Methods in Molecular Biology 591,
DOI 10.1007/978-1-60761-404-3_11, © Humana Press, a part of Springer Science+Business Media, LLC 2010

emission. As a result, only molecules that are close to the zero at the center of the "STED beam" are allowed to fluoresce and contribute to the fluorescence signal. Confining the signal to such a subdiffraction-sized spot results in higher resolution imaging when the spot is scanned through the specimen (1, 6–8). Experiments using single molecules as test objects already demonstrated a spot size as small as 16 nm in the focal plane (9, 10).

Since its first utilization for the imaging of yeast cells (7), STED microscopy has been used for a number of applications in cell biology, including the visualization of proteins at the synapse (11), cytoskeletal elements (8), mitochondrial proteins (12), the analysis of lipid rafts (13), and others (14–17). It has also been successfully used for live cell applications (7, 18, 19). As until now, most applications relied on chemically fixed cells, we present protocols that describe the preparation of immunolabeled fixed cells. In addition, we provide protocols for bead samples that may be used as standard samples to determine the performance of the instrument.

The increased resolution of STED microscopy, like any subdiffraction microscopy, may disclose shortcomings of the sample preservation which are concealed by the lower resolution of conventional (confocal) microscopy. Also, special care has to be taken to avoid spherical aberrations induced by the refractive index mismatch between the immersion system and the embedding medium of the sample, which would immediately deteriorate the obtainable optical resolution. The protocols presented here are intended as starting points for the preparation of other samples; likewise they may be used as benchmarks to test the performance of a microscope.

2. Materials

We have successfully used the reagents described below. In many cases other commercially available equivalents may also be used. If not specified otherwise, all chemicals used were of analytical grade. Standard consumables and equipment as present in most molecular biology labs is required. For imaging of the samples a custom-built STED microscope (11, 13, 14, 19) or a commercial STED microscope (Leica Microsystems, Wetzlar, Germany) can be used.

2.1. General Materials

1. Cover slips No. 1 (0.13–0.16 mm thick) (Menzel Gläser, Braunschweig, Germany) (*see* **Note 1**)

2. Microscopy slides (ISO Norm 8037/I, 26 × 76 mm, 1 mm thick (Menzel Gläser)
3. Phosphate-buffered saline (PBS, 137 mM NaCl, 3 mM KCl, 8 mM Na_2HPO_4, 1.5 mM KH_2PO_4, pH 7.4)

2.2. Materials for Embedding Media

1. Mowiol 4-88 (Calbiochem, Darmstadt, Germany)
2. Tris–HCl buffer (Tris(hydroxymethyl)aminomethane, 0.2 M, pH 8.5)
3. Glycerol
4. 1,4-Diazabicyclo[2.2.2]octan (DABCO) (Sigma-Aldrich, Saint Louis, MO)
5. 2,2′-thiodiethanol (TDE), highest purity (Sigma-Aldrich)

2.3. Materials for Bead Sample

1. Fluorescent microspheres (FluoSpheres, crimson fluorescent (625/645), 10^6 beads/ml) diameter 0.02 μm, 0.04 μm (custom-made), 0.1 μm (custom-made) or 0.2 μm (Molecular Probes/Invitrogen, Eugene, OR)
2. Absolute ethanol
3. Poly-L-lysine solution (0.1% (w/v) in H_2O (Sigma-Aldrich)
4. Embedding medium (*see* **Section 3.1**)
5. Nail polish

2.4. Materials for Immunofluorescence Labeling

1. Paraformaldehyde (PFA) (powder)
2. Absolute methanol
3. Triton X-100 solution (0.5% (v/v) in PBS)
4. Blocking solution (5% (w/v) BSA in PBS)
5. Primary antibodies (anti-Tom20, FL-145, rabbit polyclonal, Santa Cruz Biotech., Santa Cruz, CA; anti-β-tubulin, MAB3408, mouse monoclonal, Sigma-Aldrich)
6. Fluorophore-labeled secondary antibodies

3. Methods

3.1. Embedding Media

The use of immersion lenses with high numerical apertures is compromised by spherical aberrations induced by the refractive index mismatch between the immersion and the embedding medium (20, 21). A solution to this problem is to adapt the index of the embedding medium to that of the immersion medium. This is crucial for high-resolution imaging. As a beneficial secondary effect, several of the utilized mounting media reduce photobleaching (22–24).

3.1.1. Mowiol

Mowiol is widely used as embedding medium for fluorescence microscopy. For imaging of Mowiol embedded samples, oil-immersion lenses may be used. Mowiol hardens over time, alleviating the need to seal the sample. Mowiol, although convenient and sufficient for many STED applications, may be not the best choice for the most demanding applications (*see* **Note 2**).

1. Mix 6 g glycerol with 2.4 g Mowiol 4-88 in a 50 ml centrifuge tube and stir 1 h on magnetic stirrer.
2. Add 6 ml water and stir for further 2 h.
3. Add 12 ml Tris–HCl buffer (0.2 M, pH 8.4) and heat up the solution to 50°C (in a water-bath) for more than 2 h under constant agitation. Extend this step, until the Mowiol 4-88 is completely dissolved.
4. Optional: add antifading agents (e.g., 25 mg/ml DABCO); stir for more than 4 h (*see* **Note 3**).
5. Centrifuge at 7500*g* for 30 min to remove any undissolved solids.
6. Aliquot in eppendorf tubes and store at –20°C. Each aliquot can be used for several weeks if stored at 4°C.

3.1.2. TDE (2,2′-thiodiethanol)

TDE is a mounting medium allowing for the adjustment of the refractive index ranging from that of water (1.33) to that of immersion oil (1.518). It is miscible with water at any ratio (25). Because TDE does not harden, the sample has to be sealed with nail polish (*see* **Notes 4 and 5**).

1. Mix 97 ml 2,2′-thiodiethanol (TDE) with 3 ml PBS (*see* **Note 6**).
2. Stir for at least 15 min.
3. Adjust pH to 7.5 with HCl/NaOH (*see* **Note** 7).
5. Check refractive index with refractometer and adjust refractive index to 1.518 by addition of 100% TDE or PBS, if necessary (*see* **Note 8**).

3.2. Preparation of Bead Samples

One possibility to determine the resolution of a fluorescence microscope is to image fluorescent microspheres that are distinctly smaller that the resolution of the instrument. The actual physical size of the beads may vary, which could complicate the analysis of the images (*see* **Note 9**). A typical STED image of a bead sample is shown in **Fig. 11.1**.

1. Dilute the beads in ethanol. The typical dilution factor is $1{:}10^3$ for 100 nm beads and $1{:}10^5$ for 20 nm beads.
2. Sonicate the beads at least for 5 min (in ultrasonic bath) (*see* **Note 10**).

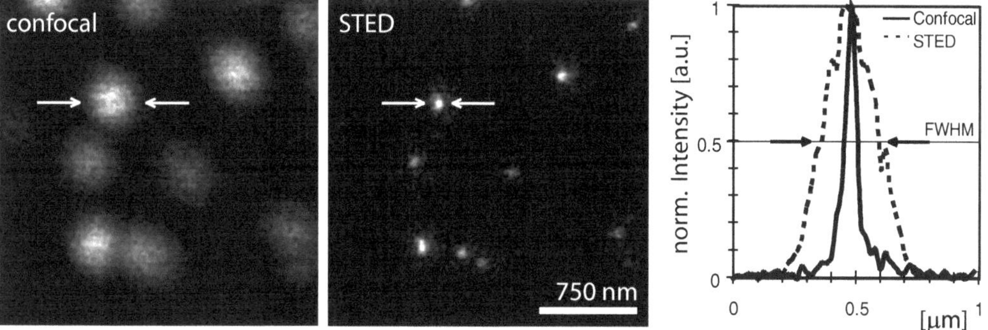

Fig. 11.1. Determination of the microscope's resolution. One way to determine the resolution of a microscope is to image fluorescent beads whose size is considerably smaller than the optical resolution of the instrument (21). Shown are dispersed fluorescent microspheres (crimson fluorescent, specified diameter: 40 nm (for details *see* **Section 2.3**)). Left: confocal image. *Right*: corresponding STED image. The *right panel* shows normalized intensity profiles through the bead marked by the arrows. The full width at half maximum (FWHM) is a measure for the optical resolution. The obtained resolution was about 250 nm (confocal) and about 60 nm (STED).

3. Clean the cover slip with ethanol.
4. Apply the bead suspension to cover slip (Use 10 µl bead suspension for a 30 mm cover slip) and allow to dry on air (*see* **Notes 10 and 11**).
5. Mount with the required embedding medium (*see* **Section 3.1**).

3.3. Immunofluorescence Labeling

Indirect immunofluorescence labeling is a widely used method to visualize specific structures in fixed cells. The same methods can be applied for STED microscopy. However, due to the higher resolution of STED microscopy, the requirements for sample preparation tend to be stricter than for conventional optical microscopy.

Most notably, inadequate sample preparation and labeling efficiencies may not be noticed with confocal microscopy, but is clearly visible with STED microscopy. Thus care has to be taken to utilize optimal fixation conditions to ensure structural preservation (*see* **Note 12**). Simultaneously, extraction conditions need to be chosen so that optimal accessibility for the antibodies to the structures is possible. The antibody concentrations, incubation times, and temperatures should be optimized to ensure optimal brightness and signal-to-noise ratios in the images (*see* **Note 13**). It may be necessary to optimize the labeling procedures for any new structure or specimen.

We present here labeling protocols for two standard samples (microtubule cytoskeleton and TOM complexes on the mitochondrial surface) which proved to be robust. Furthermore, these samples can be routinely used to determine the performance of a STED microscope. The staining procedures may be used as a starting point to optimize different samples.

3.3.1. Labeling of the Microtubule Cytoskeleton in Methanol-Fixed Mammalian Cells

Microtubules form a complex and interweaved network. Because the diameter of antibody decorated microtubules is below 70 nm (8), they are convenient test objects to monitor the performance of a STED microscope (**Fig. 11.2a**). The microtubule cytoskeleton can be fixed with ice-cold methanol (*see* **Note 12**).

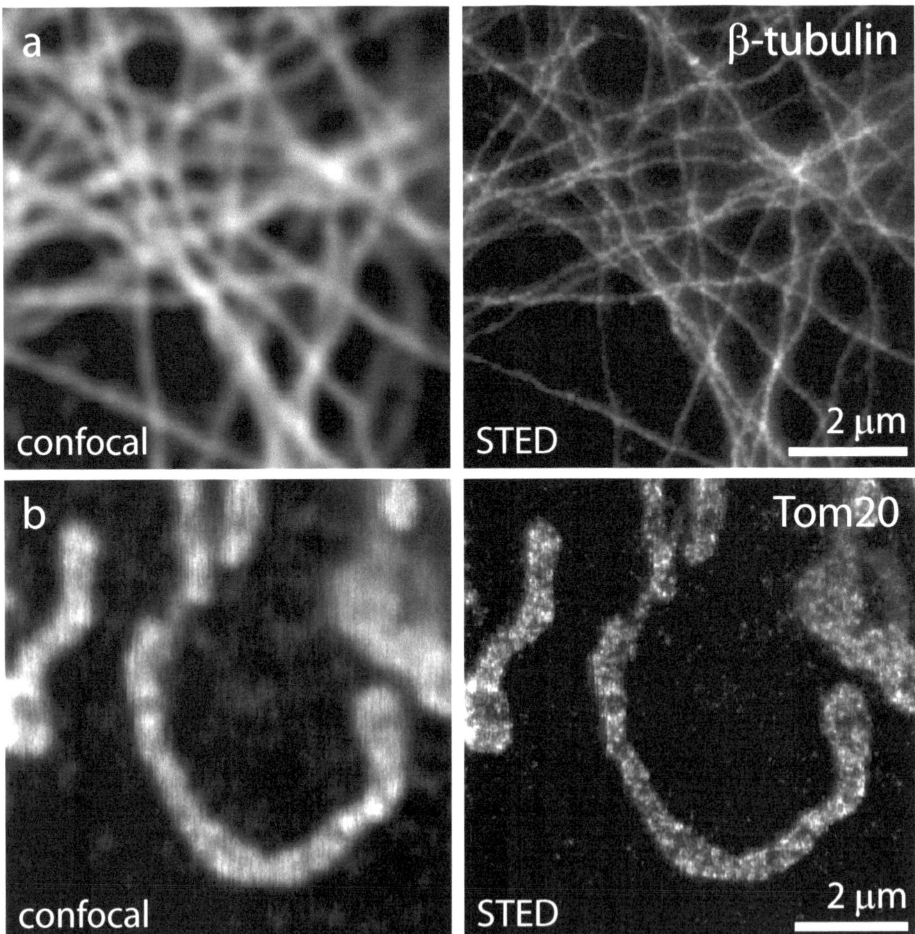

Fig. 11.2. Confocal and STED images of immunolabeled sub-cellular structures. (**a**) Microtubule cytoskeleton. PtK2 cells were fixed with ice-cold methanol and labeled with primary antibodies specific for β-tubulin. The secondary antibody was labeled with Atto647N. Left: confocal image. *Right*: corresponding STED image taken with the microscope as described in (28). (**b**) TOM (translocase of the outer mitochondrial membrane) complex. PtK2 cells were fixed with formaldehyde and labeled with primary antibodies specific for Tom20 (a subunit of the TOM complex). The secondary antibody was labeled with Atto647N. *Left*: confocal image. *Right*: corresponding STED image. In the STED image individual TOM clusters, which are concealed in the confocal image, are clearly discernible. The lateral resolution in both STED images was in the order of 50 nm.

1. Grow mammalian cells on cover slips (typically to a confluency of 50–80%) (*see* **Note 14**).
2. Fix and permeabilize the cells by incubating the cover slips in ice-cold methanol (–20°C) for 5 min (*see* **Note 12**). All subsequent steps are performed at room temperature.

3. Wash samples twice in PBS.
4. Incubate for 5 min in blocking solution (*see* **Note 15**).
5. Dilute primary antibody specific for β-tubulin to a final concentration of ~5 μg/ml in blocking solution.
6. Centrifuge antibody solution for at least 1 min at 15,000*g* (*see* **Note 16**).
7. Incubate sample for 1 h in antibody solution (*see* **Note 17**).
8. Wash cells twice in PBS (*see* **Note 18**).
9. Dilute labeled secondary antibody (anti mouse) to a final concentration of ~1–20 μg/ml in blocking solution. Optimal dilution depends on the quality of the antibody (*see* **Note 19**).
10. Centrifuge antibody solution for at least 1 min at 15,000*g*.
11. Incubate sample for 1 h in secondary antibody solution (*see* **Note 20**).
12. Wash cells in PBS for at least 5 min.
13. Mount sample on slide (*see* **Section 3.3.3.** and **Notes 21 and 22**).

3.3.2. Labeling of the TOM Complex in Formaldehyde-Fixed Mammalian Cells

The TOM (translocase of the outer membrane of mitochondria) complex is the major import pore for mitochondrial precursor proteins (26). Using conventional (confocal) microscopy the TOM complexes appear to be evenly distributed on the mitochondria. Only sub-diffraction resolution microscopy reveals that they are located in small clusters at the surface of the mitochondrial tubules (12, 14) (**Fig. 11.2b**). Due to the three-dimensional arrangement of the TOM complexes on the mitochondria and their inhomogeneous size distribution, this is a more challenging sample for high-resolution microscopy than the microtubule network. For this sample formaldehyde fixation and membrane permeabilization with detergent are required.

Preparation of 8 % (w/v) Formaldehyde Solution

1. Mix 80 ml of water and 8 g of paraformaldehyde powder.
2. Stir the suspension to disperse the powder.
3. Add 1 ml of 1 M NaOH.
4. Heat the solution to 60°C under continuous stirring until the suspension becomes clear (~15 min) (*see* **Note 23**).
5. Add 10 ml concentrated PBS (10× concentrated, pH 7.4).
6. Adjust pH to 7.4 with HCl.
7. Add water to 100 ml.
8. Store at 4°C for 1 week or at –20°C for longer time periods (*see* **Note 23**).

Immunofluorescence Labeling

1. Grow mammalian cells on cover slips (typically to a confluency of 50–80%).
2. Chemically fix cells by incubating the cover slips for 5–10 min in formaldehyde solution prewarmed to 37°C.
4. Wash cells twice in PBS.
5. Extract by incubating in 0.5 % (v/v) Triton X-100 in PBS for 5 min.
6. Incubate for 5 min in blocking solution (*see* **Note 15**).
7. Dilute primary Tom20 specific antibody to a final concentration of ~2 μg/ml in blocking solution.
8. Centrifuge antibody solution for at least 1 min at 15,000*g* (*see* **Note 16**).
9. Incubate sample for 1 h in antibody solution (*see* **Note 17**).
10. Wash cells twice in PBS (*see* **Note 18**).
11. Dilute labeled secondary antibody (anti-rabbit) to a final concentration of ~1–20 μg/ml in blocking solution. Optimal dilution depends on the quality of the antibody.
12. Centrifuge antibody solution for at least 1 min at 15,000*g*.
13. Incubate sample for 1 h in diluted secondary antibody.
14. Wash cells in PBS for more than 5 min.
15. Mount sample on slide (*see* **Section 3.3.3.** and **Notes 21** and **22**).

3.3.3. Embedding of Cell Samples

Embedding with Mowiol

1. Pipette a small drop of Mowiol (prewarmed to room temperature) onto the slide.
2. Mount cover slip with sample on slide and remove excess Mowiol with tissue paper.
3. Leave slides for several hours in the dark to allow the Mowiol to harden (*see* **Notes 2** and **3**).

Embedding with 97 % (v/v) TDE

To prevent cellular structure from distortion by osmotic shock during embedding in TDE, the exchange of water with TDE must be slow. Therefore a dilution series is required.

1. Prepare TDE dilution series with PBS: 10% (v/v) TDE, 25% (v/v) TDE, 50% (v/v) TDE, and 97% (v/v) TDE.
2. Incubate labeled sample in 10% (v/v) TDE for 10 min.
3. Incubate labeled sample in 25% (v/v) and 50% (v/v) TDE for 5 min each.
4. Incubate labeled sample in 97% (v/v) TDE twice for 5 min each.
5. Mount sample with 97% (v/v) TDE on slide.
6. Remove excess TDE.
7. Seal with nail polish (*see* **Note 5**).

4. Notes

1. Each objective lens is designed and assembled to achieve optimum performance with specific cover slips. Using a cover slip of, for example, the wrong thickness can introduce (spherical) aberrations into the imaging process which degrade resolution. Consult the manufacturer of your objective lens about which cover slip to use.
2. Mowiol is a solution of polyvinyl alcohol. It hardens over some days and can be molten again by heating the samples in buffer. The refractive index of Mowiol solutions may vary between batches. During hardening the refractive index changes from close to glycerol (~1.45) to close to immersion oil (~1.518). Some groups report shrinkage of the samples upon hardening.
3. Embedding media for fluorescence microscopy often contain chemicals that are supposed to reduce bleaching of the fluorescent probes. Prominent examples include 1,4-diazabicyclo[2.2.2]octan (DABCO, 25 mg/ml), *N*-propyl gallate (NPG, 0.5% (w/v)) and *p*-phenylene diamine (PPD, 1 mg/ml) (22–24). In rare cases these reagents reduce the dyes fluorescence intensity. The effectiveness of an antifading reagent has to be evaluated for any given fluorophore. We found that Mowiol (with DABCO) is frequently a good option. However, other (commercially available) embedding media may also be considered. Currently, TDE is, to our knowledge, the only embedding medium that allows a precise tuning of the refractive index to match that of immersion oil.
4. TDE is an inexpensive, non-toxic glycol derivative. Commercially available TDE may contain impurities. It can be further purified by distillation (boiling temperature 164–166°C at 27 hPa).

   Due to its high refractive index (~1.522) and full miscibility with water, it is suited as mounting medium with precisely adjusted refractive index (also *see* **Note 8**). Unlike Mowiol, TDE does not harden. Samples need to be sealed with nail polish.

   The properties of some fluorophores are changed in TDE: Especially the absorption and emission spectra, but also the quantum efficiency and bleaching properties may be altered when the samples are mounted in TDE.

5. Nail polish is often used to seal microscopy samples. Colored or glittering nail polish should be avoided. Certain solvents in the nail polish may quench the fluorescence.
6. Typically TDE is used in combination with a phosphate buffer, but other aqueous buffers also work.
7. Due to the viscosity of TDE, the measurement of its pH is difficult. Either a special pH electrode has to be used or, if standard electrodes are used, one has to wait for a long time until equilibrium has been reached.
8. TDE is hygroscopic. The refractive index of undiluted TDE (~1.522) may vary due to its water content.
9. Due to the production process of the beads, the variability of their diameter tends to become large for small beads (**Fig. 11.3**). If uniformly stained beads are substantially smaller than the microscope resolution, the measured brightness is proportional to the bead volume. Hence polydispersity leads to broad brightness distributions.

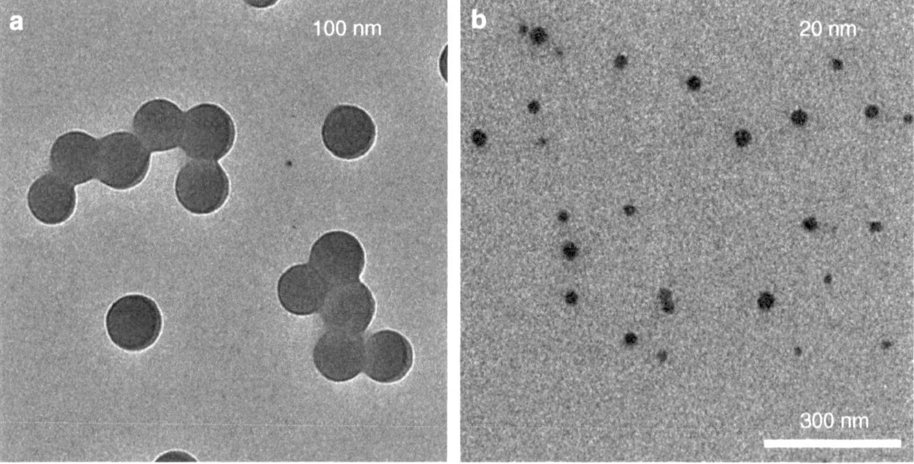

Fig. 11.3. Electron micrographs of crimson fluorescent microspheres. (**a**) 100 nm beads (*see* **Section 2.3**) exhibit a monodisperse size distribution (100 nm ± 7.5%), whereas (**b**) 20 nm beads (*see* **Section 2.3**) are more polydisperse (20 nm ± 50%). The variation in bead size has to be taken into account when determining the resolution of a microscope.

10. The concentration of purchased beads is typically very high. In suspension, they tend to aggregate, which can be alleviated by dilution and sonification before use. During drying, ethanol or surface contaminants on the cover slip may lead to droplet formation and thereby to an inhomogeneous bead distribution on the cover slip.
11. The STED beam exerts a force on particles which have a different refractive index than the surrounding medium

(optical tweezer effect). Beads may start to move or even detach from the cover slip. Increasing the stickiness of the surface by using Poly-L-lysine solution can avoid this.

12. Fixation is required to preserve cellular structures. Cross-linking fixatives like formaldehyde are advantageous for most structures. (The (microtubule) cytoskeleton may be an exception.) Due to this cross-linking activity, fixatives also may hinder antibodies from reaching their target structure. To overcome this, fixation is often combined with extraction (e.g., organic solvents like methanol, Lavdovsky's fixative, or Carnoy's fluid), which may also influence the preservation of the cellular structures. This may be a much more severe problem in STED microscopy than in conventional diffraction limited microscopy because in the latter case the fixation/extraction induced artifacts may be hidden by the lower resolution. Hence special care has to be taken to control proper preservation of cellular structures. Protocols sufficient for confocal microscopy may not be sufficient for STED microscopy.
13. Many antibody suppliers suggest performing antigen retrieval. To this end, the sample is heated (to up to 120°C) for a certain time (27). Sometimes heating is performed under high salt conditions and/or at extreme pH values. Often these procedures work astonishingly well; still special attention to the proper preservation of the analyzed structures is mandatory.
14. Immunofluorescence labeling described here was successfully performed with several mammalian cell lines, ranging from human cervix carcinoma cells like HeLa to rodent kidney cell lines like BHK-21. Information about proper cultivation conditions is available from the American Type Culture Collection (www.lgcstandards-atcc.org).
15. Several blocking agents can be used, including skimmed milk (coldwater fish) gelatin or BSA. A number of rather expensive blocking agents are on the market, including serum of the antibody producing host species or synthetic blocking agents. We recommend the use of BSA for routine labeling. For specialized applications the use of other blocking agents may be necessary.

    The blocking solution should be freshly prepared. It can be stored for longer time periods at –20°C.

    Unspecific background labeling is a common problem. Often insufficient blocking of the cells results in (high) background. To overcome this problem, the concentration of the blocking agent and/or the time of blocking may be increased. Most other problems originate from the primary and secondary antibodies. These may be solved by optimizing the antibody concentrations, incubation times, and washing conditions.

16. Many antibody-fluorophore conjugates tend to precipitate and should be centrifuged before use. This procedure will remove any aggregates that may have formed during storage.
17. After fixation, the samples must not become dry. Otherwise the specificity of the labeling is often lost. To keep the samples humid, labeling is best performed in a humidity chamber. Alternatively, the cover slips can be submersed in antibody solutions in a microtiter plate.
18. Some protocols suggest washing the sample in PBS-Triton X-100 and blocking anew with blocking solution, prior to application of the secondary antibody. In case of an unspecific background, this is recommended.
19. Several fluorophores were successfully employed for STED microscopy (*see* **Table 11.1**). The number of fluorophores suitable for STED microscopy is rapidly increasing. Currently, antibodies labeled with these fluorophores are not

Table 11.1
List of fluorophores used for STED microscopy

Dye name (manufacturer/distributor)	Excitation wavelength	STED wavelength	Reported spatial resolution (direction)	Reference(s)
ATTO532 (ATTO-Tec, Siegen, Germany)	470 nm	603–615 nm	25–72 nm (*xy*)	(11, 15, 17, 30, 31)
Chromeo 488 (Active Motif, Carlsbad, CA)	488 nm	602 nm	< 30 nm (*xy*)	(32)
YFP	490 nm	595 nm	<50 nm (*xy*)	(33)
GFP	490 nm	575 nm	~70 nm (*xy*)	(34)
ATTO565 (ATTO-Tec)	532 nm	640–660 nm	30–40 nm (*xy*)	(35)
MR 121 SE (Roche, Mannheim, Germany)	532 nm	793 nm	~50 nm (*z*)	(8)
NK51 (ATTO-Tec)	532 nm	647 nm	~50 nm (*xyz*)	(12)
RH 414 (Invitrogen Carlsbad, CA)	554 nm	745 nm	30 nm (*z*)	(36)
ATTO590 (ATTO-Tec)	570 nm	690–710 nm	30–40 nm (*xy*)	(35)
ATTO633 (ATTO-Tec)	630 nm	735–755 nm	30–40 nm (*xy*)	(35)
ATTO647N (ATTO-Tec)	635 nm	750–780 nm	~50 nm (*xy*)	(14, 19, 37)

Modified from www.nanoscopy.de (Dept. of NanoBiophotonics, Max Planck Institute for Biophysical Chemistry, February 2009). Note that the number of fluorophores that are demonstrated to be suitable for STED microscopy is steadily increasing.

always commercially available; custom labeling may be required or is advantageous to ensure a high quality of the labeled antibody conjugate. A robust protocol may be found at (29), www.probes.com or www.atto-tec.com.

20. Some antibody-fluorophore conjugates may result in an unwanted unspecific staining of cellular structures. For example, Atto647N is known to label mitochondrial membranes to some extent. Control experiments are required to determine the level of unspecific background for each secondary antibody.
21. A nuclear counterstain can be performed by mounting the samples in Mowiol or TDE containing DAPI (2 μg/ml).
22. After immunofluorescence labeling, the samples should be stored at 4°C. The samples should be imaged within the next few days.
23. Prolonged heating of the formaldehyde solution at 60°C, or too high temperatures, will induce the formation of formic acid, deteriorating the properties of the fixative. Formaldehyde is toxic. Preparation of the formaldehyde solution from solid PFA must be done under a fume hood.

    For most applications, fresh formaldehyde solutions may be kept in the refrigerator for 1 week or stored at –20°C for longer periods. Occasionally it has been reported that the use of fresh formaldehyde is superior.

Acknowledgments

We thank Sylvia Löbermann and Donald Ouw for excellent technical assistance and helpful discussion on the protocols. Furthermore we acknowledge Benjamin Harke for help with STED microscopy and Jaydev Jethwa for insightful comments. A part of the work that resulted in the protocols described in this chapter was supported by a grant from the Deutsche Forschungsgemeinschaft (JA 1129/3) to S.J.

References

1. S. W. Hell and J. Wichmann (1994) Breaking the diffraction resolution limit by stimulated emission: Stimulated emission depletion microscopy, *Opt. Lett.* **19**, 780–782.
2. S. W. Hell (2003) Toward fluorescence nanoscopy, *Nat. Biotechnol.* **21**, 1347–1355.
3. S. W. Hell, M. Dyba and S. Jakobs (2004) Concepts for nanoscale resolution in fluorescence microscopy, *Curr. Opin. Neurobiol.* **14**, 599–609.
4. S. W. Hell (2007) Far-field optical nanoscopy, *Science* **316**, 1153–1158.
5. M. Fernandez-Suarez and A. Y. Ting (2008) Fluorescent probes for super-resolution imaging in living cells, *Nat. Rev. Mol. Cell. Biol.* **9**, 929–943.
6. T. A. Klar and S. W. Hell (1999) Subdiffraction resolution in far-field fluorescence microscopy, *Opt. Lett.* **24**, 954–956.

7. T. A. Klar, S. Jakobs, M. Dyba, A. Egner and S. W. Hell (2000) Fluorescence microscopy with diffraction resolution barrier broken by stimulated emission, *Proc. Natl. Acad. Sci. USA* **97**, 8206–8210.
8. M. Dyba, S. Jakobs and S. W. Hell (2003) Immunofluorescence stimulated emission depletion microscopy, *Nat. Biotechnol.* **21**, 1303–1304.
9. V. Westphal, C. M. Blanca, M. Dyba, L. Kastrup and S. W. Hell (2003) Laser-diode-stimulated emission depletion microscopy, *Appl. Phys. Lett.* **82**, 3125–3127.
10. V. Westphal and S. W. Hell (2005) Nanoscale resolution in the focal plane of an optical microscope, *Phys. Rev. Lett.* **94**, 143903.
11. K. I. Willig, S. O. Rizzoli, V. Westphal, R. Jahn and S. W. Hell (2006) STED-microscopy reveals that synaptotagmin remains clustered after synaptic vesicle exocytosis, *Nature* **440**, 935–939.
12. R. Schmidt, C. A. Wurm, S. Jakobs, J. Engelhardt, A. Egner and S. W. Hell (2008) Spherical nanosized focal spot unravels the interior of cells, *Nat. Methods* **5**, 539–544.
13. C. Eggeling, C. Ringemann, R. Medda, G. Schwarzmann, K. Sandhoff, S. Polyakova, V. N. Belov, B. Hein, C. von Middendorff, A. Schonle and S. W. Hell (2008) Direct observation of the nanoscale dynamics of membrane lipids in a living cell, *Nature* doi:10.1038/nature07596.
14. G. Donnert, J. Keller, C. A. Wurm, S. O. Rizzoli, V. Westphal, A. Schonle, R. Jahn, S. Jakobs, C. Eggeling and S. W. Hell (2007) Two-color far-field fluorescence nanoscopy, *Biophys. J.* **92**, L67–L69.
15. R. R. Kellner, C. J. Baier, K. I. Willig, S. W. Hell and F. J. Barrantes (2007) Nanoscale organization of nicotinic acetylcholine receptors revealed by stimulated emission depletion microscopy, *Neuroscience* **144**, 135–143.
16. R. J. Kittel, C. Wichmann, T. M. Rasse, W. Fouquet, M. Schmidt, A. Schmid, D. A. Wagh, C. Pawlu, R. R. Kellner, K. I. Willig, S. W. Hell, E. Buchner, M. Heckmann and S. J. Sigrist (2006) Bruchpilot promotes active zone assembly, Ca^{2+} channel clustering, and vesicle release, *Science* **312**, 1051–1054.
17. J. J. Sieber, K. I. Willig, R. Heintzmann, S. W. Hell and T. Lang (2006) The snare motif is essential for the formation of syntaxin clusters in the plasma membrane, *Biophys. J.* **90**, 2843–2851.
18. U. V. Nagerl, K. I. Willig, B. Hein, S. W. Hell and T. Bonhoeffer (2008) Live-cell imaging of dendritic spines by sted microscopy, *Proc. Natl. Acad. Sci. USA* **105**, 18982–18987.
19. V. Westphal, S. O. Rizzoli, M. A. Lauterbach, D. Kamin, R. Jahn and S. W. Hell (2008) Video-rate far-field optical nanoscopy dissects synaptic vesicle movement, *Science* **320**, 246–249.
20. A. Diaspro (2001) Confocal and two-photon microscopy: foundations, applications and advances, Wiley-Liss, New York.
21. J. B. Pawley (2006) Handbook of biological confocal microscopy, Springer, New York.
22. A. Longin, C. Souchier, M. Ffrench and P. A. Bryon (1993) Comparison of anti-fading agents used in fluorescence microscopy: Image analysis and laser confocal microscopy study, *J. Histochem. Cytochem.* **41**, 1833–1840.
23. M. Ono, T. Murakami, A. Kudo, M. Isshiki, H. Sawada and A. Segawa (2001) Quantitative comparison of anti-fading mounting media for confocal laser scanning microscopy, *J. Histochem. Cytochem.* **49**, 305–312.
24. K. Valnes and P. Brandtzaeg (1985) Retardation of immunofluorescence fading during microscopy, *J. Histochem. Cytochem.* **33**, 755–761.
25. T. Staudt, M. C. Lang, R. Medda, J. Engelhardt and S. W. Hell (2007) 2,2′-thiodiethanol: A new water soluble mounting medium for high resolution optical microscopy, *Microsc. Res. Tech.* **70**, 1–9.
26. W. Neupert and J. M. Herrmann (2007) Translocation of proteins into mitochondria, *Annu. Rev. Biochem.* **76**, 723–749.
27. S. R. Shi, R. J. Cote and C. R. Taylor (1997) Antigen retrieval immunohistochemistry: Past, present, and future, *J. Histochem. Cytochem.* **45**, 327–343.
28. B. Harke, J. Keller, C. K. Ullal, V. Westphal, A. Schonle and S. W. Hell (2008) Resolution scaling in STED microscopy, *Opt. Expr.* **16**, 4154–4162.
29. G. T. Hermanson (2008) Bioconjugate techniques, Elsevier Acad. Press, Amsterdam [u.a.].
30. G. Donnert, J. Keller, R. Medda, M. A. Andrei, S. O. Rizzoli, R. Luhrmann, R. Jahn, C. Eggeling and S. W. Hell (2006) Macromolecular-scale resolution in biological fluorescence microscopy, *Proc. Natl. Acad. Sci. USA.* **103**, 11440–11445.
31. D. Fitzner, A. Schneider, A. Kippert, W. Mobius, K. I. Willig, S. W. Hell, G. Bunt, K. Gaus and M. Simons (2006) Myelin basic protein-dependent plasma membrane reorganization in the formation of myelin, *Embo J.* **25**, 5037–5048.
32. L. Meyer, D. Wildanger, R. Medda, A. Punge, S. O. Rizzoli, G. Donnert and S. W. Hell (2008) Dual-color STED microscopy

at 30-nm focal-plane resolution, *Small* **4**, 1095–1100.
33. B. Hein, K. Willig and S. W. Hell (2008) Stimulated emission depletion (STED) nanoscopy of a fluorescent protein labeled organelle inside a living cell, *Proc. Natl. Acad. Sci. USA* **105**, 14271–14276.
34. K. I. Willig, R. R. Kellner, R. Medda, B. Hein, S. Jakobs and S. W. Hell (2006) Nanoscale resolution in GFP-based microscopy, *Nat. Methods* **3**, 721–723.
35. D. Wildanger, E. Rittweger, L. Kastrup and S. W. Hell (2008) STED microscopy with a supercontinuum laser source, *Opt. Expr.* **16**, 9614–9621.
36. M. Dyba and S. W. Hell (2002) Focal spots of size lambda/23 open up far-field florescence microscopy at 33 nm axial resolution, *Phys. Rev. Lett.* **8816**, 163901.
37. K. I. Willig, B. Harke, R. Medda and S. W. Hell (2007) STED microscopy with continuous wave beams, *Nat. Methods* **4**, 915–918.

Chapter 12

Two-Photon Permeabilization and Calcium Measurements in Cellular Organelles

Oleg Gerasimenko and Julia Gerasimenko

Abstract

Inositol trisphosphate and cyclic ADP-ribose, main intracellular Ca^{2+} messengers, induce release from the intracellular Ca^{2+} stores via inositol trisphosphate and ryanodine receptors, respectively. Recently, studies using novel messenger nicotinic acid adenine dinucleotide phosphate (NAADP) releasing Ca^{2+} from calcium stores in organelles other than endoplasmic reticulum (ER) have been conducted. However, technical difficulties of Ca^{2+} measurements in relatively small Ca^{2+} stores prompted us to develop a new, more sensitive, and less damaging two-photon permeabilization technique. Applied to pancreatic acinar cells, this technique allowed us to show that all three messengers – IP_3, cADPR, and NAADP – release Ca^{2+} from two intracellular stores: the endoplasmic reticulum and an acidic store in the granular region. This chapter describes a detailed procedure of using this technique with pancreatic acinar cells.

Key words: Ca^{2+} stores, Ins(1,4,5)P_3, Ryanodine, Pancreas and pancreatic cells, Two-photon permeabilization, Secretory granules.

1. Introduction

Hormone induced Ca^{2+} release from intracellular stores into the cytosol (1) and intracellular Ca^{2+}-releasing messenger inositol (1,4,5-trisphosphate (IP_3) are the most important discoveries in the Ca^{2+} signaling field (2, 3). IP_3-induced Ca^{2+} release is currently accepted as the principal mechanism for generation of Ca^{2+} signals in non-excitable cells (4, 5). While it is generally accepted that the endoplasmic reticulum (ER) is the main organelle for Ca^{2+} release (6, 4, 7–9), other organelles can also serve as a possible Ca^{2+} store. Several candidates have been suggested

D.B. Papkovsky (ed.), *Live Cell Imaging*, Methods in Molecular Biology 591,
DOI 10.1007/978-1-60761-404-3_12, © Humana Press, a part of Springer Science+Business Media, LLC 2010

including the mitochondria, which play an important role in shaping cytosolic Ca^{2+} signals (10–13). Other organelles such as the nuclear envelope (14–19), the Golgi apparatus (20–22), the secretory granules (23–26), and the endosomes (27) have been reported to store and release Ca^{2+}.

In pancreatic acinar cells, the bulk of the ER Ca^{2+} store is located in the basal part of the cell (28) with thin ER projections into the secretory granule area (29) where Ca^{2+} release is usually initiated (30). Application of intracellular Ca^{2+} messenger inositol 1,4,5-trisphosphate (IP_3) or cyclic ADP ribose (cADPR) produces Ca^{2+} release specifically localized in the secretory granular area (30, 31), including novel Ca^{2+} releasing messenger nicotinic acid adenine dinucleotide phosphate (NAADP) (32–35). Several hypotheses explaining possible mechanisms of NAADP-induced Ca^{2+} release have been suggested recently (36–39).

To study effects of intracellular Ca^{2+} messengers, researchers have to use permeabilized cells which are of particular importance for studies of internal stores (2, 3). Chemical permeabilization often damages intracellular membranes and makes it difficult to detect other stores apart from the largest ER. We have tried several permeabilizing agents (digitonin, saponin, streptolysin) until we came up with the idea to use two-photon laser approach (40, 41) for permanent cellular permeabilization (42). This method is much less damaging than chemical permeabilizations. Permeabilization has been achieved by localized perforation of the membrane using two-photon (tuned to 740–750 nm) high intensity laser pulses. Two-photon light when directed to a small area of cell membrane can perforate plasma membrane allowing successful intracellular delivery of foreign DNA (41). We have modified this technique to achieve permanent permeabilization of pancreatic acinar cells (**Fig. 12.1a–d**). The result of permeabilization is the hole of approximately 2 μm in size, with most of plasma membrane and intracellular organelles intact.

We have studied Ca^{2+} release elicited by the intracellular Ca^{2+} releasing messengers NAADP, cADPR, and IP_3 and found that all three messengers can release Ca^{2+} from two separate internal stores: thapsigargin-sensitive (ER type) store and thapsigargin-insensitive acidic Ca^{2+} store (**Fig. 12.2**) (42, 43). With the help of two-photon permeabilization technique we have shown that in both stores NAADP and cADPR likely activate RyRs, whereas IP_3 activates IP_3Rs. The use of antibodies and/or large proteins also allowed us to modulate specific intracellular function(s) (**Fig. 12.3**). This technique serves as a reliable and convenient method to permeabilize cells with the minimal damage to cellular membranes and organelles.

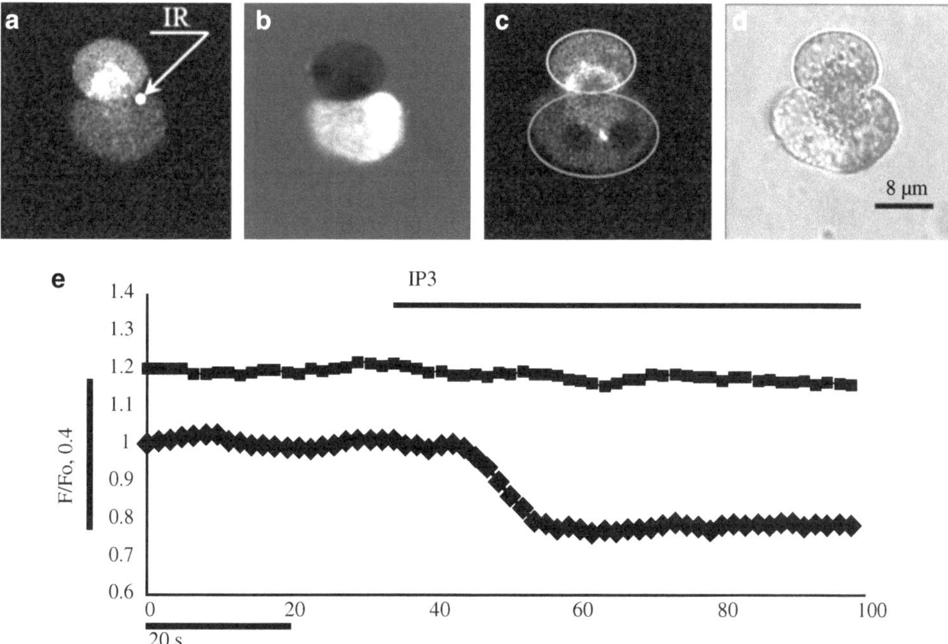

Fig. 12.1. Two-photon permeabilization of pancreatic acinar cells and the effects of Ca^{2+} releasing messengers (modified from Gerasimenko et al. (42)). **a**. A doublet of pancreatic acinar cells loaded with Fluo-5 N AM before permeabilization. Arrow shows the position of two-photon light application. **b**. Same cell doublet after permeabilization and perfusion with Texas Red dextran (3×10^3 M_r). Only the lower cell has been permeabilized and is therefore bright due to diffusion of Texas Red dextran into the cytoplasm. **c**. Same cell doublet after washing out of Texas Red dextran. Note reduced fluorescence of Fluo-5 N in the lower permeabilized cell. **d**. Transmitted light picture of the doublet (after permeabilization) shown in **a–c**. **e**. IP_3 (10 μM) was applied to the doublet shown in **a-d**. Note that IP_3 elicited a reduction in [Ca^{2+}] in the intracellular stores in the lower (permeabilized) cell, whereas there was no response in the upper (intact) cell.

2. Materials

2.1. Mouse Pancreatic Acinar Cells Preparation

1. Buffer I for cell isolation, which contains 140 mM NaCl, 4.7 mM KCl, 10 mM HEPES, 1 mM $MgCl_2$, 10 mM Glucose, 1 mM $CaCl_2$, pH 7.2 (adjusted with NaOH). Store at 4°C; pre-warm to 37°C before use.
2. Solution of collagenase (Worthington, UK; 200 u/mL; 1 mL/pancreas) in buffer I for pancreas digestion. Store at –20°C; pre-warm to 37°C before use.
3. Water-bath at 37°.
4. Centrifuge with swing rotor (Denley, UK) or any centrifuge capable of 100×*g* with swing rotor for 15 mL tubes.
5. Animal facility for mice handling (schedule 1).

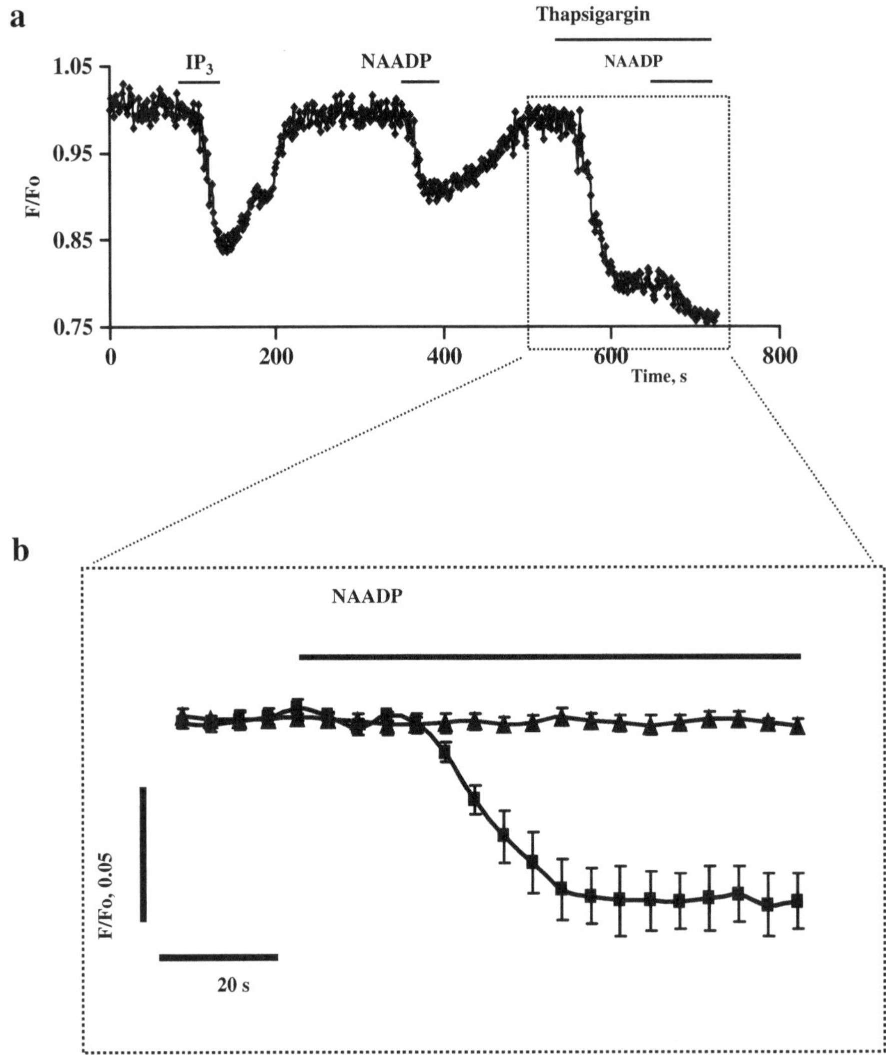

Fig. 12.2. NAADP, IP_3, or cADPR elicit Ca^{2+} release from both thapsigargin-sensitive and thapsigargin-insensitive intracellular stores (modified from Gerasimenko et al. (42)). **a**. Ins(1,4,5)P_3 and NAADP induce large Ca^{2+} responses from whole cell (before thapsigargin), while NAADP can induce only small Ca^{2+} release after high dose of thapsigargin. **b**. Same experiment as shown in 12.2a with the ROI in the granular area (*lower trace*, ■) and basal area (*upper trace*, ▲). NAADP (100 nM) induces Ca^{2+} release from the store in the secretory granule area in the presence of thapsigargin (10 μM) but not in the basal area.

2.2. Two-Photon Permeabilization

1. Buffer II for cell permeabilization, which contains 128 mM KCl, 20 mM NaCl, 10 mM HEPES, 2 mM ATP, 1 mM $MgCl_2$, 0.1 mM EGTA, 0.075 mM $CaCl_2$, pH 7.2 (adjusted with KOH). Store at 4°C. Pre-warm to 37°C before use. (see **Note 2**).

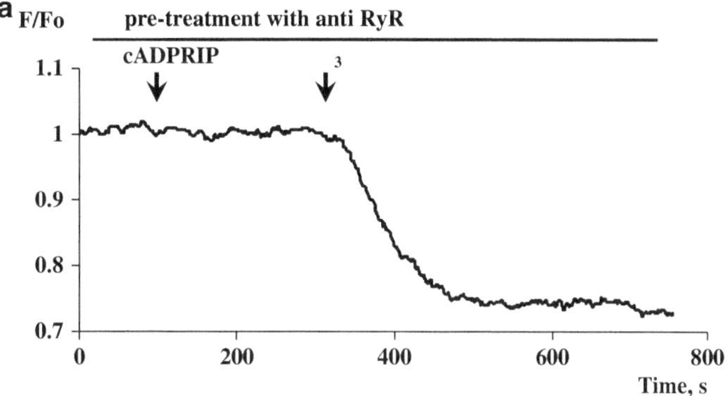

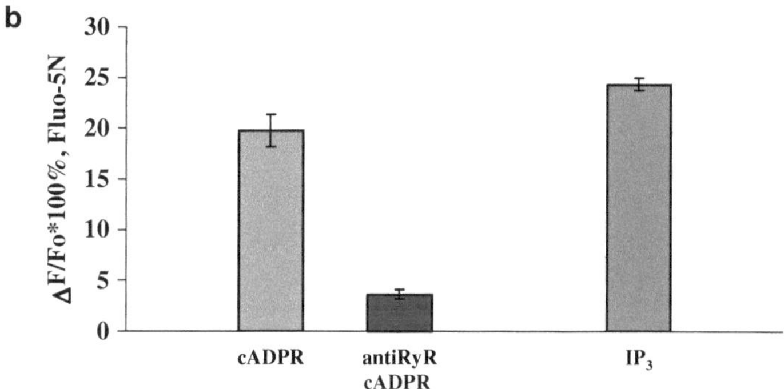

Fig. 12.3. Antibodies against RyRs completely block functional release induced by cADPR, but not by Ins(1,4,5)P_3. **a.** Permeabilized pancreatic acinar cells have been incubated with anti-RyRs antibodies (Millipore, UK) for 20 min and then cADPR has been added with no effect. After that Ins(1,4,5)P_3 induced Ca^{2+} release similar to observed in control experiments. **b**. Comparison of responses induced in the absence of antibodies to cADPR (*first column*) and in the presence of anti-RyRs antibodies to cADPR (*second column*) and to I(1,4,5)P_3 (*third column*).

2. Microscope cover slips (borosilicate glass, thickness no.1) (VWR International, UK), coated with poly-L-lysin (Sigma, UK).
3. Two-photon femtosecond laser Millennia (power 8 W) (Spectra-Physics).
4. Leica SP2 confocal two-photon microscope (similar results were also obtained using Olympus two-photon confocal microscope).
5. Perfusion chamber (self-made or from any supplier).

2.3. Fluorescent Calcium Dyes

1. Fluorescent dyes Mag-Fura-2 AM, Fluo-5 N AM, Texas Red dextran (3×10^3 M_r) (all Invitrogen, UK). Prepared as 1 mM stock solutions in DMSO, stored at –20°C.
2. 1 mM solution of the "near–membrane" Ca^{2+} indicator FFP-18 (K^+) salt (TEFLabs, USA) in buffer II, stored at –20°C.

3. Methods

3.1. Cell Preparation

1. Isolate mouse pancreatic acinar cells by collagenase digestion as described previously (29) (using schedule 1 procedure for handling CD1 mice, according to Animals (Scientific Procedures) Act). Briefly, inject extracted pancreas with 1 mL of collagenase solution, place it in 1.5 mL eppendorf tube, and incubate 15 min in a water-bath at 37°C.
2. Dissociate cells by pipetting through a large diameter tip (cut tip end if necessary).
3. Collect cells in a separate tube filled with buffer I.
4. Centrifuge cells for 1 min at 100×*g* using swing rotor at room temperature.
5. Gently resuspend the cell pellet in 2 mL of buffer I, add about 10 mL of the same buffer, and wash cells by pipetting several times.
6. Centrifuge cells for 1 min at 100×*g* using swing rotor at room temperature.
7. Resuspend cells in 2 mL of buffer I.
8. Prepare stock of indicator dye for cell loading by desiccation of Mag Fura-2AM or Fluo-5 N AM in DMSO in concentration of 2 mM according to manufacturer's instructions (Invitrogen).
9. After isolation, load cells with low affinity Ca^{2+}-sensitive dyes: Mag Fura-2 AM (5 μM) or Fluo-5 N AM (5 μM) by adding an aliquot (5 μL to 2 mL of cells) of concentrated stock and incubate for 30–45 min at 37°C in water-bath.
10. Centrifuge for 1 min at 100×*g* using swing rotor at room temperature. Resuspend cells in 2 mL of buffer I.

3.2. Two-Photon Permeabilization

1. Place 200 μL of cells on poly-L-lysine-coated cover slips attached to perfusion chamber under the microscope. Perform all experiments at room temperature. Prior to permeabilization, perfuse the cells with buffer II (K^+ rich, low Ca^{2+}, containing EGTA).
2. Stain cell membrane with the "near-membrane" Ca^{2+} indicator FFP-18 (K^+ salt) (1 μM) for 5 min to help the formation of a single, site-specific perforation in the cell membrane using the two-photon laser beam.
3. Apply a high intensity two-photon laser beam in pulse mode at 740–750 nm from Spectra-Physics (8 W Millennia femtosecond laser) to a small area of the cell membrane

(**Fig. 12.1a**) (*see* **Note 1**). This results in heating of this small membrane area and subsequent hole formation.

4. Permeabilization can be confirmed by monitoring the fluorescence of Texas Red dextran added to the extracellular medium. Upon permeabilization Texas Red dextran penetrates into the cytoplasm of targeted cell (**Fig. 12.1b**).
5. After permeabilization, perfuse cells with intracellular solution (buffer II) for 5–10 min to wash out the cytosolic component of the fluorescent dye (*see* **Note 3**): Experiments shown in **Fig. 12.1** were conducted in the same solution as above except that $CaCl_2$ was reduced to 0.05 mM.
6. Observe washing of Texas Red dextran from the extracellular solution, which confirms successful and stable permeabilization (**Fig. 12.1c**).
7. After perforation, cells should be able to respond to intracellular Ca^{2+} releasing messengers IP_3, NAADP, or cADPR, (**Fig. 12.1e–g**).

3.3. Fluorescent [Ca^{2+}] Measurements

1. Acquire fluorescent images using the Leica SP2 MP two-photon confocal microscope with objective 63× NA 1.2.
2. Excitation and emission wavelengths for Mag Fura-2 are 430 nm (2–5% power) and 460–590 nm, respectively. Alternatively, for excitation of Mag Fura-2 use the two-photon wavelength ~745 nm.
3. Excitation and emission wavelengths for Fluo-5 N are 488 nm (Argon Ion laser, 1–2% power) and 510–590 nm, respectively.
4. Collect fluorescent images with a frequency of 0.6–1.0 frame/s.
5. Excitation and emission wavelengths for Texas Red dextran are 543 nm and 580–650 nm, respectively.
6. Calculate free Ca^{2+} concentrations assuming that the K_d of Fluo-5 N for Ca^{2+} is 90 μM (*see* **Note 4**).
7. Perform the calibration procedure by applying ionomycin (10 μM) and nigericin (7 μM) with 2 mM EGTA or 10 mM $CaCl_2$.
8. Perform statistical analysis using Microsoft Excel software. Determine P values for statistical significance between sets of data using Student's t-test.

3.4. Conclusions

We compared responses to the messengers in cells permeabilized by two-photon light with responses obtained from saponin (as well as digitonin and streptolysin) permeabilized acinar cells. We found that two-photon permeabilization has two principal advantages, namely better preserved morphology (including polarity) and responsiveness. The amplitudes of the responses to IP_3 were approximately 1.5 times higher in the two-photon permeabilized cells than in cells permeabilized by saponin (even more difference for digitonin and streptolysin). Two-photon permeabilization resulted in a formation of a hole with a diameter of ~2 μm at any site selected on the surface of the cell. This hole did not close after permeabilization, as in previously published work (41), perhaps due to the larger size and/or the exposure of the cell to an intracellular solution before the two-photon pulse. The success rate with two-photon permeabilization was high (>50%) and we propose this technique as a reliable and convenient method for studies of intracellular stores and organelles.

4. Notes

1. The choice of permeabilization region on cell membrane should be preferably away from the bulk of the ER. Permeabilization can damage not only plasma membrane but also intracellular membranes and organelles which potentially can cause the dramatic loss of Ca^{2+}. In our experiments best results were obtained by placing target of laser beam near granular area (**Fig. 12.1a**).
2. Concentration of ATP can be increased to 3 mM to ensure high activity of SERCA pumps.
3. Permeabilizations were conducted at first in buffer II with calcium concentration (0.075 mM $CaCl_2$) and then before experiment calcium was reduced (by perfusion with buffer II containing 0.05 mM $CaCl_2$).
4. The pH-dependence of the K_d of Fluo-5 N for Ca^{2+} was tested. There was no significant difference between the results at pH 7.2 and at pH 6 (27).

References

1. Nielsen, S. P. and Petersen, O. H. (1972). Transport of calcium in the perfused submandibular gland of the cat. *J. Physiol.* **223**, 685–697.
2. Streb, H., Irvine, R. F., Berridge, M. J. and Schulz, I. (1983). Release of Ca^{2+} from a nonmitochondrial intracellular store in pancreatic acinar cells by inositol-1,4,5-trisphosphate. *Nature* **306**, 67–69.
3. Berridge, M. J. (1984). Inositol trisphosphate and diacylglycerol as second messengers. *Biochem. J.* **220**, 345–360.
4. Berridge, M. J. (1993). Inositol trisphosphate and calcium signalling. *Nature* **361**, 315–325.
5. Berridge, M. J., Bootman, M. D. and Roderick, H. L. (2003). Calcium signalling: dynamics, homeostasis and remodelling. *Nature Rev. Mol. Cell. Biol.* **14**, 517–529.
6. Meldolesi, J. and Pozzan, T. (1998). The endoplasmic reticulum Ca^{2+} store: a view from the lumen. *Trends. BioChem. Sci.* **23**, 10–14.

7. Petersen, O. H., Petersen, C. C. H., and Kasai, H. (1994). Calcium and hormone action. *Annu. Rev. Physiol.* **56**, 297–319.
8. Pozzan, T., Rizzuto, R., Volpe, P. and Meldolesi, J. (1994). Molecular and cellular physiology of intracellular calcium stores. *Physiol. Rev.* 74, 595–636.
9. Ashby, M. C. and Tepikin, A. V. (2002) Polarized calcium and calmodulin signaling in secretory epithelia. *Physiol. Rev.* **82**, 701–734.
10. Rizzuto, R., Brini, M., Murgia, M. and Pozzan, T. (1993). Microdomains with high Ca^{2+} close to IP_3-sensitive channels that are sensed by neighboring mitochondria. *Science* **262**, 744–747.
11. Hajnoczky, G., Robb-Gaspers, L. D., Seitz, M. B. and Thomas, A. P. (1995). Decoding of cytosolic calcium oscillations in the mitochondria. *Cell* **82**, 415–424.
12. Tinel, H., Cancela, J. M., Mogami, H., Gerasimenko, J. V., Gerasimenko, O. V., Tepikin, A. V. and Petersen, O. H. (1999). Active mitochondria surrounding the pancreatic acinar granule region prevent spreading of inositol trisphosphate-evoked local cytosolic Ca^{2+} signals. *EMBO J.* **18**, 4999–5008.
13. Pozzan, T., Magalhaes, P. and Rizzuto, R. (2000). The comeback of mitochondria to calcium signaling. *Cell. Calcium* **28**, 279–283.
14. Nicotera, P., Orrenius, S., Nilsson, T. and Berggren, P. O. (1990). An inositol 1,4,5 trisphosphate-sensitive Ca^{2+} pool in liver nuclei. *Proc. Natl. Acad. Sci. USA* **87**, 6858–6862.
15. Malviya, A. N., Rogue, P. and Vincendon, G. (1990). Stereospecific inositol 1,4,5[^{32}P] trisphosphate binding to isolated rat liver nuclei: evidence for inositol trisphosphate receptor-mediated calcium release from the nucleus. *Proc. Natl. Acad. Sci. USA* **87**, 9270–9274.
16. Gerasimenko, O. V., Gerasimenko, J. V., Tepikin, A. V. and Petersen, O. H. (1995). ATP dependent accumulation and inositol trisphosphate- or cyclic ADP-ribose-mediated release of Ca^{2+} from the nuclear envelope. *Cell* **80**, 439–444.
17. Gerasimenko, O. V., Gerasimenko, J. V., Tepikin, A. V. and Petersen, O. H. (1996a). Calcium transport pathways in the nucleus. *Pflugers Arch.* **432**, 1–6.
18. Gerasimenko, J. V., Maruyama, Y., Yano, K., Dolman, N. J., Tepikin, A. V., Petersen, O. H. and Gerasimenko, O. V. (2003). NAADP mobilizes Ca^{2+} from a thapsigargin-sensitive store in the nuclear envelope by activating ryanodine receptors. *J. Cell. Biol.* **163**, 271–282.
19. Gerasimenko, O. V. and Gerasimenko, J. V. (2004), New aspects of nuclear calcium signalling. *J. Cell. Sci.* **117**, 3087–3094.
20. Pinton, P., Pozzan, T. and Rizzuto, R. (1998). The Golgi apparatus is an inositol 1,4,5-trisphosphate-sensitive Ca^{2+} store, with functional properties distinct from those of the endoplasmic reticulum. *EMBO J.* **17**, 5298–5308.
21. Missiaen, L., Van Acker, K., Van Baelen, K., Raeymaekers, L., Wuytack, F., Parys, J. B., De Smedt, H., Vanoevelen, J., Dode, L., Rizzuto, R. and Callewaert, G. (2004). Calcium release from the Golgi apparatus and the endoplasmic reticulum in HeLa cells stably expressing targeted aequorin to these compartments. *Cell. Calcium* **36**, 479–487.
22. Dolman, N. J., Gerasimenko, J. V., Gerasimenko, O. V., Voronina, S. G., Petersen, O. H. and Tepikin, A. V. (2005). Stable Golgi-mitochondria complexes and formation of Golgi Ca^{2+} gradients in pancreatic acinar cells. *J. Biol. Chem.*, **280**, 15794–15799.
23. Fasolato, C., Zottini, M., Clementi, E., Zacchetti, D., Meldolesi, J. and Pozzan, T. (1991). Intracellular Ca^{2+} pools in PC12 cells. Three intracellular pools are distinguished by their turnover and mechanisms of Ca^{2+} accumulation, storage, and release. *J. Biol. Chem.* **266**, 20159–20167.
24. Gerasimenko, O. V., Gerasimenko, J. V., Belan, P. V. and Petersen, O. H. (1996b). Inositol trisphosphate and cyclic ADP-ribose-mediated release of Ca^{2+} from single isolated pancreatic zymogen granules. *Cell* **84**, 473–480.
25. Nguyen, T., Chin, W. C., and Verdugo, P. (1998). Role of Ca^{2+}/K^+ ion exchange in intracellular storage and release of Ca^{2+}. *Nature* **395**, 908–912.
26. Quesada, I., Chin, W. C., Steed, J., Campos-Bedolla, P., and Verdugo, P. (2001). Mouse mast cell secretory granules can function as intracellular ionic oscillators. *Biophys. J.* **80**, 2133–2139.
27. Gerasimenko, J. V., Tepikin, A. V., Petersen, O. H. and Gerasimenko, O. V. (1998). Calcium uptake via endocytosis with rapid release from acidifying endosomes. *Curr. Biol.* **8**, 1335–1338.
28. Petersen, O. H., Gerasimenko, O. V., Gerasimenko, J. V., Mogami, H. and Tepikin, A. V. (1998). The calcium store in the nuclear envelope. *Cell. Calcium* **23**, 87–90.
29. Gerasimenko, O. V., Gerasimenko, J. V., Rizzuto, R. R., Treiman, M., Tepikin, A. V. and Petersen, O. H. (2002). The distribution of

the endoplasmic reticulum in living pancreatic acinar cells. *Cell. Calcium* **32**, 261–268.
30. Thorn, P., Lawrie, A. M., Smith, P., Gallacher, D. V. and Petersen, O. H. (1993). Local and global cytosolic Ca^{2+} oscillations in exocrine cells evoked by agonists and inositol trisphosphate. *Cell* **74**, 661–668.
31. Thorn, P., Gerasimenko, O. and Petersen, O. H. (1994). Cyclic ADP-ribose regulation of ryanodine receptors involved in agonist evoked cytosolic Ca^{2+} oscillations in pancreatic acinar cells. *EMBO J.* **13**, 2038–2043.
32. Lee, H. C., Walseth, T. F., Bratt, G. T., Hayes, R. N. and Clapper, D. L. (1989). Structural determination of a cyclic metabolite of NAD^{+} with intracellular Ca^{2+}-mobilizing activity. *J. Biol. Chem.* **264**, 1608–1615.
33. Chini, E. N., Beers, K. W. and Dousa, T. P. (1995). Nicotinate adenine dinucleotide phosphate (NAADP) triggers a specific calcium release system in sea urchin eggs. *J. Biol. Chem.* **270**, 3216–3223.
34. Cancela, J. M., Churchill, G. C. and Galione, A. (1999). Coordination of agonist-induced Ca^{2+}-signalling patterns by NAADP in pancreatic acinar cells. *Nature* **398**, 74–76.
35. Cancela, J. M., Gerasimenko, O. V., Gerasimenko, J. V., Tepikin, A. V. and Petersen, O. H. (2000). Two different but converging messenger pathways to intracellular Ca^{2+} release: the roles of nicotinic acid adenine dinucleotide phosphate, cyclic ADP-ribose and inositol trisphosphate. *EMBO J.* **19**, 2549–2557.
36. Hohenegger, M., Suko, J., Gscheidlinger, R., Drobny, H. and Zidar, A. (2002). Nicotinic acid-adenine dinucleotide phosphate activates the skeletal muscle ryanodine receptor. *Biochem. J.* **367**, 423–431.
37. Galione, A., and Petersen, O. H. (2005) The NAADP receptor: new receptors or new regulation? *Mol. Interv.* **5**, 73–79.
38. Dammermann, W. and Guse, A. H. (2005). Functional ryanodine receptor expression is required for NAADP-mediated local Ca^{2+} signaling in T-lymphocytes. *J. Biol. Chem.* **280**, 21394–21399.
39. Malavasi, F., Deaglio, S., Funaro, A., Ferrero, E., Horenstein, A. L., Ortolan, E., Vaisitti, T., Aydin, S. (2008) Evolution and function of the ADP ribosyl cyclase/CD38 gene family in physiology and pathology. *Physiol. Rev.* **88**(3):841–886.
40. Piston, D. W. (1999). Imaging living cells and tissues by two-photon excitation microscopy. *Trends. Cell. Biol.* **9**, 66–69.
41. Tirlapur, U. K. and Konig, K. (2002). Targeted transfection by femtosecond laser. *Nature* **418**, 290–291.
42. Gerasimenko, J. V., Sherwood, M., Tepikin, A. V., Petersen, O. H., Gerasimenko, O. V. (2006a). NAADP, cADPR and IP3 all release Ca^{2+} from the endoplasmic reticulum and an acidic store in the secretory granule area. *J. Cell. Sci.* **119**(Pt 2):226–238.
43. Gerasimenko, J. V., Flowerdew, S. E., Voronina, S. G., Sukhomlin, T. K., Tepikin, A. V., Petersen, O. H., Gerasimenko, OV. (2006b). Bile acids induce Ca^{2+} release from both the endoplasmic reticulum and acidic intracellular calcium stores through activation of inositol trisphosphate receptors and ryanodine receptors. *J. Biol. Chem.* **281**(52): 40154–40163.

Chapter 13

Imaging and Analysis of Three-Dimensional Cell Culture Models

Benedikt W. Graf and Stephen A. Boppart

Abstract

Three-dimensional (3D) cell cultures are important tools in cell biology research and tissue engineering because they more closely resemble the architectural microenvironment of natural tissue, compared to standard two-dimensional cultures. Microscopy techniques that function well for thin, optically transparent cultures, however, are poorly suited for imaging 3D cell cultures. Three-dimensional cultures may be thick and highly scattering, preventing light from penetrating without significant distortion. Techniques that can image thicker biological specimens at high resolution include confocal microscopy, multiphoton microscopy, and optical coherence tomography. In this chapter, these three imaging modalities are described and demonstrated in the assessment of functional and structural features of 3D chitosin scaffolds, 3D micro-topographic substrates from poly-dimethyl siloxane molds, and 3D Matrigel cultures. Using these techniques, dynamic changes to cells in 3D microenvironments can be non-destructively assessed repeatedly over time.

Key words: 3D culture, optical coherence tomography, multiphoton microscopy.

1. Introduction

Culturing cells on two-dimensional (2D) substrates has been a standard technique in cell biology research for decades. However, the culture flasks and dishes typically used do not replicate the natural microenvironment of cells in tissue. Research has shown that the extracellular environment has a profound impact on cell biology, including differentiation of stem cells (1, 2). As a result, there has been an increased interest in developing culturing techniques that more closely resemble natural tissue. Many different types of three-dimensional (3D) cultures and 3D scaffold materials have

D.B. Papkovsky (ed.), *Live Cell Imaging*, Methods in Molecular Biology 591,
DOI 10.1007/978-1-60761-404-3_13, © Humana Press, a part of Springer Science+Business Media, LLC 2010

been developed. A commonly used substance, Matrigel, is derived from the extra cellular matrix (ECM) of mouse tumor (sarcoma). Other techniques are based on seeding cells in porous 3D scaffolds constructed with synthetic materials. Three-dimensional cell cultures have become important for both research and clinical applications, with one goal being the development of advanced biocompatible engineered tissues (3, 4).

While 3D cell cultures more accurately model the natural environment of cells, their use also presents new challenges. Many of the tools used to visualize live cells in 2D cell cultures such as bright field and phase microscopy rely on light transmitted through the sample. These approaches are impractical for viewing cells in 3D cultures since the whole sample may be too thick for light to effectively pass through. For visualizing 3D cultures and thick tissue samples non-destructively, imaging techniques that rely on epi-illumination (with light collected in the backward direction) are more suitable. Different epi-illumination imaging techniques exist and can be based on the detection of fluorescence as well as backscattered light. Fluorescence-based techniques are most commonly used to visualize a fluorescent marker that has been targeted to a specific area or molecule of interest. Alternatively, the autofluorescent properties of cells and tissues can frequently provide sufficient contrast to identify structural and biochemical components. Imaging techniques based on the scattering of light are commonly used to observe the structural and dimensional characteristics of a sample. For example, by measuring the wavelength-dependent scattering properties of the sample, one can infer the sizes of the scatterers present, such as the size of individual cells or nuclei.

Confocal microscopy (CM) is an imaging technique that is capable of high resolution optical sectioning in relatively thick samples and can be used both in fluorescence as well as reflectance mode (*see* **Note 1**). The penetration depth of confocal microscopy is limited to roughly less than 100 μm (5). Deeper penetration depths for fluorescence imaging can be achieved by using multiphoton microscopy (MPM). This technique relies on non-linear optical effects and can perform high resolution optical sectioning in samples up to a millimeter thick, depending on tissue type (6). Scattering-based imaging with penetration depths of up to several millimeters can be achieved by using optical coherence tomography (OCT) (7). OCT constructs depth-resolved images by using interferometric techniques to measure the time-of-flight of scattered photons. Determining which imaging technique is appropriate for a specific 3D cell culture depends on the optical properties, type, and thickness of the culture, as well as the features of interest in the culture. Some of the imaging techniques described here can be used simultaneously, which provides a more comprehensive view of a sample based on differing contrast-enhancing mechanisms. Since these imaging techniques have little effect on the viability of living cells, they are well-suited for observing dynamic changes in 3D cell cultures.

This chapter will focus on imaging techniques for assessing cells grown in different 3D cultures. Optical coherence tomography, multiphoton microscopy, and confocal microscopy are used to visualize structural and functional properties of the samples. OCT and CM are used to observe the location and functional activity of cells over a period of several days in a chitosin scaffold. OCT and MPM are used to obtain high-resolution images of cells seeded on 3D micro-topographic substrates, as well as in 3D cultures using Matrigel as a scaffold material. The effect of mechanical stimuli on the cells in these cultures is also observed.

Understanding the principles and limitations of confocal microscopy, multiphoton microscopy, and optical coherence tomography is essential when choosing the appropriate technique for imaging a specific type of 3D culture, scaffold, or substrate. The most important parameters to consider for imaging 3D cultures are the optical contrast mechanism (fluorescence or scattering), the resolution, and the penetration depth. CM, MPM, and OCT instruments are all commercially available, but can also be constructed in a lab that is equipped with sufficient optical hardware components.

1.1. Confocal Microscopy

Confocal microscopy is a high resolution imaging technique that enables optical sectioning of non-transparent samples. There are several types of CM but the technique that has gained the most widespread use in the life sciences is confocal laser scanning microscopy. This technique is based on point illumination of the sample with a laser, and spatial filtering of the returning beam with a pinhole to block light from outside the focus. The beam is scanned across the sample and an image is constructed. CM can be used in both fluorescence mode as well as reflectance mode.

Spatial resolution in CM depends on the numerical aperture (NA) of the objective, the pinhole size, and the wavelength of the light. Sub-micrometer resolution can be achieved with commercial systems allowing even the smallest cells and sub-cellular features to be visualized. Penetration depth in thick samples is limited by the fact that scattering in the sample causes the illuminating beam to defocus. This will decrease the amount of light that passes through the pinhole as the imaging depth becomes deeper, effectively limiting the imaging depth. The penetration depth also depends on the optical properties of the sample, the NA of the objective, and wavelength of the light source. Generally, longer wavelength light penetrates deeper in a sample due to less absorption and scattering. Fluorescence CM typically uses shorter wavelength (visible) light to excite commercial fluorophores through absorption. Reflectance CM does not have stringent wavelength requirements and can readily be performed with longer wavelength light sources to improve penetration depth (8). Ongoing research is developing newer fluorescence-based probes or dyes that are excited by near-infrared wavelengths. The deepest penetration depth achievable with CM is typically limited to less than 100 μm (9, 10).

1.2. Multiphoton Microscopy

Multiphoton microscopy is a high-resolution fluorescence imaging technique that has at least a twofold improvement in penetration depth over confocal microscopy (11). It is based on nonlinear optical processes that absorb two or more near-infrared photons and subsequently emit a single photon in the visible range. It is most commonly used to excite exogenous fluorescent dyes or probes through two-photon absorption. This process requires a high intensity of photons in both time and position (*see* **Note 2**). This is typically accomplished using a laser with ultra-short pulse duration (such as a titanium-sapphire laser) and a high numerical aperture objective. The high intensity requirement restricts the two-photon absorption to within the focal volume. This results in a high spatial resolution without the need for spatial filtering. In addition, MPM reduces photobleaching since fluorescence excitation is limited to a small volume (12). Because near-infrared wavelengths are used to excite the fluorescent molecules, imaging penetration is significantly better than CM, and there is little absorption or thermal effects. As a result, MPM is preferable to CM for imaging 3D cell cultures or tissue when they are thick or highly scattering.

1.3. Optical Coherence Tomography

Optical coherence tomography is a technique which is capable of obtaining high resolution scattering-based images of thick tissue samples (7, 13). OCT operates by focusing light from a broad bandwidth light source into a sample and determining the time it takes for the light to return to a detector. The echo time-of-flight information is used to determine the depth in the sample from where light was scattered. OCT is similar in principle to ultrasound imaging except that near-infrared light is used instead of acoustic waves. Compared to clinical ultrasound, OCT has a higher resolution but a shallower penetration depth. Because the speed of light is significantly faster than the speed of acoustic waves, the time information must be measured indirectly using interferometry. Light scattered from the sample is combined with a reference beam and the resulting interference pattern is measured. The interference pattern contains the time-of-flight information of the light in the sample arm. Light scattered from different depths can be distinguished because of the low coherence property of the laser or light source. Cross-sectional OCT images are constructed by scanning the laser beam across the sample and acquiring depth-dependent scattering profiles at each position.

The setup of an OCT system can be divided into four parts: source, sample arm, reference arm, and detector arm. The source is commonly a near-infrared laser or superluminescent diode with a broad spectral bandwidth. The beam from the source is split by a beam splitter into the reference arm and the sample arm. Light reflected from the sample and the reference arm recombine at the beam splitter and is directed to the detector. The sample arm consists of focusing optics (which also function as collecting optics), the sample, and a mechanism for scanning the beam or the sample. The arrangement of the reference arm and the detection arm is dependent on the detection scheme being used. There are two basic schemes, time-domain OCT and frequency-domain OCT. Time-domain OCT consists of a single detector and mechanism for rapidly scanning the delay in the reference arm. At each delay of the reference arm, scattering from a specific depth in the sample can be determined. Frequency-domain OCT is performed by using a fixed reference arm path and measuring the spectral interference pattern in the detector arm. The entire depth-scattering profile can be obtained by taking one measurement of the spectrum and computing the inverse Fourier transform. Frequency-domain OCT does not require any mechanical scanning in the reference arm, which results in significantly improved speed and sensitivity (14). Frequency-domain OCT is further sub-divided into two categories: spectral (or Fourier) domain OCT and swept-source OCT (*see* **Note 3**). In spectral-domain OCT the spectrum of the broad bandwidth light source is measured using a spectrometer (15). In swept-source OCT the wavelength of a tunable laser is rapidly swept over the bandwidth (16). The broad spectrum is thus encoded in time and a single detector is used instead of a spectrometer. A simplified schematic of a spectral-domain OCT system is shown in **Fig. 13.1** to better demonstrate the principles of OCT.

Resolution in the axial and transverse directions is decoupled in OCT. Axial resolution is governed by the coherence length of the light source which is inversely dependent on the spectral bandwidth. Axial resolution in a typical OCT system is less than 10 microns. Transverse resolution is dictated by the optics used to focus the beam onto the sample. Transverse resolution for a typical OCT system is around 10 μm as well, so a large confocal parameter, or depth-of-focus, can be achieved. Just as with CM and MPM, imaging depth for OCT is highly dependent on the optical properties of the sample as well as the wavelength of the source. Penetration depth can be up to several millimeters, even in highly scattering samples, because OCT not only spatially filters out-of-focus light as in CM but also coherently rejects photons that are not singly backscattered from points in the tissue and

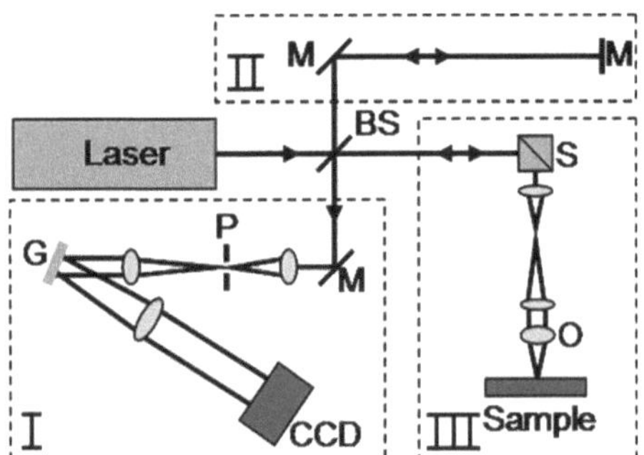

Fig. 13.1. Schematic of a spectral/Fourier-domain OCT system. The beam from the broad bandwidth light source is split into the reference arm (II) and the sample arm (III) using a beam splitter. The beam in the sample arm is focused onto a sample and scanned to acquire images. Light backscattered from the sample and the reference end mirror is recombined at the beam splitter and sent to the detector arm (I). The detector arm is a spectrometer that measures the spectral interference pattern. Scanning of the beam and acquisition from the charge-coupled device (CCD) camera is synchronized using a personal computer. Abbreviations: BS, beamsplitter; G, grating; M, mirror; O, objective; P, pinhole; S, beam scanners.

therefore no longer remain coherent with the light reflected back in the reference arm (17).

A variation of OCT, called optical coherence microscopy (OCM), utilizes a high NA objective to achieve similar transverse spatial resolution as reflectance CM, at the expense of a short depth-of-focus (18). In this case, the axial resolution may then be determined by the focusing optics instead of the source bandwidth. Optical sectioning is achieved with both spatial and coherence gating, instead of spatial filtering alone. The advantage of OCM is that it has a greater penetration depth than CM. Because of the high NA and the narrow beam waist, the short focal region is effectively limited to a single *en face* plane in the sample. As a result, the beam in OCM is typically scanned in two lateral dimensions to acquire an *en face* image, as opposed to a cross-sectional (depth-resolved) image as in OCT. Because OCM uses a high NA objective and can use the same laser source (titanium:sapphire) as in MPM, it is possible to perform OCM at the same time as MPM to collect image data based on multiple contrast mechanisms (19–22). This enables simultaneous acquisition of high resolution scattering and fluorescence information from thick, highly scattering samples. OCT has also been used simultaneously with fluorescence CM to provide structural and functional information in engineered tissues (23).

Additional variations of OCT and OCM exist that enable imaging based of different contrast mechanisms. Spectroscopic OCT (24) and OCM (25) measure the depth-resolved spectrum of backscattered light, which allows wavelength-dependent properties of the sample to be assessed. Biomechanical properties can be spatially resolved using optical coherence elastography (26, 27). Doppler OCT is a technique which measures frequency shifts of backscattered light to measure dynamic events such as blood flow *in vivo* (28).

2. Materials

2.1. Chitosin Scaffold Culture

1. NIH 3T3 fibroblast cells (American Type Culture Collection, Manassas, VA).
2. 4 μg of GFP-vinculin plasmid (provided by Dr. Benjamin Geiger, Weizmann Institute of Science, Israel) diluted with 50 μL of FreeStyle 293 expression (Invitrogen).
3. 2 μL of lipofectamine 2000 reagent (Invitrogen) diluted with 50 μL of FreeStyle 293 expression.
4. 2% (wt%) chitosan flakes (Sigma-Aldrich) dissolved in a 0.2 M acetic acid aqueous solvent.
5. Phosphate-buffered saline (PBS, 1×) solutions containing 70, 50, and 0% of ethanol.
6. Plastic 8 mm diameter and 3 mm depth cylindrical molds (cap of eppendorf tube).
7. Portable microincubator, LU-CPC (Harvard Apparatus, Holliston, MA).

2.2. Matrigel and Micro-topographic Culture

1. Matrigel solution (BD Bioscience, Bedford, MA).
2. Microtextured poly(dimethly-siloxane) (PDMS) substrates with 3D topographic features (provided by James Norman, Stanford University).
3. Hoechst 33342 nuclear dye (Invitrogen).
4. 3D reconstruction software, Analyze 5.0 (Mayo Clinic, Rochester, MN).
5. Mechanical stimulation system Tissue Train and StageFlexer (Flexcell, NC).

2.3. Optical Coherence Tomography Instrument

The OCT setup used in this experiment is a custom built, fiber-based, time-domain OCT system. The axial and lateral resolutions of the system are 3 and 10 μm, respectively. The major components are as follows:

1. Titanium:sapphire laser (Kapteyn-Murnane Laboratories) pumped by a Nd:YVO_4 laser (Coherent) that produces ~90-fs pulses with an 80-MHz repetition rate. The center wavelength is 800 nm with a bandwidth of 20 nm or greater.

2. Ultrahigh numerical aperture fiber UHNA4 (Thorlabs, Karslfeld, Germany) to broaden the spectrum from 20 to 100 nm.
3. Galvanometer-driven retroreflector delay line in reference arm operating at 30 Hz.
4. Galvanometer-controlled mirrors for scanning of beam across the sample.
5. Achromatic lens with 20 mm focal length and 12.5 mm diameter (Thorlabs).
6. Photodetector (Newport).

2.4. Integrated Optical Coherence and Multiphoton Microscope

The integrated OCM and MPM microscope is a custom-built system that uses free space optics (19). The OCM modality has a spectral-domain detection scheme. The axial and lateral MPM resolutions are 0.8 and 0.5 μm, respectively, while the axial and lateral OCM resolutions are 2.2 and 0.9 μm, respectively. The major components of the system are as follows:

1. Titanium:sapphire laser (Kapteyn-Murnane Laboratories) pumped by a Nd:YVO_4 laser (Coherent) with center wavelength at 800 nm with a bandwidth of 120 nm.
2. Galvanometer-controlled mirrors for scanning of beam across the sample.
3. Microscope objective, 20×, 0.95 NA, water-immersion (Olympus).
4. Cold Mirror dichroic mirror (CVI Laser, Livermore, CA) and H7421-40 photo-multiplier tube (Hamamatsu, Inc.) for detection of multiphoton signal.
5. Emission filters to detect fluorescence from GFP (FF01-520/35-25, Semrock) and from Hoechst nuclear dye (BG39, CVILaser).
6. Custom-built spectrometer for detection of OCM signal. Consists of collimating optics, a blazed diffraction grating having 830.3 grooves/mm and a line-scan camera (L104k–2 k, Basler, Inc.) that contains a 2048-element CCD array maximum readout rate of 29 kHz.

3. Methods

3.1. Transfecting 3T3 Cell Line to Co-express GFP-Vinculin

1. Seed 7×10^5 3T3 fibroblast cells on each well of a six-well plate.
2. Mix 4 μg of diluted DNA and 2 μL of diluted Lipofectamine 2000 and add to each of the wells.
3. Incubate cells for 20 min then clean with PBS.

4. Resulting cells are a stable cell line of cells that express GFP-vinculin.

3.2. Chitosin Scaffold Cell Culture Preparation

1. Filter chitosan solution and transfer to cylindrical molds.
2. Freeze cylinders at –20°C and lyophilize for 4 days until solvent is completely removed (*see* **Note 4**).
3. Sterilize scaffold with ethanol/PBS solutions.
4. Hydrate scaffolds in cell culture media for 12 h before seeding cells.
5. Seed cells at a concentration of 5×10^6 cells/cm^3.
6. Place scaffold into micro-incubator for imaging.

3.3. Matrigel and Micro-Topographic Culture Preparation

1. Stain GFP-transfected 3T3 fibroblast cells with Hoechst nuclear dye and incubate for 30 min.
2. Mix solution with 7×10^4 GFP-transfected 3T3 cells with thawed Matrigel solution at a 1:1 volume ratio.
3. Solidify the mixture in a 37°C incubator.
4. Place Matrigel culture into micro-incubator for imaging.
5. Seed GFP-transfected 3T3 cells at concentration of 5×10^5 cell/ml on a micro-topographic PDMS substrate with 10 μm height and diameter cylindrical pegs.

3.4. OCT and CM Imaging of Cell Dynamics in Chitosin Scaffold Culture

OCT and CM images of the chitosin scaffold cell culture were taken at several time points after seeding to observe dynamic changes to the distribution of the cell population (29, 30). OCT is used to visualize changes to the cell population that occur throughout the entire volume of the 3D chitosin scaffold culture. CM, in comparison, provides cellular-level resolution and functional information near the surface of the cell culture.

1. The micro-incubator is placed in the sample arm of the OCT system allowing the chitosin scaffold to be imaged while maintaining the sterile environment. Three-dimensional OCT images of the same chitosin scaffold are taken at 1, 3, 5, 7, and 9 days after seeding (*see* **Note 5**).
2. To verify the dynamic changes observed in the OCT images taken on different days, different identical samples are histologically prepared after being incubated for the same durations (*see* **Note 6**). The histological samples are viewed with a standard light microscope and compared to cross-sectional OCT images (*see* **Fig. 13.2**).
3. To more effectively visualize the whole cell population, 3D OCT data images are reconstructed and viewed from different angles (*see* **Fig. 13.3**).

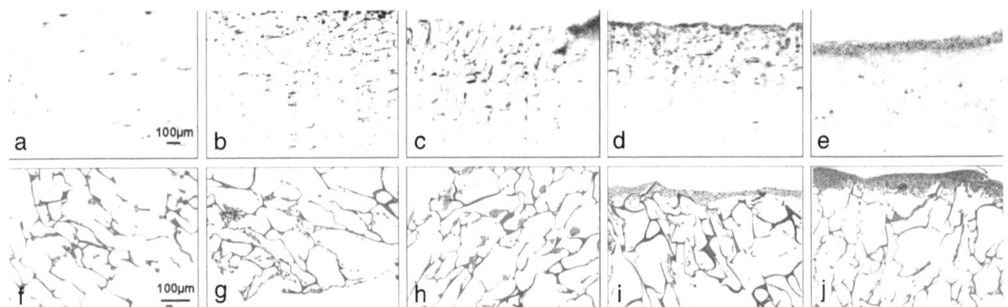

Fig. 13.2. Cross-sectional OCT (**a**–**e**) and corresponding hematoxylin and eosin-stained histology (**f**–**j**) images of chitosin scaffold cultures at day 1, 3, 5, 7, and 9, respectively. Cells are relatively evenly distributed throughout the scaffold after seeding. After several days, it is evident that the cells have migrated to the surface of the culture and have formed a dense, highly scattering layer. This demonstrates the ability of OCT to observe morphological features in a 3D cell culture at different times during the development of the 3D culture. Note the difference in image scale between OCT and histology images. Modified figure used with permission (20).

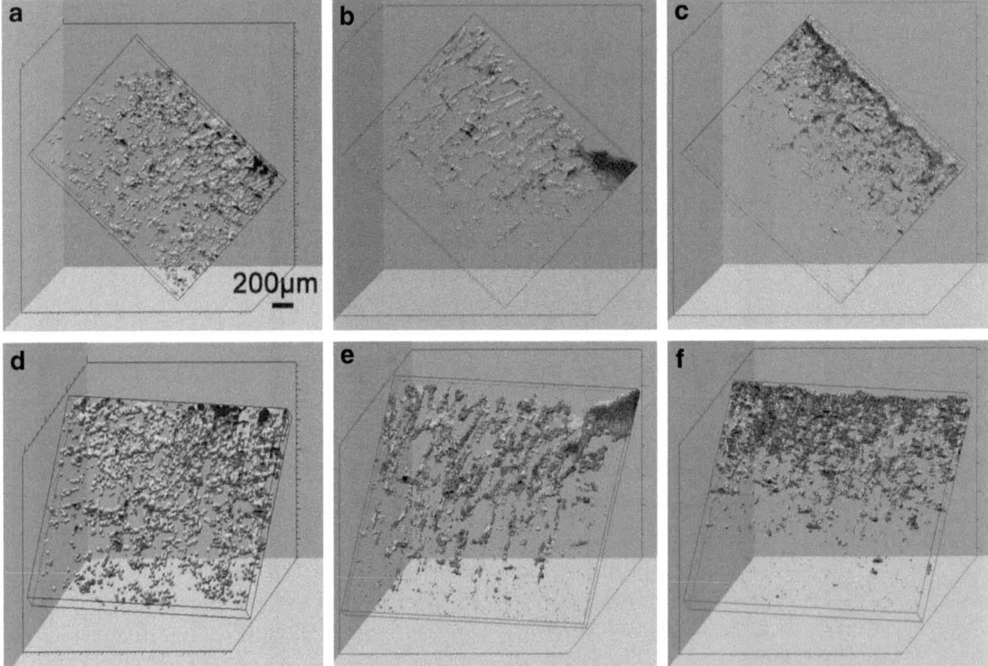

Fig. 13.3. Three-dimensional OCT reconstructions of chitosin scaffolds from two different viewing angles taken at 3 (**a**, **d**), 5 (**b**, **e**), and 7 (**c**, **f**) days. Rotational angles of image **a**–**c** are 40° (*y*), 20° (*z*), 70° (*x*) while **d**–**f** are 0° (*y*), 10° (*z*), 50° (*x*). Viewing a 3D reconstruction from many angles, and computationally sectioning out planes from arbitrary angles, enables a more complete view and assessment of the 3D characteristics of a sample, compared to single cross-sectional images. Modified figure used with permission (20).

4. To visualize the culture at cellular resolution and to observe functional characteristics (*see* **Note 7**), CM images of the superficial layers of the chitosin scaffold are taken in fluorescence mode using a commercial system, DM-IRE (Leica Microsystems, Bensheim, Germany). CM images are taken at 1, 3, 5, and 7 days to observe dynamic changes to the culture (*see* **Fig. 13.4**).

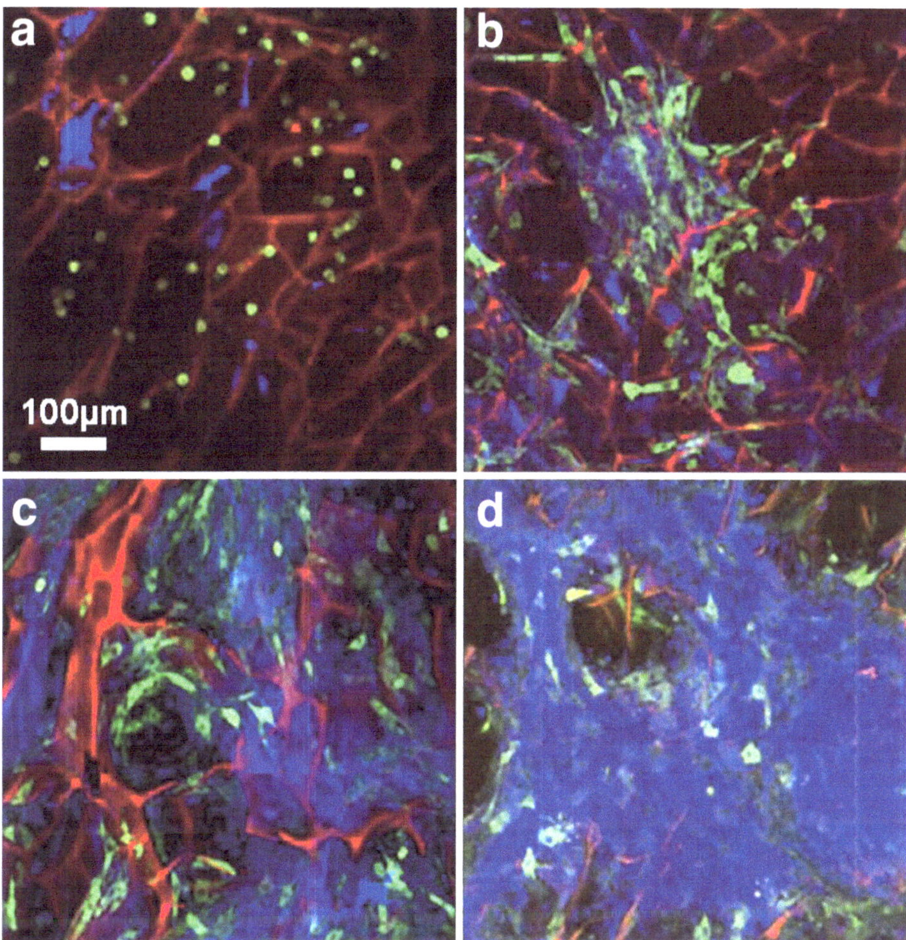

Fig. 13.4. Fluorescence CM images of GFP-transfected fibroblasts in a 3D chitosin scaffold culture taken at day 1, 3, 5, and 7 (**a**–**d**, respectively). The *red channel* is the autofluorescence signal from the scaffold, the *green channel* is the GFP fluorescence from the cells, and the *blue channel* is the backscattering signal predominantly from the deposited extracellular matrix and over time, the higher scattering cell density near the surface of the scaffold. One day after seeding, the fibroblasts are spherical in shape and have a low GFP signal. As the cells grow and attach to the substrate, there is an increase in expression of the GFP-vinculin gene. The increase in the backscattering signal by day 7 verifies that the cells have formed a dense, highly scattering layer near the surface over a period of several days. Modified figure used with permission (20).

3.5. Imaging Effects of Mechanical Stimuli on 3D Cultures with OCM and MPM

Mechanical forces are known to affect morphology and genetic expression in cells. To observe the effects of mechanical forces on cells in 3D cultures both micro-topographic substrates and Matrigel cultures (*see* **Note 8**) were imaged before and after being subjected to uniaxial stretching (31). Integrated OCM and MPM allows simultaneous acquisition of fluorescence and scattering-based images. For these samples, this enables visualization of individual cells (via MPM) and the local structural microenvironment of the substrates (via OCM).

1. *En face* OCM and MPM images of the 3D micro-topographic substrate are taken 3 days after seeding (*see* **Fig. 13.5a**). Two MPM channels are obtained by acquiring the same image twice using different emission filters for fluorescence detection (*see* **Note 9**).
2. The culture is mechanically stimulated with the FlexCell apparatus with 5% cyclic, 1 Hz equibiaxial sinusoidal stretching for a period of 18 h.
3. *En face* OCM and MPM images are again taken to observe the changes to cell morphology caused by the mechanical stimulation (*see* **Fig. 13.5b**).
4. 3D OCM and MPM images of the Matrigel culture are taken prior to mechanical stimulation. Three-dimensional data sets are obtained by acquiring a series of *en face* planes at different depths in the sample. Using the Analyze software, *en face* planes are computationally extracted from the 3D data to observe the structure (*see* **Fig. 13.6a**) and functional (*see* **Fig. 13.6b**) properties of cells in the culture.
5. The Matrigel culture is subjected to the same mechanical strain as in step 2.

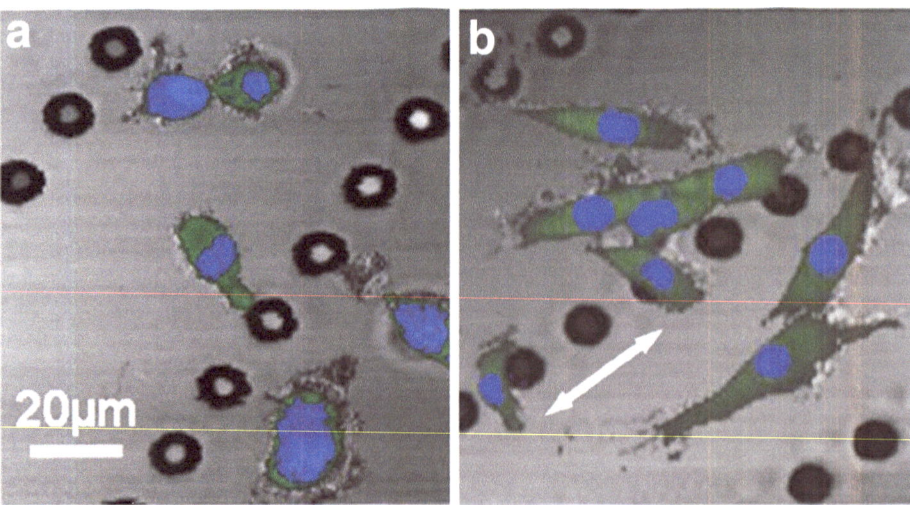

Fig. 13.5. OCM/MPM images of GFP-vinculin fibroblast cells on a micro-topographic substrate (linear arrays of micropegs) before (**a**) and after (**b**) mechanical stimulation. OCM shows the features of the substrate (*grey*) while MPM images show the GFP fluorescence (*green channel*) and the nuclear stain (*blue channel*). Visualizing the nucleus enables individual cells to be clearly identified while the GFP signal shows the extent of the cell body, the level of adhesion of the cell to the substrate, and interactions with adjacent cells. The flexible polymer (PDMS) substrate was repeatedly stretched in the direction of the *white arrow* in (**b**). Mechanical stimulation results in an elongation of the fibroblast cells, roughly along the direction of stretch. The stretching causes an elongation of the cells and an increase in the distribution of the GFP signal indicating that there is an increase in expression of vinculin. Modified figure used with permission (31).

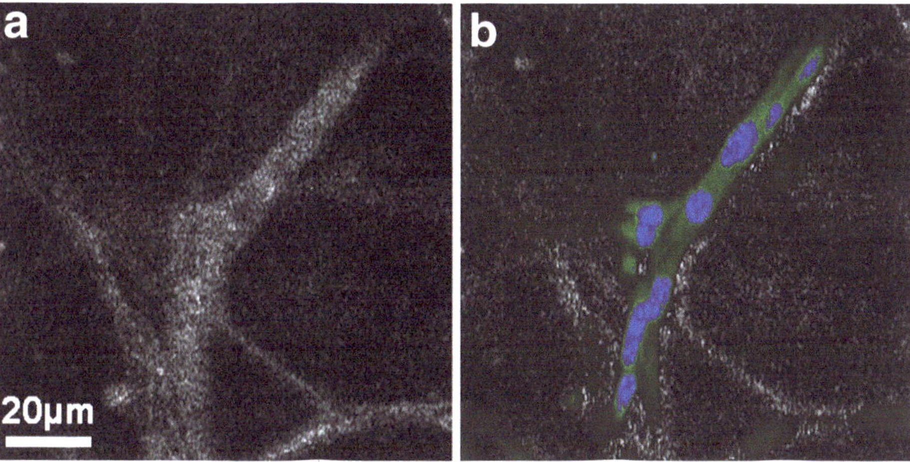

Fig. 13.6. OCM (**a**) and MPM overlaid on OCM (**b**) images from a 3D Matrigel scaffold seeded with GFP-vinculin fibroblast cells stained with a nuclear dye prior to mechanical stimulation. The green signal in the MPM image is from GFP and indicates the relative degree of cell adhesion. The *blue signal* is the nuclear stain which identifies individual cells. Scattering in this culture, as seen in the OCM image, is caused by both the cells and the extra-cellular environment (collagen secreted by the cells). It is observed that the cells have organized into an interconnected architecture along the fibrous collagen structure of the Matrigel. Modified figure used with permission (31).

6. 3D OCM and MPM images are taken after stimulation for comparison. Different projections of the overlaid 3D data set are viewed using the Analyze software (*see* **Fig. 13.7**).
7. Analysis of the before and after OCM and MPM images allowed the effects of mechanical stimulation on the function and morphology of cells in these two types of cultures to be observed (*see* **Note 10**).

4. Notes

1. Traditional microscopy requires that a specimen be fixed and cut into thin slices and placed on a glass slide for imaging. Optical sectioning refers to imaging thin sections of the intact specimen. In addition to enabling visualization of samples in 3D, optical sectioning has the advantage of eliminating the need for the harsh processing steps that physically destroy the specimen and may alter some of its structural properties. The ability to obtain optical sections in live specimens allows one to observe dynamic events in biology. Instead of viewing different specimens at fixed points in time, single specimens can be observed longitudinally over time with minimal disturbance. The optical sectioning techniques discussed in this chapter are particularly

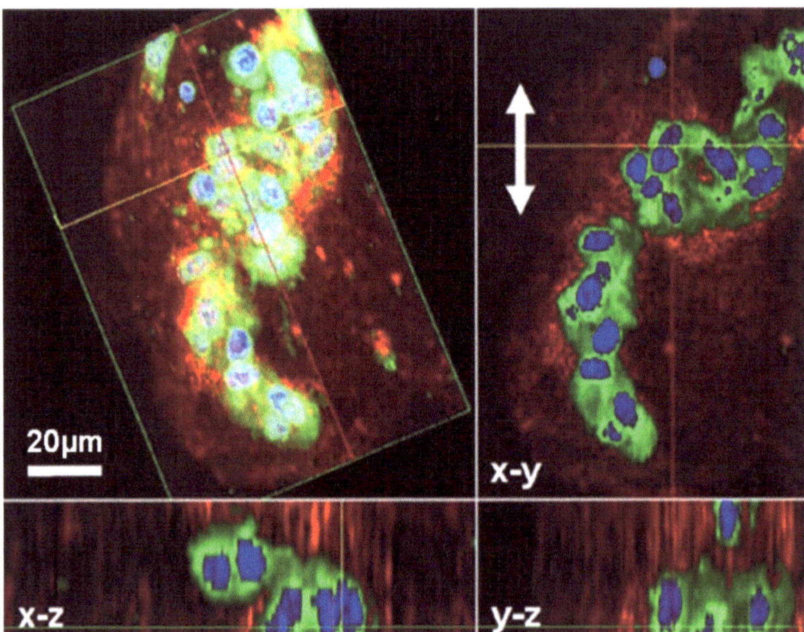

Fig. 13.7. Multimodality 3D OCM/MPM image of GFP-vinculin fibroblasts in a 3D Matrigel scaffold acquired after mechanical stimulation. The 3D reconstruction is shown (*top left*) as well as projections along different axes giving a comprehensive view of this cluster of cells. The red channel corresponds to the scattering from the sample (OCM) while the green and blue channels are fluorescence images (MPM). Three-dimensional reconstruction, rotation, and computational sectioning at arbitrary plane angles allow for a more comprehensive view of a 3D cell culture. Mechanical stimulation in the direction of the arrows in the *x-y* plane has induced the cells to become more spherical and form clusters, potentially indicating that adhesion sites with the scaffold were broken during mechanical stimulation. Modified figure used with permission (31).

well-suited for imaging thick, highly scattering samples that do not allow light to be transmitted.

2. MPM is typically less harmful to biological specimen than fluorescence mode CM because it minimizes photobleaching and photodamage (32). However, MPM requires focusing ultrashort laser pulses into a sub-femtoliter volume of a sample. Although power levels needed for MPM are moderate (2–20 mW), the high instantaneous intensity of light could possibly damage a specimen through different mechanisms. Exposure to ultrashort pulses of light has been shown to cause cell death through oxidative photodamage (33, 34). In practice, damage can be minimized by limiting the power levels and reducing the exposure time of the sample to the laser beam. The later can be achieved by rapidly scanning the beam or blocking the beam path to the sample when the beam is not scanning.
3. A key advantage of swept-source OCT over spectral-domain OCT is the imaging speed. Imaging speed for spectral-domain OCT is limited by the read-out speed

of the line camera in the spectrometer. Currently, maximum readout rates are ~100 kHz, or ~100,000 columns in OCT or points in OCM, per second). Swept-source OCT using Fourier-domain mode-locked lasers can achieve speeds of several hundred kHz (16). However, these lasers currently operate in the 1.3 μm wavelength range. Spectral-domain OCT can be performed with a broad bandwidth laser source centered at any wavelength; however, line cameras sensitive to wavelengths greater than 1000 nm use indium-gallium-arsenide detection elements and can be expensive. Very high imaging speeds are needed when imaging *in vivo* or when observing fast dynamic events.

4. Resulting porous scaffold should have an interconnected porosity of >80% with an average pore size of 100 μm.
5. With this OCT system, the resolution is sufficient so that the location and structure of the cell population can be visualized, even though individual cellular features cannot be identified. The chitosin scaffold itself is not visible due to the low scattering properties of the material. OCT was capable of imaging throughout the depth of the sample (~1 mm).
6. Histological preparation included fixing samples with 3.7% formaldehyde, embedding in paraffin, cutting into 5 μm slices with a microtome, and staining with hematoxylin and eosin.
7. These fibroblast cells are transfected to co-express GFP with vinculin. Vinculin is a surface adhesion protein and thus the presence of a GFP signal correlates to cell–cell and cell–substrate adhesion.
8. Although cells in 3D micro-topographic cultures are located on a surface of the polymer substrate, these substrates have defined 3D features and characteristics, compared to flat planar cultures, and have been shown to affect cellular organization (35). Cells in the Matrigel culture are in a true 3D microenvironment.
9. All fluorescent dyes used with single photon excitation can be used for MPM. The emission spectrum of a dye excited with either one photon or with two simultaneous photons is similar and the same filters can often be used. However, the two-photon excitation spectrum of a fluorescent molecule is not easily predicted from its one-photon spectrum and must therefore be determined experimentally in many cases. The two-photon emission spectra for many commonly used dyes and biomolecules have been characterized (36, 37).

10. The response of the cells in the Marigel culture to stretching is opposite to what was observed with the cells on the micro-topographic substrates, where the fibroblasts were initially more spherical, and became elongated following mechanical stimulation. The different responses could potentially be explained by differences in adhesion strength to the PDMS and to the Matrigel, or to the differences in strain experienced at the cellular level in each type of 3D culture. These images demonstrate that OCM and MPM can enable the observation of the complex cell-scaffold interactions in 3D cultures. With this integrated imaging technique, 3D multi-modality information from cultures can be acquired, providing a comprehensive view of the structural and functional characteristics of the sample.

References

1. Weaver, V. M., Petersen, O. W., Wang, F., Larabell, C. A., Briand, P., Damsky, C. and Bissell, M. J. (1997) Reversion of the malignant phenotype of human breast cells in three-dimensional culture and *in vivo* by integrin blocking antibodies. *J. Cell Biol.* **137**, 231–245.
2. Fuchs, E., Tumbar, T. and Guasch, G. (2004) Socializing with the neighbors: stem cells and their niche. *Cell* **116**, 769–778.
3. Friedrich, M. J. (2003) Studying cancer in 3 dimensions: 3D models foster new insights into tumorigenesis. *J. Am. Med. Assoc.* **290**, 1977–1979.
4. Lee, J., Cuddihy, M. J. and Kotov, N. A. (2008) Three-dimensional cell culture matrices: state of the art. *Tissue Eng. Part B* **14**, 61–86.
5. Pawley, J. (ed.) (1995) *Handbook of Biological Confocal Microscopy.* Springer, New York, NY.
6. Helmchen, F. and Denk, W. (2005) Deep tissue two-photon microscopy. *Nat. Methods* **2**, 932–940.
7. Schmitt, J. M. (1999) Optical coherence tomography (OCT): a review. *IEEE J. Sel. Top. Quantum Electron.* **5**, 1205–1215.
8. Rajadhyaksha, M., Grossman, M., Esterowitz, D., Webb, R. H. and Anderson, R. R., (1995) *In vivo* confocal scanning laser microscopy of human skin: melanin provides strong contrast. *J. Invest. Dermatol.* **104**, 946–952.
9. Schmitt, J. M., Knüttel, A. and Yadlowsky, M. (1994) Confocal microscopy in turbid media. *J. Opt. Soc. Am. A* **11**, 2226–2235.
10. Smithpeter, C. L., Dunn, A. K., Welch, A. J. and Richards-Kortum, R. (1998) Penetration depth limits of *in vivo* confocal reflectance imaging. *Appl. Opt.* **37**, 2749–2754.
11. Centonze, V. E. and White, J. G. (1998) Multiphoton excitation provides optical sectionsfrom deeper within scattering specimens than confocal imaging. *Biophys. J.* **75**, 2015–2024.
12. Rubart, M. (2004) Two-photon microscopy of cells and tissue. *Circ. Res.* **95**, 1154–1166.
13. Huang, D., Swanson, E. A., Lin, C. P., Schuman, J. S., Stinson, W. G., Chang, W., Hee, M. R., Flotte, T., Gregory, K., Puliafito, C. A. and Fujimoto, J. G. (1991) Optical coherence tomography. *Science* **254**, 1178–1181.
14. Leitgeb, R. Hitzenberger, C. K. and Fercher, A. F. (2003) Performance of Fourier-domain vs. time-domain optical coherence tomography. *Opt. Express* **11**, 889–894.
15. Nassif, N., Cense, B., Park, B. H., Yun, S. H., Chen, T. C., Bouma, B. E., Tearney, G. J. and de Boer, J. F. (2004) *In vivo* human retinal imaging by ultrahigh-speed spectral domain optical coherence tomography. *Opt. Lett.* **29**, 480–482.
16. Huang, S. W., Aguirre, A. D., Huber, R. A., Adler, D. C. and Fujimoto, J. G. (2007) Swept source optical coherence microscopy using a Fourier domain mode-locked laser. *Opt. Express* **15**, 6210–6217.
17. Welzel, J. (2008) Optical coherence tomography in dermatology: a review. *Skin Res. Tech.* **7**, 1–9.
18. Aguirre, A. D., Hsiung, P., Ko, T. H., Hartl, I. and Fujimoto, J. G. (2003) High-resolution optical coherence microscopy for high-speed, *in vivo* cellular imaging. *Opt. Lett.* **28**, 2064–2066.

19. Vinegoni, C., Ralston, T., Tan, W., Luo, W., Marks, D. L. and Boppart, S. A. (2006) Integrated structural and functional optical imaging combining spectral-domain optical coherence and multiphoton microscopy. *Appl. Phys. Lett.* **88**, 053901.
20. Tang, S., Sun, C. H., Krasieva, T. B., Chen, Z. and Tromberg, B. J. (2007) Imaging subcellular scattering contrast by using combined optical coherence and multiphoton microscopy. *Opt. Lett.* **32**, 503–505.
21. Tang, S., Krasieva, T. B., Chen, Z. and Tromberg, B. J. (2006) Combined multiphoton microscopy and optical coherence tomography using a 12-fs broadband source. *J. Biomed. Opt.* **11**, 020502.
22. Beaurepaire, E., Moreaux, L., Amblard, F. and Mertz, J. (1999) Combined scanning optical coherence and two-photon excited fluorescence microscopy. *Opt. Lett.* **24**, 969–971.
23. Dunkers, J., Cicerone, M. and Washburn, N. (2003) Collinear optical coherence and confocal fluorescence microscopies for tissue engineering. *Opt. Express* **11**, 3074–3079.
24. Morgner, U., Drexler, W., Kärtner, F. X., Li, X. D., Pitris, C., Ippen, E. P., and Fujimoto, J. G. (2000) Spectroscopic optical coherence tomography. *Opt. Lett.* **25**, 111–113.
25. Xu, C., Vinegoni, C., Ralston, T., Luo, W., Tan, W. and Boppart, S. (2006) Spectroscopic spectral-domain optical coherence microscopy. *Opt. Lett.* **31**, 1079–1081.
26. Schmitt, J. (1998) OCT elastography: imaging microscopic deformation and strain of tissue. *Opt. Express* **3**, 199–211.
27. Liang, X., Oldenburg, A. L. Crecea, V., Chaney, E. J. and Boppart, S. A. (2008) Optical micro-scale mapping of dynamic biomechanical tissue properties. *Opt. Express* **16**, 11052–11065.
28. Milner, T. E., Srinivas, S., Wang, X., Malekafzali, A., Van Gemort, M. J., Nelson, J. S., Chen, Z. (1997) Noninvasive imaging of *in vivo* blood flow velocity using optical Doppler tomography. *Opt. Lett.* **22**, 1119–1121.
29. Tan, W., Sendemir-Urkmez, A., Fahrner, L. J., Jamison, R., Leckband, D. and Boppart, S. A. (2004) Structural and functional optical imaging of three-dimensional engineered tissue development. *Tissue Eng.*, **10**, 1747–1756.
30. Tan, W., Oldenburg, A. L., Norman, J. J., Desai, T. A. and Boppart, S. A. (2006) Optical coherence tomography of cell dynamics in three-dimensional tissue models. *Opt. Express* **14**, 7159–7171.
31. Tan, W., Vinegoni, C., Norman, J. J., Desai, T. A. and Boppart, S. A. (2007) Imaging cellular responses to mechanical stimuli within three-dimensional tissue constructs. *Microsc. Res. Tech.* **70**, 361–371
32. Squirrell, J. M., Wokosin, D. L., White, J. G. and Bavister, B. D., (1999) Long-term two-photon fluorescence imaging of mammalian embryos without compromising viability. *Nat. Biotechnol.* **17**, 763–767.
33. Tirlapur, U. K., König, K., Peuckert, C., Krieg, R. and Halbhuber, K. J. (2001) Femtosecond near-infrared laser pulses elicit generation of reactive oxygen species in mammalian cells leading to apoptosis-like death. *Exp. Cell. Res.* **263**, 88–97.
34. Sun, C., Chen, C., Chu, S., Tsai, T., Chen, Y. and Lin, B. (2003) Multiharmonic-generation biopsy of skin. Opt. Lett. **28**, 2488–2490.
35. Norman, J. J. and Desai, T. A. (2005) Control of cellular organization in three dimensions using a microfabricated polydimethylsiloxane-collagen composite tissue scaffold. *Tissue Eng.* **11**, 378–386.
36. Bestvater, F., Spiess, E., Stobrawa, G., Hacker, M., Feurer, T., Porwol, T., Berchner-Pfannschmidt, U., Wotzlaw, C. and Acker, H. (2002) Two-photon fluorescence absorption and emission spectra of dyes relevant for cell imaging. *J. Microsc.* **208**, 108–115.
37. Zipfel, W. R., Williams, R. M., Christie, R., Nikitin, A. Y., Hyman, B. T. and Webb, W. W. (2003) Live tissue intrinsic emission microscopy using multiphoton-excited native fluorescence and second harmonic generation. *PNAS* **100**, 7075–7080.

Chapter 14

Long-Term Imaging in Microfluidic Devices

Gilles Charvin, Catherine Oikonomou, and Frederick Cross

Abstract

During the past 10 years, major developments in live-cell imaging methods have accompanied growing interest in the application of microfluidic techniques to biological imaging. The broad design possibilities of microfabrication and its relative ease of implementation have led to the development of a number of powerful imaging assays. Specifically, there has been great interest in the development of devices in which single cells can be followed in real-time over the course of several generations while the growth environment is changed. With standard perfusion chambers, the duration of a typical experiment is limited to one cell generation time. Using microfluidics, however, long-term imaging setups have been developed which can measure the effects of temporally controlled gene expression or pathway activation while tracking individual cells over the course of many generations. In this paper, we describe the details of fabricating such a microfluidic device for the purpose of long-term imaging of proliferating cells, the assembly of its individual components into a complete device, and then we give an example of how to use such a device to monitor real-time changes in gene expression in budding yeast. Our goal is to make this technique accessible to cell biology researchers without prior experience with microfluidic systems.

Key words: Microfluidic devices, long-term imaging, live-cell imaging, time-lapse fluorescence microscopy, PDMS microfabrication, temporally controlled gene expression.

1. Introduction

In the past 15 years, the first use of the green fluorescent protein (GFP) as a marker for gene expression (1) and the subsequent development of numerous GFP variants and detection schemes have greatly increased the potential and utility of live-cell imaging (2, 3). These discoveries have triggered the development of new imaging techniques to expand the applications of live-cell

D.B. Papkovsky (ed.), *Live Cell Imaging*, Methods in Molecular Biology 591,
DOI 10.1007/978-1-60761-404-3_14, © Humana Press, a part of Springer Science+Business Media, LLC 2010

fluorescence microscopy (4). Notably, confocal microscopy and other high-resolution techniques have considerably improved our capacity to precisely monitor, in real-time, various biological processes in single cells or in tissues (4).

In parallel with these developments in fluorescence imaging techniques, a growing number of microfluidics techniques have been developed and applied to biology, allowing greater flexibility in experimental design (5–7). In particular, microfluidic devices have allowed researchers to apply time-dependent stimuli and monitor single cell responses to changing conditions (8–11). The ability to microfabricate versatile custom-designed flow chambers using standard, and relatively inexpensive, photolithography techniques (12) and the increased throughput of microfluidic assays (compared to experiments involving standard perfusion chambers) have contributed to the success of microfluidics. However, the majority of such techniques, which are usually developed in bioengineering or biophysics departments, have not yet been widely adopted by cell biology laboratories.

An inherent problem with standard perfusion chambers is the impossibility of tracking a large number of successive divisions of unicellular organisms such as bacteria or yeast under the microscope. In these setups, the majority of the progeny of a dividing cell tend to be washed away as medium is changed or an inducer is added. Consequently, imaging is usually limited to one or two cell divisions. Two main designs have been developed to overcome this issue so that cell proliferation can be monitored over several generations (typically 8–10), while controlling the environment. In the "flow aside" design (*see* **Fig. 14.1a**), cells (budding yeast are shown) are loaded into a sub-compartment (cell trap) of the microfluidic device, which is located to the side of the main flow channel (13, 14). The small height of the trap (typically 3–5 μm) ensures that the cells proliferate in two dimensions, in the same focal plane. In the "flow above" design (*see* **Fig. 14.1b**), a physical barrier – a diffusive cellulose membrane – separates the cells, which sit on a coverslip, from the main flow passing through a microfluidic chamber placed on top of the membrane (8, 15). In this case, the membrane allows both media and inducers to diffuse freely in and out of the cell environment, ensuring good control of growth conditions. For both types of devices, the time required to change conditions is limited by diffusion. For small molecules (such as those in media, whose molecular weight is much smaller than the 14 kDa membrane cutoff), the cellulose mesh of the membrane does not affect molecules' diffusivity. Therefore, for a typical distance of 30 μm between the cells and the main channel, the diffusion time is no greater than 1 min for both the "flow above" and the "flow aside" designs.

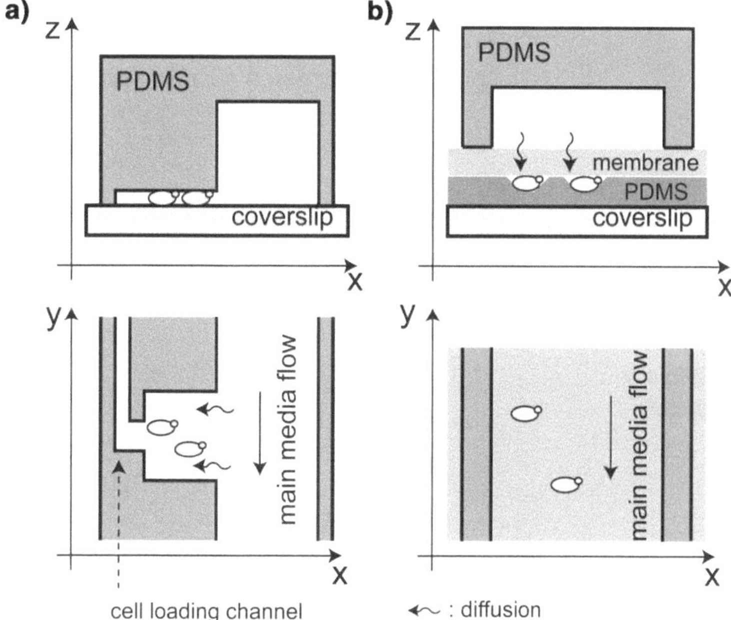

Fig. 14.1. Common designs of microfluidic devices for multigenerational imaging. (**a**) "Flow aside" geometry. *Top panel* shows side view of a PDMS chip with a glass coverslip. *Bottom panel* shows top view of cell trap compartment of the device, along with the small connecting channel used to load cells. (**b**) "Flow above" geometry, with cells physically separated from the main flow by a diffusive membrane. *Top panel* shows side view and *bottom panel* shows top view.

An advantage of the "flow above" design is that it requires less skill to fabricate, as all the necessary parts can be made without the use of photolithography techniques and without the additional constraint of the height of the cell trap, which complicates alternative designs. Thus, it is easier to implement in cell biology labs, most of which have limited or no access to microfabrication facilities. This setup has been extensively used in experiments with *Saccharomyces cerevisiae* (and to a lesser extent with *Escherichia coli)*, and it should be suitable for many other unicellular organisms such as *S. pombe*. The setup can also be used in many experimental paradigms. In the context of cell cycle studies, our main focus has been to induce temporally controlled expression of a gene placed under the control of a regulatable promoter, such as *MET3*pr or *GAL1*pr (15). Another current application of this device is to monitor the output of signal transduction pathways, such as the pheromone response pathway in budding yeast (9).

In this paper, we provide a protocol which allows the user to build and use a "flow above" microfluidic device. As an alternative to fabricating the components, a commercial version of this setup, which includes all parts described in **Section 3** (product reference YC-1, Warner Instruments Inc, Hamden, CT) may be used (this product is scheduled for release in mid-2009).

2. Materials

2.1. Cellulose Membranes

1. Dialysis tubing Cellusep T3 (Fisher Scientific), a roll with flat width 33 mm, thickness 23 μm, and molecular weight cut-off 12–14 kDa (*see* **Notes 1** and **2**).
2. 200 mL 10 mM Tris–EDTA (TE) buffer, pH8. Store at room temperature.
3. 250 mL TE buffer supplemented with 2% w/v Na_2CO_3 (TEC). Store at room temperature.
4. Several standard petri dishes.
5. Polydimethylsiloxane (PDMS) and curing agent Sylgard 184 (Dow Corning, Midland, MI). Store at room temperature.

2.2. PDMS Coated Coverslips

1. Coverslips, gauge 1.5, 24 × 50 mm (Fisher Scientific).
2. Clean, plain, circular silicon wafer, 100 mm diameter (Silicon Valley Microelectronics, Santa Clara, CA) or a flat plastic surface (e.g., a transparency film for laser printers).
3. Silanization agent such as trimethylchlorosilane (TMCS) (Sigma-Aldrich) – only needed if using a silicon wafer. Store at room temperature.
4. Large petri dish (diameter >100 mm) – only needed if using a silicon wafer.

2.3. PDMS Flow Chamber

1. Mould to cast PDMS flow chambers. We recommend polyoxymethylene material, which is easy to machine, brand name Delrin (McMaster-Carr, Atlanta, GA).
2. PrecisionGlide needles, gauge 23 (Becton-Dickinson, Franklin Lakes, NJ).

2.4. Assembly of the Microfluidic Device

1. 5-mm-thick cover for the flow cell (*see* **Section 3.4**), made of acrylic glass (*see* **Section 3.3**).
2. PrecisionGlide needles, gauge 21 (Becton-Dickinson, Franklin Lakes, NJ).
3. Formulation S-54-HL Tygon microbore tubing, 0.5 mm internal diameter (US Plastic Corp., Lima, OH).
4. 1 mL syringe.

5. 5 mL growth medium for yeast, e.g., synthetic complete yeast medium (the carbon source used depends on desired experimental conditions).
6. 50 mL 20% w/v solution of Triton X-100 in water.
7. Desired strain of *S. cerevisiae*.
8. An epifluorescence inverted microscope and dedicated time-lapse acquisition software.
9. *Optional* – peristaltic pump to drive flow through the device, e.g., Ismatec IPC (Ismatec, Glattbrugg, Switzerland).
10. *Optional* – aquarium pump and bubblers to oxygenate media.

3. Methods

The flow cell setup consists of the following parts (*see* **Fig. 14.2**):

- diffusive cellulose membrane (**Fig. 14.2-1**) that ensures diffusion of media, inducers, and other molecules smaller than ~14 kDa from the main flow chamber to the cell environment.
- coverslip coated with PDMS (**Fig. 14.2-2**). The PDMS coating acts as a soft cushion for the cells to prevent damage resulting from constraining the cells with the membrane.
- main flow chamber (**Fig. 14.2-3**) through which the exchange medium flows.
- metal coverslip holder (**Fig. 14.2-4**) and acrylic glass cover (**Fig. 14.2-5**), which sandwich and clamp the elements listed above in order to ensure the mechanical integrity of the setup and to prevent media leakages. Four screws (*see* holes on **Fig. 14.2-6**) are used for this purpose.
- inlet and outlet formed with blunted needles (**Fig. 14.2-7**) connected to transparent Tygon tubing (**Fig. 14.2-8**).

Preparation of these parts is described below.

3.1. Preparation of Membranes

1. Unroll the dialysis tubing on a glass rectangle or other clean hard surface. Using a scalpel, cut a piece of dialysis tubing of the desired size. For example, for use with a 50 × 24 mm coverslip, a size of 20 × 49 mm is optimal (*see* **Note 3**).
2. Fill two petri dishes with water.
3. Place the cut section of tubing into the first petri dish (it may quickly curl up and then uncurl), then pick it up with

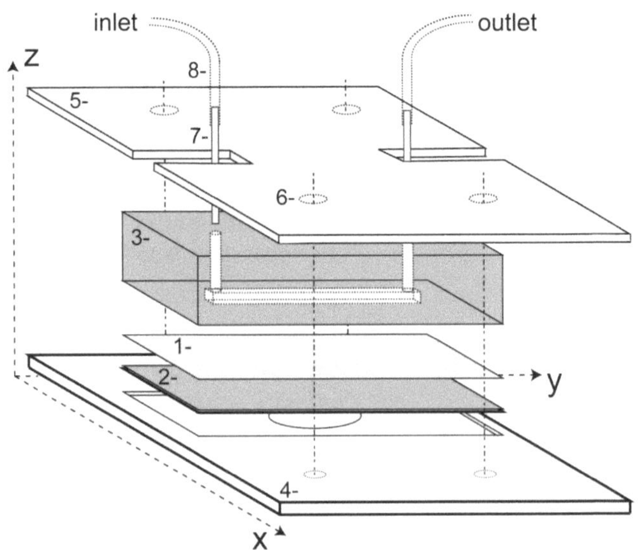

Fig. 14.2. Design of microfluidic flow cell used in (15). *1* – cellulose membrane; *2* – PDMS-coated coverslip; *3* – PDMS flow chamber; *4* – coverslip holder; *5* – acrylic glass cover; *6* – screw holes for clamping apparatus; *7* – blunt needle; *8* – connecting tubing.

forceps, hold the short end between finger and thumb, and gently wrinkle this end until you can take the two pieces apart with forceps (*see* **Note 4**). Place the membranes in the second petri dish, trying not to fold them.

4. Fill a large beaker with 200 mL TEC buffer, transfer the membranes, and loosely cover the beaker with a plastic lid. Boil on a hot plate for 20 min. The solution should become yellowish (*see* **Note 5**).
5. Take out the membranes and put them in a petri dish with water. Rinse the beaker, fill with 200 mL of TE buffer, transfer the membranes back to the beaker, and boil for 30 min. If the buffer is yellowish, discard it and repeat this step.
6. Let the beaker cool down on the bench, then place the membranes in a fresh petri dish with TE buffer, where the membranes can be stored for months at 4°C (*see* **Note 6**).

3.2. Preparation of PDMS-Coated Coverslips

1. Heat an oven to 80°C.
2. If you are using a plastic film as your flat surface, go to step 4. If you are using a silicon wafer, clean the surface of the wafer using a KimWipe soaked in acetone. Remove all dust and traces of PDMS from previous use. Rinse it with ethanol and dry under a stream of air.
3. Coat the silicon wafer with TMCS (**caution:** toxic and corrosive) in order to passivate the surface (*see* **Note 7**). Under

the hood, place the wafer in a large petri dish, add 500 μL TMCS to the center of the wafer, and put the lid on the dish. Tilt to spread the liquid evenly and let the wafer sit for 15 min until the TMCS has evaporated completely (*see* **Note 8**).

4. Prepare PDMS mixture using a small petri dish: add 30 g PDMS, then 3 g of curing agent (i.e., 10:1 ratio) (*see* **Note 9**). Vigorously stir the mixture with a P1000 pipette tip (bend at the end for best results) for several minutes until it resembles dense snow. Wait about 30 min for bubbles to disappear. (The process can be sped up by using a vacuum chamber.) This gives enough material to coat several coverslips.
5. Place the passivated wafer or plastic sheet on the bench. Using a P200 pipette and a tip with a cut end, dispense 70 μL of PDMS mixture onto the wafer (or plastic sheet) as a thin line. Place a 24 × 50 mm coverslip on top of this line (*see* **Notes 10** and **11**). Repeat two more times in order to have three coverslips on the wafer or film. Incubate for about 30 min until the PDMS has spread evenly across the coverslip (*see* **Note 12**), and then transfer the wafer (or plastic film) to the oven and bake for 45 min.
6. Remove the wafer (or plastic film) from the oven and peel off the coverslips carefully with a scalpel. Coated coverslips can be stored in clean dry petri dishes for further use.

3.3. Preparation of Flow Chambers

In this protocol, a mould is used to cast a PDMS flow chamber. The mould for a simple microfluidic chip with one channel does *not* require advanced microfabrication techniques such as photolithography. For the design and dimensions of our mould (*see* **Fig. 14.3a**), the use of Delrin and a simple milling machine is sufficient. Internal dimensions of the mould are approximately 48 × 20 mm. Small holes at both ends of the channel are used to insert blunt needles (gauge 23), thus producing the inlet and outlet of the chip.

1. Preheat oven to 80°C and prepare about 30 g of PDMS mixture as described in **Section 3.2**, step 4.
2. Insert needles into the mould as shown (*see* **Fig. 14.3b**).
3. Using a standard disposable plastic pipette, add approximately 5 mL of PDMS mixture to the mould. The total height of the PDMS chip should be approximately 5 mm, and the amount of PDMS material deposited should be adjusted accordingly (**Fig. 14.3c**).
4. Allow the PDMS to settle for 30 min, then bake in the 80°C oven for 2 h.
5. Rotate the needles on the mould to free them and pull them up out of the chip. Flip the mould and lift up the bottom

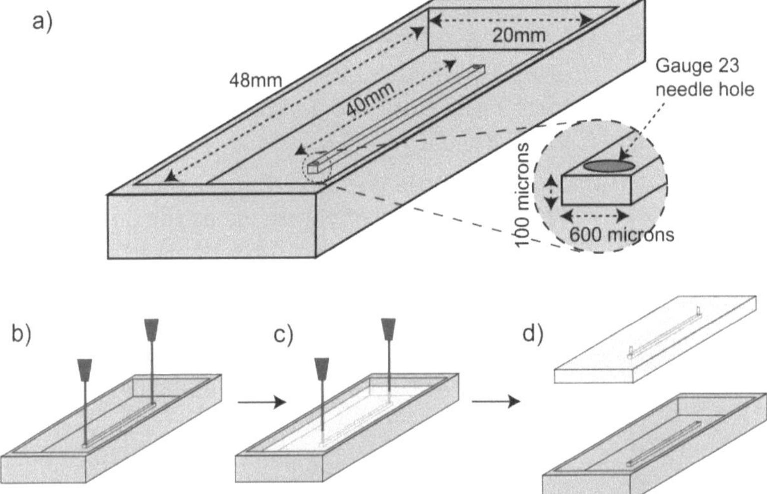

Fig. 14.3. Preparation of PDMS flow chamber. (**a**) Plastic mould used to cast PDMS. Precision machining generates an emboss 100 μm high, 600 μm wide, and 40 mm long. (**b**) Small holes are drilled for needles prior to PDMS addition (**c**). (**d**) After baking, PDMS chip is easily removed with scalpel.

carefully with a scalpel, a little at a time on each edge. Once the bottom is removed, the flow cells should come off easily (**Fig. 14.3d**).

6. Verify that the needle holes are connected to the channel (*see* **Note 13**).
7. *Optional*: if you have access to an air plasma cleaner, treat the flow chamber for about 30 s. This makes the surface of the PDMS hydrophilic, therefore preventing unwanted air bubbles from appearing in the channel during use (*see also* **Note 14**).

3.4. Assembling the Microfluidic Device

1. Rinse the PDMS-coated coverslip with ethanol, then with water, and dry it under a stream of air. Identify the side which has the PDMS coating (*see* **Note 11**) and place the coverslip on a flat surface with this side up.
2. Sonicate log-phase yeast cells of your desired strain and measure their optical density at 660 nm (OD_{660}) using an absorption spectrometer. Dilute to $OD_{660} = 0.02$ in an eppendorf tube in approximately 100 μL volume.
3. Pipette 20 μL of cells onto the middle of the coverslip.
4. Using forceps, center and place a membrane (prepared in **Section 3.3**) on top of the coverslip (*see* **Note 15**).
5. Let the membrane dry for about 15 min until its entire surface acquires a mottled bluish-gray color (*see* **Note 16**).

6. While the membrane is drying, prepare two connecting needles (gauge 21). Cut each needle to make it blunt on both ends (*see* **Fig. 14.2**-7), and attach a piece of tubing to one end (*see* **Fig. 14.2-8**). The tubing should be long enough to connect the chip to a syringe or pump.
7. When the membrane is ready, place the coverslip inside its holder (**Fig. 14.2-4**).
8. Place the flow cell (**Fig. 14.2-3**) and its cover (**Fig. 14.2-5**), with the flow channel facing down, on top of the coverslip. Applying pressure to the assembled unit with a finger, insert the four screws (**Fig. 14.2-6**) and tighten them each a little at a time until moderate resistance is felt. Do not overtighten.
9. Plug the needles of the connecting tubes (**Fig. 14.2**-7) into the inlet and outlet of the flow chamber (**Fig. 14.2-3**).
10. Connect the inlet tubing to a syringe filled with medium. *Very slowly* (no more than 100 μL/min) load approximately 100 μL of medium into the chamber of the chip. Make sure that medium flows smoothly and that all air bubbles are removed from the main channel (*see* **Note 17**).
11. Place the device on the stage of an inverted microscope (*see* **Note 18**) and connect it to a peristaltic pump, syringe pump, or gravitational flow (*see* **Note 19**).

3.5. Using the Microfluidic Device for Imaging

1. With medium flowing at a constant rate, find cells underneath the channel using transmission (bright field, phase contrast, or DIC) mode. If the cells are moving, the membrane was not dry enough and the apparatus should be reassembled (go back to **Section 3.4**, step 1). You may use the same flow cell, coverslip, and membrane. If cells are static and medium flows well, check that there is no excessive pressure in the chamber by analyzing dynamic changes in the focus (*see* **Note 20**)
2. Use time-lapse microscope software to record images over the course of your experiment. As an example, a sequence of phase contrast (top) and fluorescent images (bottom) is provided (*see* **Fig. 14.4**), which was acquired in a typical long-term imaging experiment (*see* **Notes 21** and **22**).
3. Once the experiment is finished, wash the flow chamber and tubing with a 50% ethanol solution in water for 30 min using a 100 μL/min flow rate. Do not use bleach as it affects PDMS transparency and takes a long time to remove. Rinse with DI water for a further 30 min.
4. Disconnect tubing and carefully disassemble the device. Discard the membrane. Rinse the coverslip with 95% ethanol to remove immersion oil, then with water, and place in a

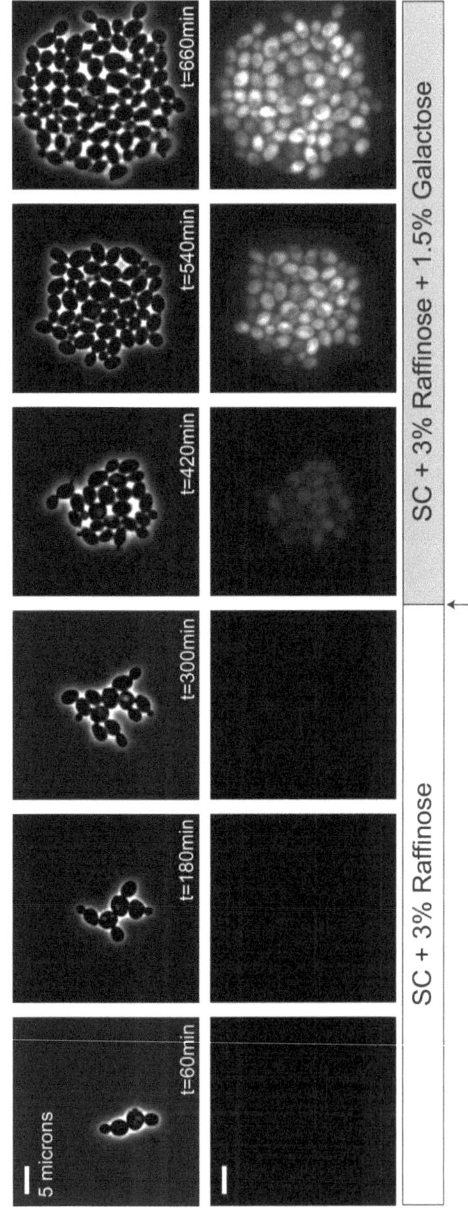

Fig. 14.4. Example of time-lapse experiment conducted with flow cell: monitoring turn-on of the *GAL1* promoter. *Top series* of images (phase contrast) shows exponential growth of *GAL1pr-Venus-degron* cells. *Bottom series* shows corresponding fluorescent images. 1.5% galactose was added to SCR medium at t=360 min.

petri dish containing 20% Triton solution. After a few minutes, rinse with water and examine the channel for visible cell clumps. If any are present, use a P200 pipette tip to clear the channel. Place the flow cell in the same Triton-containing petri dish and perform the cleaning procedure as described for the coverslip. Store dry at room temperature for further use (*see* **Note 23**).

4. Notes

1. Dialysis membranes with lower molecular weight cutoffs (e.g., 6 or 8 kDa) can also be used. However, 14 kDa membranes allow for better diffusion of larger molecules such as yeast alpha-factor pheromone.
2. To ensure rapid diffusion of medium and inducers across the membrane, its thickness should not exceed 40 μm.
3. It is easy to prepare several membranes at a time. To do this, fold the tubing end-over-end to produce a stack of tubing. Put a coverslip on top of the stack to guide you and use a scalpel to cut the membranes to the appropriate dimensions.
4. Once cut, dialysis tubing comprises two equal sides which must be separated to provide two membranes.
5. Cellulose membranes are conditioned in glycerol. This compound must be removed from the pores or diffusion through the membrane is greatly reduced.
6. Never allow the membranes to dry out.
7. Passivation of the wafer is necessary to be able to peel off the coated coverslips easily after baking.
8. The TMCS will leave colored traces on the surface of the wafer.
9. PDMS is a viscous silicon polymer which turns into a solid elastomer in the presence of curing agent and heat.
10. When applying the coverslip, small bubbles may appear in the middle of the PDMS material. Use a scalpel to shift the coverslip upward so that these bubbles are no longer in the middle and will not cause optical distortion during imaging.
11. It will be important to know which side of the coverslip has the PDMS coating. At this stage of fabrication, you may use a lab marker to mark the coverslip in an unambiguous way before placing it, marked side down,

onto the PDMS. This protects the ink from the ethanol used during the cleaning process. Alternatively, after the coverslip is prepared, tap both sides with a pair of forceps. PDMS has a duller, plasticky sound.

12. Some PDMS material may expand beyond the edges of coverslip. Measure the coverslip thickness after baking to ensure that it is compatible with the working distance of the microscope objective. If not, then reduce the amount of deposited PDMS material.
13. In standard microfluidic devices, connecting holes are usually generated after the curing procedure by punching the chip with a sharpened blunt needle. Our method was found to be easier and more reliable.
14. Microfluidic chips can be reused many times. After each experiment, they should be cleaned with detergent (20% Triton) and ethanol. Biofilms tend to form in the channel over time, which make it less transparent and more difficult to clean. Regular air plasma treatments ensure that the channel remains clean and hydrophilic for a longer period of time.
15. Dialysis membrane is a tube that has a polarity. The outer side has scratches which may compromise the quality of phase contrast images if they appear in the same focal plane as imaged cells. Use the slight curvature of the membrane along its length to determine the inside of the tube and place this concave side face-down on the coverslip.
16. Medium from the cells and water from the soaked membrane provide excess liquid which prevents good contact between the membrane and coverslip. This liquid must evaporate for cells to be properly constrained by the membrane. As the membrane dries, it begins to stick to the coverslip and becomes optically clear (it is opaque when wet). However, if the membrane becomes completely dry it will peel back from the coverslip. Drying is non-uniform, as edges tend to dry faster and must be rewetted by adding a drop of water and blotting the excess liquid with a KimWipe.
17. There should not be any leakage from the PDMS flow chamber. If this occurs, it is usually due to clogging of either the inlet or outlet tubing. Overtightened screws can also block the channel and prevent flow. If the flow is not steady, adjust the screws and check again.
18. A standard inverted microscope with epifluorescence capabilities can be used. Budding yeast cells are best imaged using a 63× or 100× magnification objective. High numerical aperture (NA = 1.4) is recommended for fluorescence imaging.

19. Peristaltic pumps and syringe pumps may be programmed for automated media switching.
20. To check how sensitive the focal plane is to changes in flow conditions, defocus the image by a few microns while looking at cells and see whether the halos surrounding the cells are undulating. If so, tighten or loosen screws to improve flow. This can be done without removing the flow cell from the stage. Ideally, cells should stay in focus when the flow is changed or stopped.
21. Yeast cells carrying a *GAL1-Venus-degron* construct (Venus is a variant of yellow fluorescent protein and the degron is a small sequence that destabilizes the protein to which it is attached (15)) were introduced into the microfluidic device with synthetic complete + 3% raffinose (SCR) medium and imaged every 3 min for 11 h. After 6 h, medium was switched from SCR to SCR + 1.5% galactose in order to activate transcription downstream of the *GAL1*pr (see the fluorescence rise in the bottom series of images). Doubling time for yeast under both conditions is about 2 h.
22. We recommend maintaining a constant flow rate, typically 50 μL/min, throughout the experiment, even without media changes. Aquarium pumps and bubblers can be used to aerate media during the assay. In the absence of flow or proper oxygenation, the level of fluorescence of constitutive markers is seen to decrease dramatically over the course of a typical (12 h) experiment. This is likely due to a decrease in the concentration of oxygen, which is required for maturation of fluorescent proteins (2).
23. Coverslips can be reused several times. The hydrophobicity of the PDMS surface promotes cell adhesion, so coverslips should be discarded when they become too hydrophilic.

Acknowledgments

G.C. is supported by the Human Frontier Science Program Organization. C.O. and F.R. are supported by the National Institutes of Health.

References

1. Chalfie M., Tu Y., Euskirchen G., Ward W.W. and Prasher D.C. (1994) Green fluorescent protein as a marker for gene expression. *Science* **263**:802–5.
2. Tsien R.Y. (1998) The green fluorescent protein. *Annu Rev Biochem* **67**:509–44.
3. Shaner N.C., Campbell R.E., Steinbach P.A., Giepmans B.N., Palmer A.E. and Tsien R.Y. (2004) Improved monomeric red, orange and yellow fluorescent proteins derived from Discosoma sp. red fluorescent protein. *Nat Biotechnol* **22**:1567–72.

4. Goldman R.D. and Spector D.L. (2004) Live Cell Imaging. Cold Spring Harbour Laboratory Press, Woodbury, NY.
5. Voldman J., Gray M.L. and Schmidt M.A. (1999) Microfabrication in biology and medicine. *Annu Rev Biomed Eng* **1**: 401–25.
6. Whitesides G.M, Ostuni E., Takayama S., Jiang X. and Ingber D. (2001) Soft lithography in biology and biochemistry. *Annu Rev Biomed Eng* **3**:335–73.
7. Sia S.K., Whitesides G.M. (2003) Microfluidic devices fabricated in poly(dimethylsiloxane) for biological studies. *Electrophoresis* **24**(21): 3563–76.
8. Balaban N.Q., Merrin J., Chait R., Kowalik L. and Leibler S. (2004) Bacterial persistence as a phenotypic switch. *Science* **300**(5690): 1622–5.
9. Paliwal S., Iglesias P.A., Campbell K., Hilioti Z., Groisman A. and Levchenko A. (2007) MAPK-mediated bimodal gene expression and adaptive gradient sensing in yeast. *Nature* **446**(7131): 46–51.
10. Hersen P., McClean M.N., Mahadevan L., Ramanathan S. (2008) Signal processing by the HOG MAP kinase pathway. *Proc Natl Acad Sci U S A* **105**(20):7165–70.
11. Bennett M.R., Pang W.L., Ostroff N.A., Baumgartner B.L., Nayak S., Tsimring L.S., Hasty J. (2008). Metabolic gene regulation in a dynamically changing environment. *Nature* **454**(7208):1119–22.
12. McDonald J.C., Duffy D.C., Anderson J.R., Chiu J.T., Wu H., Schueller O.J.A., Whitesides G.M. (2000) Fabrication of microfluidic systems in poly(dimethylsiloxane). *Electrophoresis* **21**:27–40.
13. Cookson S., Ostroff N., Pang W.L., Volfson D. and Hasty J. (2005) Monitoring dynamics of single-cell gene expression over multiple cell cycles. *Mol Syst Biol* **1**: 00024.
14. Lee P.J., Helman N.C., Lim W.A., and Hung P.J. (2008) A microfluidic system for dynamic yeast cell imaging. *BioTechniques* **44**(1): 91–5
15. Charvin G., Cross F.R., Siggia E.D. (2008) A microfluidic device for temporally controlled gene expression and long-term fluorescent imaging in unperturbed dividing yeast cells. *PLoS ONE* **3**:e1468.

Chapter 15

Monitoring of Cellular Responses to Hypoxia

Christoph Wotzlaw and Joachim Fandrey

Abstract

Oxygen is essential for survival of aerobic organisms. Sensing changes in the environmental oxygen concentration and appropriate adaptation to such changes are essential for organisms to survive. Hypoxia inducible factor 1 (HIF-1) is the key transcription factor in controlling the expression of oxygen-dependent genes required for this adaptation. HIF-1 is a heterodimer of an oxygen dependent α-subunit and constitutive β-subunit. Abundance and activity of HIF-1 is controlled by post-translational hydroxylation. Microscopic analysis of the assembly and activation process of HIF-1 has become an important tool to better understand HIF-1 regulation. Confocal laser microscopy provides exact images of HIF-1α that is translocated into the nucleus under hypoxia and its disappearance upon reoxygenation. To exactly follow the protein–protein interaction during the assembly process of HIF-1, both subunits were labeled by fusing them to fluorescent proteins. Fluorescence resonance energy transfer (FRET) was used to determine the interaction of both subunits in living cells by confocal microscopy.

Key words: Oxygen sensing, hypoxia inducible factor 1, HIF-1, protein–protein interactions, fluorescence resonance energy transfer (FRET), confocal microscopy.

1. Introduction

When oxygen supply to the tissue is reduced during ischemia or impaired respiration, a state of hypoxia develops. Organisms have to adapt to hypoxia to avoid severe damage or even death of respective tissues or the whole organism. Common to all countermeasures to ameliorate hypoxia is the ability of cells to rapidly detect changes in oxygen concentration (1). The adaptive response that is then initiated may include changes in oxygen capacity of the blood, improved blood circulation and ventilation or a switch to anaerobic metabolism. These responses depend on

D.B. Papkovsky (ed.), *Live Cell Imaging*, Methods in Molecular Biology 591,
DOI 10.1007/978-1-60761-404-3_15,

acute changes of different ion channel conductivities and subsequently cell membrane potential but also on long term changes in expression of a considerable number of genes. Hypoxia inducible factor 1 (HIF-1) is the transcriptional factor that coordinates central response of cells and organs to hypoxia (2). HIF-1 is composed of an oxygen labile α-subunit (HIF-α) and a constitutively expressed nuclear β-subunit (HIF-1β). Regulation of HIF-1α abundance is achieved through post-translational hydroxylation of specific proline residues within this subunit by oxygen-dependent prolyl hydroxylases, which thus act as bona fide cellular oxygen sensors. While hydroxylation initiates proteasomal degradation of HIF-1α under normal oxygen concentrations, reduced activity of oxygen sensor hydroxylases under hypoxia allows HIF-1α to evade destruction and form the HIF-1 dimer with its β-subunit (3). Similarly, oxygen-dependent hydroxylation of an asparagine residue within the HIF-1α protein controls transcriptional activity of the HIF-1 complex (3). One of the many target genes activated by HIF-1 under hypoxia is erythropoietin, the key regulator of red blood cell production (4). Erythropoietin is the paradigm of a hypoxia-induced gene which had helped to decipher HIF-1 activation and cellular oxygen sensing. However, it has not been resolved yet how stabilized HIF-1α interacts with its partners, both HIF-1β to form the HIF-1 dimer and also coactivators which are required for transcriptional activity. Recently, we have introduced a fluorescence energy transfer technique to visualize for the first time the interaction between the two subunits of HIF-1 and monitor the assembly of HIF-1 complex in living cells (5). In addition, we have developed and optimized scanning procedures, specialized software and evaluation processes that can also be applied to the studies of protein–protein interactions in living cells and some other models (6–8). In this paper we describe confocal imaging of HIF-1α and its nuclear distribution. In a specialized chamber which allows to study living cells under well-defined hypoxia through the microscope we follow the dimerization of HIF-1α and HIF-1β.

2. Materials

2.1. Cell Culture

1. Dulbecco's modified Eagle's medium (DMEM with 4.5 g/l glucose, L-glutamine, and pyruvate; all from Invitrogen GmbH, Karlsruhe, Germany) supplemented with penicillin (100 U/ml), streptomycin (100 μg/ml; Invitrogen GmbH, Karlsruhe, Germany) and 10% fetal bovine serum (FBS; Biocompare, South San Francisco, CA).

2. DMEM without FBS for plasmid transfection.
3. 10× stock of phosphate buffered saline (PBS), which contains 1.37 M NaCl, 27 mM KCl, 100 mM Na_2HPO_4, 18 mM KH_2PO_4, pH 7.4 (adjusted with HCl), autoclaved. PBS is prepared by 1:10 dilution of stock with distilled water.
4. Solution of trypsin (0.25%) and ethylenediaminetetraacetic acid (EDTA) (1 mM).
5. Cell culture flask (75 cm^2 bottom area) (Greiner Bio-One GmbH, Solingen, Germany).
6. Freely available software "ImageJ" from the National Institutes of Health, USA (http://rsbweb.nih.gov/ij).

2.2. Immunofluorescence of Hypoxia Inducible Factor 1α

1. 24-well polystyrene plates with 1.9 cm^2 growth area per well (Greiner Bio-One GmbH, Solingen, Germany).
2. Glass cover slips with a diameter of 1.2 cm (Glaswarenfabrik Karl Hecht KG "Assistent," Sondheim/Rhön, Germany).
3. 10× stock of PBS (*see* **Section 2.1**, Step 3). PBS is prepared by 1:10 dilution of stock with distilled water.
4. Fixation solution: 1:1 (v/v) methanol/acetone, stored at –20°C.
5. Permeabilization solution: 0.5% (v/v) Triton X-100 (Invitrogen GmbH, Karlsruhe, Germany) in PBS.
6. Mouse monoclonal anti-HIF-1α antibody (Transduction Lab., Lexington, KY).
7. Secondary goat anti-mouse IgG conjugated to Alexa Fluor 568 (Invitrogen GmbH, Karlsruhe, Germany). Caution: the conjugate is light sensitive.
8. Nuclear stain Hoechst33342 (Invitrogen GmbH, Karlsruhe, Germany), 5 μM solution in water. Caution: the fluorophore is light sensitive.
9. Antibody dilution buffer: 3% (w/v) bovine serum albumin (BSA) in PBS.
10. Anti-fading and mounting reagent Prolong Gold (Invitrogen GmbH, Karlsruhe, Germany).
11. Standard inverted fluorescence confocal microscope, one or two photon excitation.
12. Standard incubator (Heraeus) for cell culture, set at 37°C, 5% CO_2, and 100% humidity.
13. Hypoxia incubator (Heraeus) with adjustable oxygen concentration, down to 1% O_2.

2.3. Microscopy Study of HIF-1α Under Defined Gas Composition and Interaction Analysis Between HIF-1 Subunits Using Sensitized FRET Technique

1. Fugene6 transfection reagent (Roche Diagnostics GmbH, Mannheim, Germany).
2. Fusion proteins: HIF-1α fused to enhanced cyan fluorescent protein (ECFP) (Clontech GmbH, Germany) and HIF-1α fused to enhanced yellow fluorescent protein (EYFP) (Clontech GmbH, Germany) – ECFP-HIF-1α and EYFP-HIF-1β, respectively.
3. Glass-bottom cell culture dishes, 3.5 cm in diameter (WillcoWells BV, Amsterdam, the Netherlands). Physical characteristics of central glass insert allow all standard formats of microscopy including UV light excitation.
4. Chamber for cell culture dishes (Luigs & Neumann, Ratingen, Germany) for observation of living cells on microscope stage under the conditions of controlled temperature, humidity and gas composition, e.g., O_2 and CO_2 (5). Gas concentrations are individually adjustable by a gas mixing device (Newport Spectra-Physics GmbH, Darmstadt, Germany) connected to the chamber.
5. For microscopy of HIF-1α under different oxygenation conditions, a standard inverted confocal microscope with objective lenses plan apochromat 40× (Nikon GmbH, Düsseldorf, Germany) is used.
6. For sensitized FRET analysis of HIF-1α-HIF-1β interaction, a confocal microscope is used with one excitation laser line between wavelengths of 405 and 444 nm for ECFP (termed "donor" for FRET measurements) and a second laser between wavelengths of 514 nm and 532 nm for EYFP (termed "acceptor" for FRET) excitation.
7. Recommended filters FRET: Two band pass emission filters – 480/30 and 545/40 nm (AHF AG, Tübingen, Germany).

3. Methods

3.1. Cell Culture

Procedures described below are performed on human osteosarcoma cell line U2OS. Other cell lines available from the American Type Culture Collection (ATCC, Manassas, VA) such as hepatoma (HepG2) or hepatocellular carcinoma cells (Hep3B), and human embryonic kidney 293 cells containing the SV40 large T-antigen (293T) can also be used.

In samples with high cell density hypoxia at the bottom of cell culture flask may develop during experiment if oxygen supply to the cells is limited by diffusion (9). Therefore, for flasks with 75 cm^2 bottom area we recommend a moderate density of cells not exceeding 80% confluency. For immunofluorescence experiments cells are grown on glass cover slips in 24-well plates. Cover slips are sterilized by dipping them in 95% ethanol.

Cells in a flask are detached using the addition of 2 ml trypsin/EDTA. Two minutes later, trypsin is inactivated with 8 ml DMEM-FBS, and 30 μl of the suspension are transferred with a pipette into the wells containing 500 μl DMEM-FBS. The plate is steadily shaken to obtain a uniform cell spreading.

For experiments, 130 μl of cell suspension are taken and pipetted into sterile 3.5-cm dishes containing 1 ml DMEM–FBS. The dish is steadily shaken to obtain uniform cell spreading. The plate or dishes with cells are then put in CO_2 incubator for 3–6 h to allow the cells to attach. Immunofluorescence staining can be done 24–48 h later.

3.2. Immunofluorescence of HIF-1α

1. Seed cells and grow them on sterile cover slips in 24-well plates as described in **Section 3.1** (*see* **Note 1**).
2. Expose cells for 4 h to four different conditions: (i) ambient O_2 concentration (normoxia); (ii) low O_2 concentration, 1% O_2 (hypoxia); (iii) 100 μM of $CoCl_2$ in DMEM-FBS at normoxia – a common mimetic of hypoxia (4); (iv) hypoxia and subsequent reoxygenation (*see* **Note 2**).
3. After incubation, take all plates out and quickly rinse twice with ice cold PBS, except for plates subject to reoxygenation. These plates are placed into normoxia incubator for 5 min and then rinsed with ice cold PBS twice.
4. Add ice cold methanol/acetone (1:1) and incubate for 5 min to fix the cells. Then discard the solution, wash the cells briefly with PBS, and discard PBS.
5. Permeabilize cells by incubating the samples in Triton X-100/PBS for 5 min at room temperature, then wash with PBS again. For some cells such as U2OS this step is unnecessary.
6. Block samples for 15 min in antibody dilution buffer (3% (w/v) bovine serum albumin in PBS); discard buffer afterwards.
7. Incubate samples at room temperature in a volume of 200 μl with mouse monoclonal anti-HIF-1α antibody diluted 1:50 in buffer for 60 min.
8. Wash samples three times for 5 min with PBS.
9. Add secondary antibody diluted 1:400 in antibody dilution buffer (volume: 200 μl) and incubate for 30–60 min at room temperature (*see* **Note 3**). Discard secondary antibody.

10. For nuclear staining (optional), add 200 μl of 5 μM Hoechst33342 solution and incubate for 5–10 min at room temperature.
11. Wash samples three times for 5 min with PBS and dry.
12. Add 2.5 μl of anti-fading and mounting medium to samples on cover slips which are then put on microscopy slide. Dry slides in the dark at room temperature for at least 3 h before measurements.
13. View slides under a wide-field microscope with selected excitation and emission wavelengths or on a confocal microscope with laser light excitation. Alexa Fluor568 is excitable at wavelengths 500–600 nm and emits fluorescence at 580–700 nm. Hoechst33342 is excitable between 300 and 400 nm and emits in the range of 400–600 nm. Due to the distinct spectra of the fluorophores, no crosstalk perturbs images when localizing HIF-1α relative to the nucleus (*see* **Fig. 15.1**).

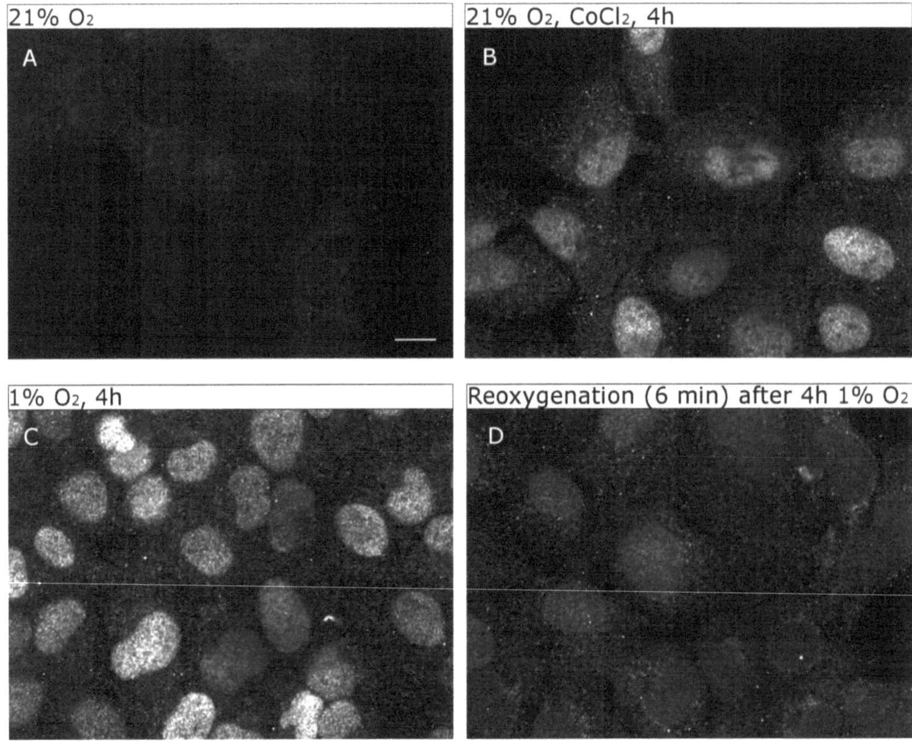

Fig. 15.1. Endogenous HIF-1α detected by immunohistochemistry. **(a)** Only weak signals from small amounts of HIF-1α protein are detected under normoxic conditions (21% O_2) due to oxygen dependent degradation of HIF-1α. **(b)** Nuclear accumulation of HIF-1α after treatment with hypoxia mimetic $CoCl_2$, which inhibits degradation of HIF-1α. **(c)** Hypoxia induced accumulation of HIF-1α in the nucleus due to reduced degradation. Oxygen serves as substrate for prolyl hydroxylases, which act as oxygen sensor through initiation of normoxic degradation of HIF-1α. **(d)** Rapid decrease of cellular HIF-1α protein upon reoxygenation. The return of oxygen after a hypoxic period enhances prolyl hydroxylase activity and initiates the fast degradation of HIF-1α within a few minutes.

3.3. Microscopic Study of HIF-1α Under Defined Gas Composition

1. Seed and grow cells onto sterile 3.5 cm dishes as described in **Section 3.1**.
2. Mix in eppendorf tube 4.5 μl of Fugene6 with 150 μl DMEM *without* serum. After 5 min, add 1.5 μg plasmid DNA encoding ECFP-HIF-1α, gently shake, and incubate for 30–40 min at room temperature.
3. Add this solution to the dish with cells, gently shake, and incubate for 6 h to allow transfection of cells with HIF-1α fusion protein.
4. Replace medium with fresh DMEM-FBS (*see* **Note 4**). Transfection is now completed.
5. Incubate cells for further 24–48 h. This allows synthesis of fusion proteins from plasmids in sufficient quantities so that fluorescence signal is bright enough for microscopic analysis.
6. Place one dish on microscope stage into the chamber adjusted to desired temperature and gas composition. Equilibrate cells for at least 15 min to these conditions.
7. Scan a stack of 6–8 fluorescent images of the middle part of the cell which shows high fluorescence intensity. The distance between sections/planes in *z*-direction should be kept between 500 and 700 nm. This allows evaluation of experimental data even if the cell moves along the optical axis during scanning (*see* **Note 5**).
8. Switch O_2 concentration to 1% and after 2 h take a new stack of 3D image.
9. Open the first 3D stack in image analysis software (*see* **Note 6**). Select the image with highest HIF-1α fluorescence and, after subtracting background value, calculate mean fluorescence. Process the second 3D stack similarly. Present results as a diagram.

3.4. Analysis of Interactions Between HIF-1 Subunits Using Sensitized FRET Technique

According to the theory of fluorescence resonance energy transfer (FRET), energy from the donor molecule is transferred non-radiatively to the acceptor if the two molecules come closer than a critical distance value. For the majority of fluorophore pairs this value is near 10 nm. In this case, when exciting the donor, fluorescence of the acceptor can be recorded (*see* **Fig. 15.2**).

Microscopic study of interaction between proteins requires initial calibration procedures, in which the main characteristics of the imaging system and fluorophores and "spectral bleed-through" values are determined. Protein–protein interaction analysis under the microscope requires labeling of the two proteins with different fluorophores. In the case of FRET, excitation of the donor molecule will initiate acceptor fluorescence if the distance between the two protein–fluorophore molecules is less than

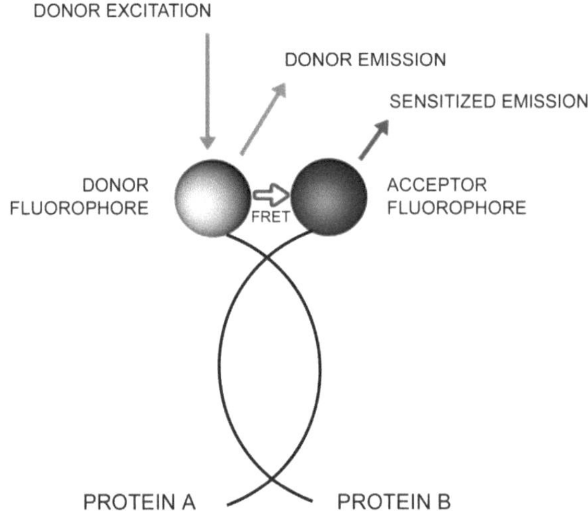

Fig. 15.2. Schematic description of fluorescence resonance energy transfer (FRET) to detect protein–protein interaction in involved living cells. Interaction of proteins requires close neighborhood of the molecules. Proteins can be labeled in living cells by generating fusion proteins containing a fluorophore and target protein. If the two fluorophore-labeled proteins come closer than 10 nm to each other due to their interaction, energy from the donor fluorophore which has been excited by laser light can be non-radiatively transferred to the acceptor fluorophore causing sensitized emission of light at the emission wavelength of the acceptor. The closer the two proteins, the greater the proportion of energy transferred to the acceptor and the smaller the proportion of residual donor emission.

10 nm. Since common fluorophores show broad excitation and emission bands, the acceptor can also be excited directly by donor excitation light, and donor emission may be collected in the acceptor emission channel. This effect, called bleed-through from one fluorophore channel to the other, has to be determined individually for each microscopy system and fluorophore combination. Bleed-through values are needed for correct quantification of non-radiative energy transfer from donor to acceptor.

1. Seed and grow cells on sterile 3.5 cm dishes as described in **Section 3.1**.
2. Perform transfection with following fusion protein constructs: (a) pECFP-HIF-1α only, (b) pEYFP-HIF-1β only, (c) pECFP-HIF-1α and pEYFP-HIF-1β, (d) pECFP and pEYFP. Add to each dish 4.5 μl of Fugene6 mixed in eppendorf tube with 150 μl of DMEM *without* FBS. After 5 min add 1.5 μg plasmid DNA of each construct, gently shake the tube with the transfection solution and incubate for 30–40 min at room temperature. Then add the transfection solution to the cell culture dish and gently shake the dish. Incubate cells for 6 h, then change medium for fresh DMEM-FBS.

3. Grow cells for further 24–48 h after transfection to allow synthesis of fusion proteins from plasmids and sufficiently high fluorescent signals for microscopic analysis.
4. Place dish into microscope chamber set to 21% O_2, 79% N_2, and 5% CO_2 (normoxia) and 37°C. Adapt cells to these conditions for at least 15 min before starting microscopic analysis.
5. Perform calibration process for each protein–protein interaction analysis as follows (*see* **Note 7**). Determine the degree by which donor fluorescence is detected in the acceptor channel (relative to donor fluorescence in the donor channel) using appropriate excitation and emission filters for the donor. Image five cells transfected only with ECFP-HIF-1α donor by exciting them with a laser line between 405 and 444 nm (optimal for ECFP) and collecting donor fluorescence with (i) the acceptor filter (band pass emission filter: 545/40 nm, AHF AG, Tübingen, Germany) to produce an image called *BTfd*; (ii) with the donor filter (band pass emission filter: 480/30 nm, AHF AG, Tübingen, Germany) to produce an image called *BTd*).
6. Determine the degree by which acceptor fluorescence is detected in the acceptor channel under donor excitation (relative to acceptor fluorescence in the acceptor channel). Image five cells transfected with EYFP-HIF-1β acceptor only as follows: (i) excite them with the donor laser and collect fluorescence with the acceptor filter to produce an image called *BTfa*; (ii) excite with a laser line between 511 and 532 nm (optimum for EYFP) and collect fluorescence with the acceptor filter to produce an image called *BTa*. This measurement takes into account that EYFP molecules are excitable to a certain degree by the laser used for donor excitation.
7. Determine background by scanning an area outside fluorescent cells. Background values are subtracted from every image.
8. Calculate calibration parameters:

$$BTfd/BTd = BTD\,(\text{spectral bleed-through of the donor}) \quad [1]$$

$$BTfa/BTa = BTA\,(\text{spectral bleed-through of the acceptor}) \quad [2]$$

   Mean BTD and BTA values in cells are calculated.
9. For the analysis of protein–protein interaction, image at least 20 cells transfected with both fluorescent donor ECFP-HIF-1α and acceptor EYFP-HIF-1β as follows: (i) excite with the donor laser and collect fluorescence with

the donor filter, to produce an image called *donor image*, *D*; (ii) excite with the donor laser and collect fluorescence with the acceptor filter, to produce an image called *FRET image*, *F*; (iii) excite with the acceptor laser and collect fluorescence with the acceptor filter to produce an image called *acceptor image*, *A*.

10. Determine background values by scanning areas outside fluorescent cells and subtract them from every image. Then calculate mean fluorescent signals for each cell.

11. Evaluate images of co-transfected cells. In the case of FRET from ECFP donor to EYFP acceptor due to close proximity of the two HIF-1 subunits ($< \sim 10$ nm), the percentage of transferred energy relative to the energy absorbed by the donor determines *FRET efficiency*. It can be calculated for every pixel of a FRET image using the following formula:

$$\text{FRET efficiency (\%)} = [1 - [D/(D + F - D \times BTD - A \times BTA))] * 100, \quad [3]$$

where *D* is donor fluorescence in the donor channel under donor laser excitation; *F* is fluorescence in the acceptor channel under donor laser excitation; *A* is acceptor fluorescence in the acceptor channel under acceptor laser excitation; *BTD* and *BTA* are spectral bleed-through values of the donor and acceptor, respectively; and $(F - D \times BTD - A \times BTA)$ is the energy transferred from ECFP to EYFP detected as EYFP fluorescence (*see* **Note 8**).

12. Calculate mean FRET efficiency values for fluorescent nuclei of co-transfected cells. FRET efficiency is also influenced by the number of acceptor molecules surrounding a donor molecule (8). To take this fact into account the system specific ratio of acceptor to donor molecules is determined as the ratio of the fluorescence from acceptor molecules to that of the donor molecules; FRET efficiencies are presented as values over this ratio. **Figure 15.3a** shows the example of HIF-1α/β FRET in U2OS cells, where the image represents typical distribution of fluorescent signals within the cell and the graph represents regression curve of mean FRET values from ~20 cells with different acceptor/donor molecule ratios.

13. Account for *random FRET*, which is caused by random collisions of fluorophores during the time of imaging. In addition, some fluorophores derived from *green fluorescent protein* (GFP) such as ECFP and EYFP tend to form dimers causing false-positive signals (*see* **Note 9**). Therefore, control measurements need to be done with cells cotransfected

with pECFP and pEYFP fluorophore plasmids, rather than fusion proteins of HIF-1 subunits. Experimental procedure is the same as in **Section 3.4**. Mean FRET efficiency values of cotransfected cells are calculated and presented as dependence of the ratio of acceptor and donor fluorescence (*see* **Fig. 15.3**).

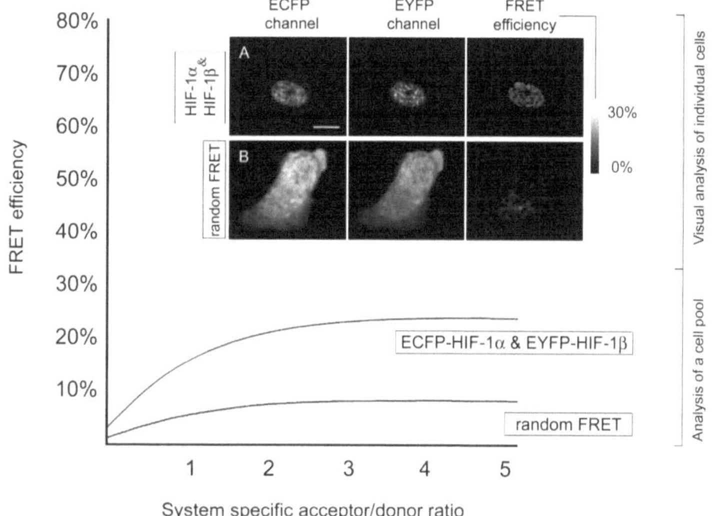

Fig. 15.3. Microscopic evaluation of FRET between HIF-1α and HIF-1β. FRET efficiency is a function of the ratio of fluorescent acceptor to donor molecules determined by proximity of the two proteins. The *inserts* show examples of microscopic scans of a single osteosarcoma cell expressing (**a**) ECFP-HIF-1α and EYFP-HIF-1β fusion proteins or (**b**) ECFP and EYFP fluorophores only. The latter experiment is used to determine FRET caused by random collision and/or dimerization of the fluorophores (random FRET) which has to be taken into account when trying to prove protein–protein interaction.

4. Notes

1. To avoid hypoxic conditions due to high oxygen consumption by the cell layer, it is recommended to work with cells having confluency below 80%, i.e., within 24–36 h after seeding, before culture density becomes too high.
2. For each condition the experiments are conducted in triplicate. Use two incubators: one for normoxia (21% O_2) and the other for hypoxia (1% O_2).
3. For this and the following step, protect samples with aluminium foil to avoid photodamage.
4. This step is recommended for sensitive cell lines but not necessary for U2OS cells.

5. To reduce destruction of fluorophores during imaging experiments, one can reduce laser power or shorten scanning periods by lowering image resolution. However, reduced laser power requires higher detector sensitivity to balance the loss of signal. This can cause higher background and noise.
6. A number of plug-ins for the software "ImageJ" is available, which helps evaluate images taken for microscopic protein–protein interaction analysis (6, 7).
7. With respect to FRET analysis, ECFP-fusion proteins represent *donor*s and EYFP-fusion proteins *acceptors*. During this step imaging characteristics of the microscopic system have to be standardized.
8. In this equation it is not taken into account that during the non-radiative resonance energy transfer other relaxation processes may take place. Their amount should be negligible.
9. New developments in the field of fluorophore design promise to reduce the amount of false-positive FRET signal. New mutants of existing GFP variants eliminate hydrophobic interactions between them.

Acknowledgments

The authors thank Professor Helmut Acker for critically reading the manuscript. This work was supported by the European Commission under the 6th Framework Programme (Contract No: LSHM-CT-2005-018725, PULMOTENSION) to Joachim Fandrey. This publication reflects only the authors' views and the European Community is in no way liable for any use that may be made of the information contained therein.

References

1. Acker, T., Fandrey, J., and Acker, H. (2006) The good, the bad and the ugly in oxygen-sensing: ROS, cytochromes and prolyl-hydroxylases. *Cardiovasc. Res.* **71**, 195–207
2. Semenza, G. L. (2000) HIF-1: mediator of physiological and pathophysiological responses to hypoxia. *J. Appl. Physiol* **88**, 1474–1480
3. Fandrey, J., Gorr, T. A., and Gassmann, M. (2006) Regulating cellular oxygen sensing by hydroxylation. *Cardiovasc. Res.* **71**, 642–651
4. Fandrey, J. (2004) Oxygen-dependent and tissue-specific regulation of erythropoietin gene expression. *Am. J. Physiol. Regul. Integr. Comp. Physiol.* **286**, R977–R988
5. Wotzlaw, C., Otto, T., Berchner-Pfannschmidt, U., Metzen, E., Acker, H., and Fandrey, J. (2007) Optical analysis of the HIF-1 complex in living cells by FRET and FRAP. *FASEB J.* **21**, 700–707
6. Feige, J. N., Sage, D., Wahli, W., Desvergne, B., and Gelman, L. (2005) PixFRET, an ImageJ plug-in for FRET calculation that can accommodate variations in spectral bleed-throughs. *Microsc. Res. Tech.* **68**, 51–58
7. Hachet-Haas, M., Converset, N., Marchal, O., Matthes, H., Gioria, S., Galzi, J. L., and Lecat, S. (2006) FRET and colocalization analyzer – a method to validate

measurements of sensitized emission FRET acquired by confocal microscopy and available as an ImageJ plug-in. *Microsc. Res. Tech.* **69**, 941–956

8. Zacharias, D. A., Violin, J. D., Newton, A. C., and Tsien, R. Y. (2002) Partitioning of lipid-modified monomeric GFPs into membrane microdomains of live cells. *Science* **296**, 913–916
9. Wolff, M., Fandrey, J., and Jelkmann, W. (1993) Microelectrode measurements of pericellular PO_2 in erythropoietin-producing human hepatoma cell cultures. *Am. J. Physiol.* **265**, C1266–C1270

Chapter 16

Imaging of Cellular Oxygen and Analysis of Metabolic Responses of Mammalian Cells

Andreas Fercher, Tomas C. O'Riordan, Alexander V. Zhdanov, Ruslan I. Dmitriev, and Dmitri B. Papkovsky

Abstract

Many parameters reflecting mitochondrial function and metabolic status of the cell, including the mitochondrial membrane potential, reactive oxygen species, ATP, NADH, ion gradients, and ion fluxes (Ca^{2+}, H^+), are amenable for analysis by live cell imaging and are widely used in many labs. However, one key metabolite – cellular oxygen – is currently not analyzed routinely. Here we present several imaging techniques that use the phosphorescent oxygen-sensitive probes loaded intracellularly and which allow real-time monitoring of O_2 in live respiring cells and metabolic responses to cell stimulation. The techniques include conventional wide-field fluorescence microscopy to monitor relative changes in cell respiration, microsecond FLIM format which provides quantitative readout of O_2 concentration within/near the cells, and live cell array devices for the monitoring of metabolic responses of individual suspension cells. Step by step procedures of typical experiments for each of these applications and troubleshooting guide are given.

Key words: Cell metabolism, mitochondrial function, cellular oxygen, phosphorescent oxygen-sensitive probe, phosphorescence quenching, live cell imaging, microsecond FLIM.

1. Introduction

Molecular oxygen (O_2) is the key substrate of aerobic organisms and the terminal acceptor of the electron transport chain, which can serve as an informative marker of cell metabolism and mitochondrial function (1–6). The development of a number of fluorescence and phosphorescence based O_2-sensitive probes and assays (7–10) has provided relatively simple, non-chemical and

D.B. Papkovsky (ed.), *Live Cell Imaging*, Methods in Molecular Biology 591,
DOI 10.1007/978-1-60761-404-3_16, © Humana Press, a part of Springer Science+Business Media, LLC 2010

non-invasive means to measure O_2 in samples containing respiring cells. Based on the collisional (i.e., non-chemical) quenching by O_2 of certain photoluminescent structures (8), these probes combine good selectivity and reliable work across the whole physiological range of O_2 (0–250 μM) and particularly at low O_2 concentrations. They allow contact-less "sensing" of O_2 in complex biological samples and particularly in samples containing respiring mammalian cells. Using extracellular O_2-sensitive probes, optical O_2 sensing and respirometry were initially applied to the imaging of tissue oxygenation (9) and analysis of bulk consumption of O_2 by populations of cells. A panel of high-throughput O_2 consumption assays performed in standard microtiter plates with detection on a fluorescent plate reader have been developed and successfully used in a number of useful applications (10, 11–15).

More recently, O_2 probes suitable for intracellular use have been developed (16, 17–22), which enable the analysis of localized O_2 gradients within the cells and a number of additional measurement tasks. In particular, real-time monitoring of fast, transient respiratory responses of individual cells and cell populations (16, 20, 23, 24), which thus far was outside the scope of traditional O_2-sensing probes and techniques, has been achieved. Some of these probes are compatible for fluorescence microscopy imaging. It has been shown that using these intracellular O_2 probes and relatively simple cell-handling procedures, sensing of intracellular O_2, and real-time monitoring of metabolic and respiratory responses of live mammalian cells and tissue can be performed (15, 21, 23). Several different cell types, particularly PC12, HepG2, HCT116, HeLa, SH-SY5Y, and A549 cells were examined so far by O_2 imaging, which has been shown to provide information-rich data about cell metabolism and good spatial and temporal resolution (16, 22).

Application of the O_2 imaging approach to suspension cells has been limited due to cell mobility and difficulties with sample manipulation and effector addition. These limitations have been overcome by the use of Live Cell Array™ (LCA) – a convenient platform developed for imaging of suspension cells (24). LCA provides capturing of cells within an array of specially designed microwells, whereby each well retains a single cell while maintaining cell viability. LCAs facilitate high content image analysis of large number of individual suspension cells with effector treatments.

In this chapter we describe the use of intracellular O_2-sensing probes based on phosphorescent platinum(II)-porphyrin dyes, in conjunction with fluorescence imaging for the assessment of mitochondrial function and real-time monitoring of responses of mammalian cells to metabolic effectors. These probes have been developed in our lab and commercialized by Luxcel Biosciences (Cork, Ireland). Standard wide-field fluorescence intensity imaging in time-lapse mode, as well as microsecond FLIM with pulsed LED excitation are described and applied to commonly used cell models including the adherent HCT116 cells, differentiated PC12 cells, and suspension Jurkat T cells.

2. Materials

1. HCT116 human colorectal carcinoma, PC12 rat pheochromocytoma, and Jurkat human T-lymphoma cell lines (American Tissue Culture Collection, ATCC) (*see* **Note 1**).
2. Cell culture media: McCoy's 5A modified medium (McCoy), Roswell Park Memorial Institute 1640 (RPMI) (Sigma-Aldrich), Dulbecco's modified Eagle's medium (DMEM) (Invitrogen) (*see* **Note 2**).
3. Fetal bovine serum (FBS) and horse serum (HS) (Sigma).
4. Tissue culture grade dimethylsulfoxide (DMSO) and ethanol.
5. Penicillin/streptomycin (antibiotic/antimycotic) stock solution (Sigma)
6. Collagen IV and poly-D-lysine (Sigma) – prepared as 5 μg/mL solution in growth medium for coating the dishes for cell adhesion.
7. Nerve growth factor (NGF) (Sigma) for differentiation of PC12 cells.
8. MitoXpress™ and IC60N phosphorescent oxygen-sensitive probes (Luxcel Biosciences, Ireland) (*see* **Note 3**).
9. EndoPorter™ reagent (Gene Tools, USA) for probe transfection – prepared as 1 mM stock in DMSO (*see* **Note 4**).
10. Glass-bottom imaging dishes (MatTek, USA).
11. Custom-made glass cylinders, approximately 9 mm outer diameter, 7 mm inner diameter, 15 mm long – one per dish (*see* **Note 5**).
12. Vacuum grease Apiezon L (Apiezon Corporation, USA).
13. Metabolic effectors (Sigma): mitochondrial uncoupler carbonylcyanide-*p*-trifluoromethoxyphenylhydrazone (FCCP) and complex III inhibitor antimycin A – prepared as 1 mM stock solutions in DMSO; extracellular Ca^{2+} chelator ethylene glycol-bis(2-aminoethylether)-N,N,N′,N′-tetraacetic acid (EGTA) – prepared as 0.5 M solution in water, pH 7.4.

14. Live Cell Array™ devices (Nunc) for imaging of suspension cells.
15. Standard tissue culture equipment: CO_2 incubator, laminar flow cabinet, Gilson pipettes, centrifuge, vacuum aspirator, inverted transmission microscope, hemocytometer.

2.1. Imaging Equipment

1. Standard wide-field fluorescent microscope Axiovert 200 with heated stage and objective (Carl Zeiss, Germany).
2. Fast gated CCD camera (LaVision, Germany) coupled to the microscope.
3. Custom-made LED excitation module (LaVision) coupled to the microscope, synchronized with CCD camera and controlled by the microscope software (LaVision). The module contains LED driver and a set of bright 397, 470, and 590 LEDs which are selected manually.
4. Filter cubes for the O_2-probe: 395 nm bandpass excitation, 450 nm dichroic mirror, 650 nm longpass emission (Carl Zeiss) (*see* **Note 6**). Filter cubes for the other fluorescent probes (e.g., 488/525, 535/550 nm – Carl Zeiss), as required.
5. Incubation system for cell culture (PeCon, Germany) consisting of a heating insert mounted on the microscope stage, low-volume incubation chamber, temperature, and O_2 and CO_2 controllers. This system allows rapid adjustment of temperature, gas composition (O_2, CO_2), and humidity in the sample compartment and changing of O_2 concentration.
6. Perfusion pump (Ismatec SA, Glattbrug, Switzerland) with 0.02 mm i.d. tubing (Gradko International Limited, Winchester, England) and a standard 22.5G hypodermic needle which are connected to the incubation chamber (*see* details in **Section 16.3**). The set-up provides effector addition during time-lapse imaging experiments, without disturbing the sample, gas composition, temperature, and focus of the region of interest (ROI).
7. ImSpector software (LaVision) for instrument control, FLIM and time-lapse measurements, image analysis, and data processing.
8. Wide-field fluorescence microscope IX51 (Olympus) equipped with a long distance 40 × objective LUCPFLN (Olympus), climate control chamber for imaging in Live Cell Array devices. Equipped with a 395 nm excitation filter and a 600 nm cut-off emission filter.

3. Methods

3.1. Culturing and Preparing the Cells for Imaging Experiments

1. All cell culture work is performed under a biological safety cabinet. HCT116 cells are cultured as adherent cells; PC12 cells are initially cultured in suspension and then adhered on pre-coated surfaces and differentiated (*see* **Notes 1** and **2**). Jurkat cells are cultured in suspension.
2. HCT116 cells are cultured in 75 cm^2 cell culture flasks with 0.2 μm vent cap in McCoys 5A medium containing 2 mM L-glutamine, 100 U/mL penicillin, 100 μg/mL streptomycin (P/S), 10% FBS, maintaining them in CO_2 incubator at 37°C and 5% CO_2. Cells are grown to a density of 1–2 × 10^6 cells/mL, regularly split, and reseeded at 1.5 × 10^5 cells/mL in 10 mL of medium. Cells are harvested from the flask using trypsin/EDTA, counted on a haemocytometer, and plated on imaging dishes by dispensing 0.2 mL of cell suspension (6 × 10^4 cells/mL) in the medium. Cells are then grown in minidishes to a confluence of approximately 90% (*see* **Note** 7) changing growth medium every 2 days.
3. PC12 cells are cultured in 75-cm^2 flasks in RPMI medium supplemented with 2 mM L-glutamine, 10% horse serum (HS), 5% fetal bovine serum (FBS), P/S at 5% CO_2 with regular passages in 30 mL at 0.5–1 × 10^5 cells/mL. For seeding, cells are harvested, centrifuged at 200*g* for 5 min, separated from the supernatant, washed in PBS, centrifuged again, and resuspended in 1 mL of trypsin/EDTA solution. Cells are then incubated for 90 s at 37°C, diluted with 10 mL of RPMI supplemented with 1% HS, and gently passed ten times through 23G needle to separate individual cells. The cells are counted and seeded in MatTek dishes pre-coated with a mixture of collagen IV (0.007%) and poly-D-lysine (0.003%) at 2.5 × 10^4 cells/dish. When the cells reach the confluence of ~50%, they are differentiated for 3–5 days in RPMI supplemented with 1% horse serum, P/S, and 100 ng/mL NGF. After differentiation cell layer becomes confluent and cells remain active for 2–4 days (*see* **Notes 7** and **8**).
4. Jurkat cells are cultured in 75-cm^2 flasks in RPMI containing 10% FBS, P/S to a density of 1–2 × 10^6 cells/mL at 37°C and 5% CO_2. Cells are split regularly and reseeded at 1.50 × 10^5 cells/mL in 20 mL of medium (*see* **Note 8**). Cells are then harvested from the flask, counted on a hemocytometer, centrifuged at 200*g*, resuspended in medium, adjusted to a concentration 1 × 10^6 cells/mL, and kept at 37°C.

3.2. Imaging of Adherent HCT116 Cells Loaded with MitoXpress Probe

1. Reconstitute a 1× package of MitoXpress probe (dry sample in a plastic vial) in 0.1 mL of growth medium with additives used to culture HCT116 cells, to produce a 10-μM probe stock (*see* **Note 9**).
2. Pipette the required amount of medium (0.2 mL per dish) to a glass or plastic vial, add stock solutions of probe (100 μL/mL) and Endoporter (6 μL/mL), and incubate the resulting probe loading solution for 10–15 min at 37°C.
3. Take the dishes with HCT116 cells prepared for imaging experiments as described in **Section 3.1**, step 2, and transfer 200 μL of probe loading solution to each dish.
4. Incubate the dishes in CO_2 incubator at 37°C for 24–30 h (*see* **Note 10**).
5. After the incubation, carefully aspirate probe loading solution from the dish and discard it. Add 1 mL of fresh medium, incubate for 5 min, then aspirate and discard. Repeat washing of the cells two more times and finally add 0.2 mL of medium (*see* **Note 11**).
6. Switch on the microscope and cell incubation system and allow them to warm up and equilibrate (T, CO_2, O_2) for at least 30 min. Set up the software for image acquisition and time-lapse experiments and the required filter combination (395 nm excitation and >650 nm emission).
7. Take clean glass cylinder and apply a thin, uniform layer of vacuum grease to one of its edges.
8. Take the imaging dish with cells (prepared in **Section 3.1**), aspirate medium, and then firmly attach the cylinder with its greased side to the glass bottom of the dish where the cells are grown (*see* **Note 12**). Add 180 μL of medium to the cap and 1 mL of medium to the outer space and place the measurement unit inside the heating insert. The assembled unit with glass cap is shown in **Fig. 16.1**.
9. Assemble the incubation chamber and allow the dish with the cap and cells to equilibrate for about 20 min (*see* **Note 13**).
10. Using a 100-μL Hamilton syringe, load the solution of effector into the tubing and pump it inside the chamber to allow temperature and gas equilibration.
11. Switch the microscope in transmission mode; focus on cells inside the cap area.
12. Switch the microscope to fluorescence mode. Select ROI where the cells have relatively uniform density and good loading with probe. Representative bright field and

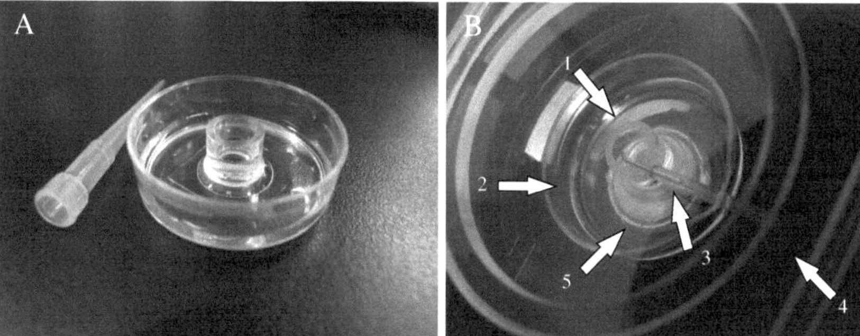

Fig. 16.1. Photographs of the MatTek dish with cap used for O_2 imaging of adherent cells. (**a**) General view of 35-mm imaging dish with attached glass cylinder mounted using vacuum grease. (**b**) Incubation chamber of the microscope with a sample prepared for O_2 measurements: 1 – glass cylinder; 2 – imaging dish; 3 – syringe needle connected to the pump for application of effector solution; 4 – heating insert with sample holder of the incubation system; 5 – microscope objective. Measurement requires restricted surface area to reduce convection and mass exchange within the sample, which is provided by a cylindrical cap attached to the bottom of the dish with cells. Effector addition should not disturb the sample. Small volume of effector stock solution is applied on top of the medium by means of perfusion pump with bended needle or gently applied onto the wall of the cap with a micropipette.

fluorescent images of HCT116 cells loaded with probe are shown in **Fig. 16.2a, b**.

13. Initiate image acquisition in time-lapse mode and monitor basal fluorescent signal of the cells under resting conditions.
14. If the signals are sufficiently high (compared to blanks outside the cells) and stable, proceed to the experiments with effector treatments (*see* **Note 14**). Initially, it is recommended to test cellular response by treating the cells with uncoupler FCCP and/or with mitochondrial inhibitor Antimycin A (*see* **Note 15**).
15. Move the microscope stage to a new field with cells within the dish, adjust the focus, and initiate image acquisition in time-lapse mode.
16. After recording the basal signal of resting cells for 5–15 min, make the addition of FCCP between two measurements. We normally add 1/10 of the total sample volume in the cap (i.e., 20 μL–180 μL) using the lowest possible flow rate of the pump (*see* **Note 16**). Continue image acquisition for 10–30 min after the addition of FCCP. When the response is finished, stop the acquisition and analyze signal profile(s).
17. Interpretation of signal profiles of the intracellular probe. Basal fluorescent signal prior to effector addition reflects O_2 concentration within the cell which corresponds to the resting level of respiration. An increase in probe signal after FCCP addition reflects increased respiration which

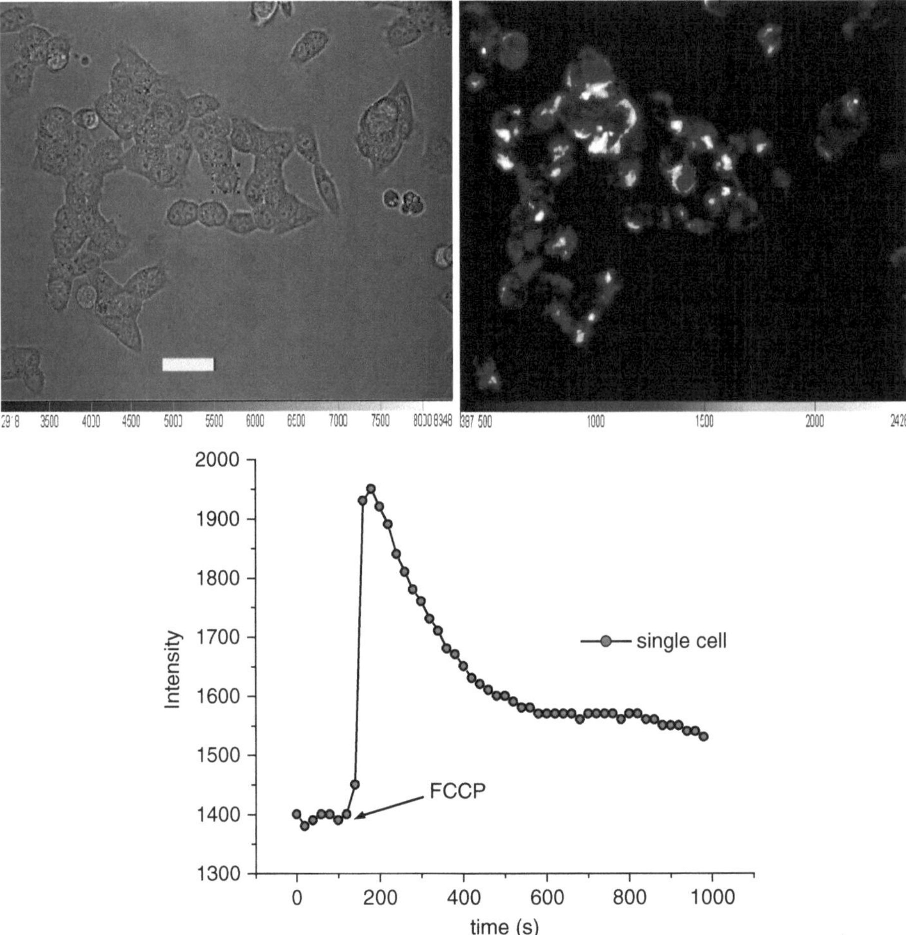

Fig. 16.2. Bright field (**a**) and fluorescent (**b**) images of HCT116 cells prepared for imaging in a MatTek dish and loaded with IC60 probe. (**c**) Respiratory response of HCT116 cells to the addition of FCCP measured by conventional imaging (390 nm bandpass/650 nm longpass). The increase in fluorescent signal reflects increased respiration due to mitochondrial uncoupling.

causes a reduction in cellular O_2. Conversely, Antimycin A is expected to inhibit cell respiration and elevate cellular O_2 to the levels present in bulk medium (i.e., air-saturated), thus bringing probe signal to a lower level (*see* **Note 17**). A characteristic profile obtained with HCT116 cells and FCCP is shown in **Fig. 16.2c**.

18. If the responses to FCCP and Antimycin A treatments are clearly seen, the samples are prepared properly, and the cells are in good condition, one can proceed to the experiments with stimulants having unknown effects on cell respiration.

19. In this case, take a new dish with cells loaded with probe and follow steps 11–17 above. Make effector addition and

monitor relative changes in intracellular O_2, which reflect the changes in cellular respiration. If required, repeat treatment.

20. Quantitative assessment. Plot signal profile for ROI with cells; adjust intensity scale as required. Select ROI outside the cells and plot the profile of blank signal. Subtract blank profile from the cell signal profile to produce the profile of probe fluorescence (*see* **Note 18**).

3.3. FLIM Analysis of Responses of Differentiated PC12 Cells Loaded with IC60 Probe

1. Prepare differentiated PC12 cells in imaging dishes as described in **Section 3.1**, step 3.
2. Take the vial containing IC60 probe, prepare stock solution and then load the cells with this probe using the same procedure as described above in **Section 3.2**, steps 1–5, for the MitoXpress probe.
3. Prepare samples and imaging system as described in **Section 3.2**, steps 6–10.
4. Start with a bright field image of the sample for pre-focusing, using repetition mode with an exposure time of 5 ms.
5. Change to fluorescence mode. Using steady-state mode (continuous excitation) and exposure time of 300–1000 ms, take one fluorescence image of the sample. Fluorescent signal received by the camera should be in the region of 1000–3000 counts per pixel (*see* **Note 19**). Make adjustments of the focus and sensitivity/scale. Repeat the adjustment taking more images if required.
6. Program the software for time-lapse experiments in the FLIM mode. Using ImSpector software, set up the parameters of phosphorescence lifetime measurements: excitation pulse width, camera modulation, gate time, and modulation frequency (ratio of the gate time and repetition time) (*see* **Note 20**). General scheme of phosphorescence decay measurement implemented in our system is shown in **Fig. 16.3**. Typical settings for the IC60 probe are given in **Table 16.1**.
7. Perform one set of measurements to generate phosphorescence decay curve for the ROI containing loaded cells. Make final adjustment of the sensitivity and intensity signals generated by FLIM. Intensity at the first delay time, which is a function of the exposure time, excitation pulse length, and x-y binning, should be in the range 1000–3000 counts (*see* **Note 21**). Signal intensity can be increased by binning; however, this will decrease image resolution.
8. Move the sample to a new field with cells and commence FLIM measurements. Each time measure

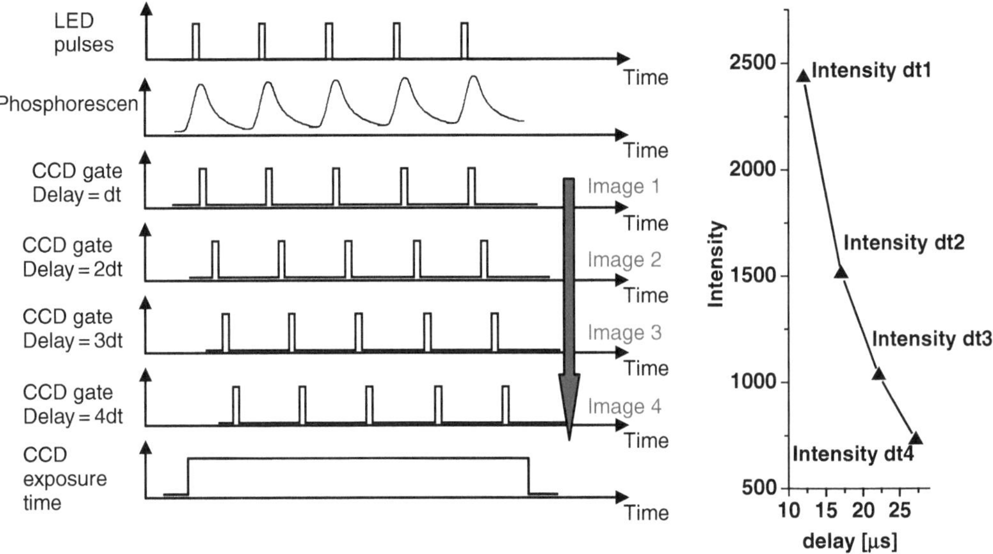

Fig. 16.3. Schematic description of FLIM measurements and the main parameters used: LED pulsing rate, gate time on the CCD camera, CCD camera exposure time, delay time, and the number of points on the curve (*left panel*). Measured phosphorescence decay curve (*right panel*).

Table 16.1
Typical settings for FLIM measurements with the O_2 probe

Parameter	Setting
Exposure time	1–3 s
Pulse length	10 μs
Integration start time	2 μs after the pulse
Integration end time	Typically 60 μs
Number of measurement points	10–20 points
Gate time	10–30 μs
Cycle time	100 μs

phosphorescence decay of the whole image, repeating this every 2–3 min.

9. Between two measurements of the phosphorescence decay make effector additions (refer to **Section 3.2**, steps 16 and 17 for details) and continue acquisition of decay curves. After the required number of measurements, stop the acquisition.
10. Define ROIs within the sample which contain cells loaded with probe, and generate phosphorescence decay curves for each ROI and for each time point of kinetic measurement. Extrapolate these decay curves with single-exponential fits

and determine phosphorescence lifetime value for each measurement.

11. Plot lifetime values as a function of time for each of the ROI and analyze the resulting profiles. Examples of FLIM images and respiratory response of PC12 cells to the depletion of extracellular Ca^{2+} are shown in **Fig. 16.4**. Intensity profiles can also be generated from the images (*see* **Fig. 16.4b**). If required, phosphorescence lifetime profiles produced by FLIM can be converted into O_2 concentration scale, using probe calibration which is determined separately (*see* steps 12–16 below).

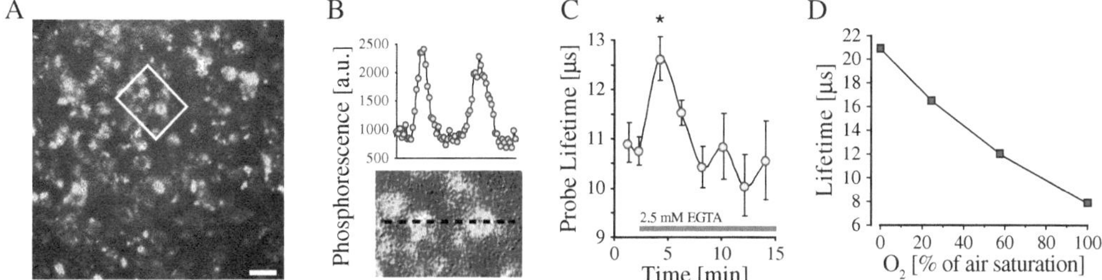

Fig. 16.4. Respiratory response of differentiated PC12 cells loaded with IC60 probe to the depletion of extracellular Ca^{2+} measured by FLIM. (**a**) Distribution of the probe intensity signal and selected ROI (*inset*). (**b**) Profile of phosphorescence intensity within ROI showing signal-to-noise ratio. (**c**) The response to the addition of EGTA (2.5 mM) with a transient increase in probe lifetime reflecting partial deoxygenation of the cell due to increased respiration ($p<0.01$ at the peak of the response). (**d**) The relationship between probe lifetime and O_2 concentration (calibration).

12. Take a fresh imaging dish (without glass cap) containing cells pre-loaded with the probe and 1–2 mL of medium and add Antimycin A to the medium at a final concentration of 10 μM. This should block cellular respiration and eliminate O_2 gradients in the sample.

13. Insert the dish in the incubation chamber with pre-set pO_2 (e.g., 3%) and incubate it for about 30 min to achieve temperature and gas equilibrium.

14. Prepare the microscope for FLIM measurements as described above (*see* steps 4–7 above). When the sample is ready, measure phosphorescence decay curve. Calculate the lifetime as described in step 10 above.

15. Change the incubation system to a new pO_2 setting and re-equilibrate the sample by incubating it for ~30 min. Move the sample/microscope stage to a new field with cells and perform lifetime measurement again. Repeat this at 3–4 other pO_2 settings.

16. Plot the relationship between phosphorescence lifetime of the probe loaded intracellularly and O_2 concentration. Adjust it to the desired O_2 scale: at 37°C and normal

ambient pressure, 20.5% O_2 in the chamber corresponds to 100% of air saturation or 200 μM O_2. Use the resulting relationship to convert measured phosphorescent lifetime profiles into profiles of cellular O_2 concentration.

3.4. Imaging in Live Cell Arrays of Suspension Jurkat Cells Loaded with the IC60 Probe

1. Reconstitute 3 × package of IC60 probe (300 nmoles dried in a vial) in 0.3 mL of medium to produce a 10-μM probe stock.
2. Take in a plastic vial 200 μL of suspension of Jurkat cells (1×10^6 cells/mL) and add 20 μL of probe and 1.2 μL of Endoporter stock solutions. Incubate at 37°C, 5% CO_2 for 24 h (*see* **Note 10**).
3. After incubation, wash the cells three times by centrifuging them at 200*g*, removing supernatant, and finally resuspend the cells in 0.1 mL of medium. Keep this suspension (approximately $\sim 2 \times 10^6$ cells/mL) at 37°C, 5% CO_2 for further use.
4. Switch on the Olympus IX51 microscope. Take a new LCA device, place it on the stage of the Olympus microscope, and allow to warm up and equilibrate at 37°C for about 20 min (*see* **Note 13**).
5. Using a P20 pipette load the LCA with 2.5 μL of suspension of cells prepared in step 3 above. Incubate for 5 min at 37°C and then flush with 15 μL of fresh medium.
6. Using transmission mode, focus on the microwells with cells within the LCA. Ensure that the device is loaded properly and the majority of the microwells contain trapped cells (*see* **Note 22**). Sample images of the LCA with cells are shown in **Fig. 16.5a, b**.
7. Switch the microscope to fluorescence mode, adjust the focus and initiate image acquisition in time-lapse mode. Take images every 20–60 s for about 10 min (*see* **Note 14**), using excitation at 395 nm and collection of emission with a 600 nm cut-off filter. Make sure that cell loading with probe is sufficient, adjust the sensitivity and scale as required.
8. After basal respiration (resting cells) is recorded, pause monitoring and apply effector treatment. Take solution of effector in medium (diluted to the required final concentration) and apply with a Gilson P20 pipette 15 μL of this solution to the loading bay of the LCA. After the effector is loaded in the LCA, resume monitoring for the required period of time (*see* **Note 23**).
9. Select several ROIs within the LCA – controls with cells – and analyze measured profiles and respiratory responses. If required, use the well without cells as a reference (blank).

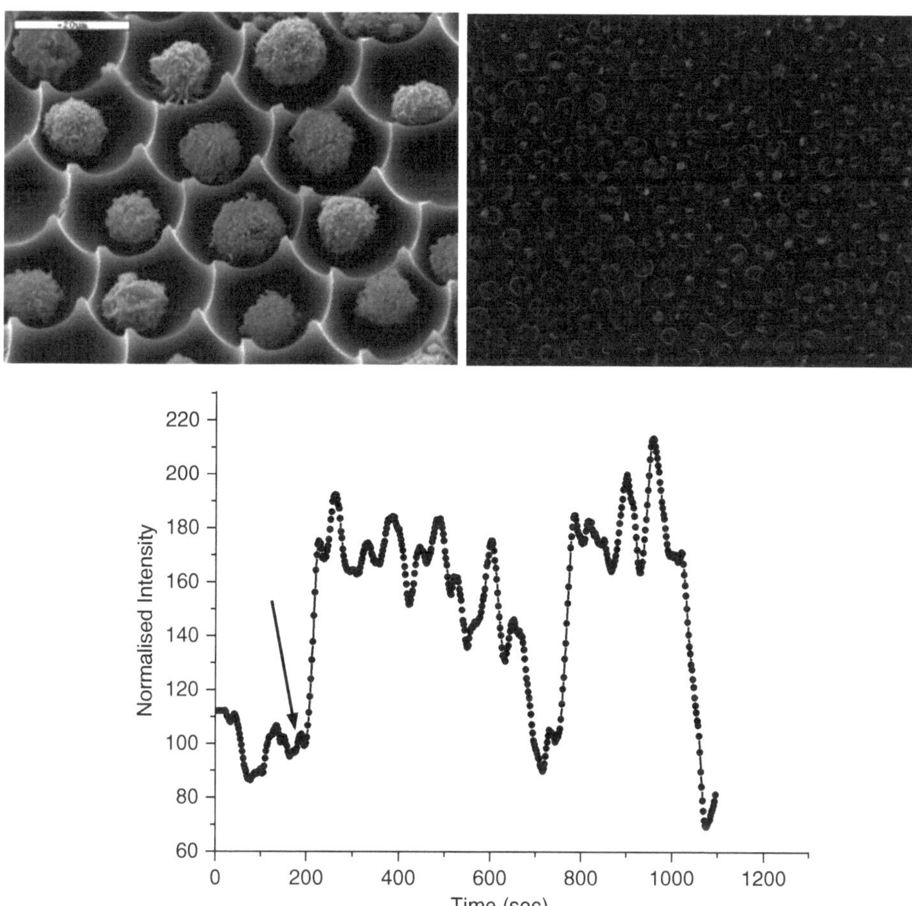

Fig. 16.5. (**a**) Photograph of a working section of Live Cell Array unit used for imaging of suspension cells (reproduced from (24) with permission of the Royal Society of Chemistry). (**b**) Fluorescent image showing individual Jurkat cells loaded with the O_2 probe trapped in the cavities of LCA. (**c**) Metabolic response of individual Jurkat cell to the addition of mitochondrial uncoupler FCCP (1 μM), measured using the LCA.

For interpretation of measured signal profiles and respiratory responses of cells, refer to **Section 3.2**, step 18. Example of response of individual Jurkat cell to FCCP treatment is shown in **Fig. 16.5c**.

4. Notes

1. A range of common cell lines can be used for imaging intracellular O_2; however, probe loading efficiency and metabolic activity may vary greatly for different cell types. Cells having well-developed mitochondrial machinery and

high rates of O_2 consumption (e.g., primary cells) are the preferred choice.

2. Growth medium, additives, and optimal culturing conditions may vary for each particular cell type.
3. Other O_2 probes for intracellular use, including those which do not require transfection reagents, are also available or in development. Contact corresponding suppliers (e.g., Luxcel Biosciences, www.luxcel.com) and research labs to get advice on the most suitable probe for your particular imaging system, measurement task, cell type.
4. In our experience, probe transfection with EndoPorter works well with many common cell types. Other loading reagents and procedures may be efficient as well. Generally, probe loading method and procedure need to be optimized for each cell type and medium.
5. Instead of the imaging dishes with cylinders, glass-bottom 96-well imaging microplates, multiwell flexiPERM inserts (Greiner Bio, Germany) can also be used. Devices with larger well size (e.g., 24-well plates) are less preferred.
6. Equivalent live cell fluorescent imaging systems, which are spectrally compatible and sensitive enough with the O_2 probes, can also be used. The MitoXpress™ probe is excitable at either 370–395 or 525–545 nm, and emits around 630–700 nm (maximum at 650 nm). The IC60 probe is excitable at 380–400 as well as at 580–600 nm, and it emits in the near infrared range 720–850 nm (maximum at 760 nm). IC60 probe is more photostable; however, its very longwave emission and lower brightness (compared to the MitoXpress™) can be problematic for some imaging systems.
7. Relatively high surface density of cells and high respiratory activity are the main requirements of a successful metabolic assay using the intracellular O_2 probes. Cells having high density of mitochondria and high levels of respiration/oxidative phosphorylation such as PC12 and HCT116 produce robust responses to stimulation with various metabolic effectors, whereas slowly respiring and highly glycolytic cells would be difficult to analyze by this method.
8. Optimal conditions for culturing depend on particular cell type. Differentiation is important for many cells to display a range of functional characteristics observed in vivo.
9. Probe loading by facilitated endocytosis in the presence of Endoporter reagent does not impose special requirements on the medium used. Other loading reagents and methods may require serum-free conditions.

10. Probe loading with EndoPorter is relatively slow; however, it works well in different media and cell types and produces high fluorescent signals which are easy to measure.
11. Extreme care should be taken not to damage the cell layer and wash the cells off the surface.
12. The cylinder is required to stabilize local oxygen gradients at cell layer by reducing surface area of the medium which in turn reduces convection and mass exchange within the sample. Based on our experience, this set-up produces better quality results.
13. Efficient temperature control of the measurement system and equilibration of the sample are very important. All O_2 probes produce fluorescence response to temperature fluctuations.
14. During these preparations, sample exposure to excitation light should be kept to a minimum. Photostability of the MitoXpress probe is moderate, whereas the IC60 probe is more photostable but less bright.
15. Optimal effector concentrations which produce maximal metabolic response depend on the cell type. We normally use 0.5–5 μM for FCCP and 10 μM for Antimycin A (in the well).
16. Mixing and disturbing the sample during the addition should be avoided as this will affect local O_2 gradients at cell monolayer and, hence, the probe signal.
17. Compared to mitochondrial inhibitors, uncouplers and activators of metabolism produce strong and positive optical response which is easier to measure. If basal respiration of cell monolayer is low and unable to create local O_2 gradient, metabolic responses may not be detectable, even though they are present. This is determined by the principle of operation of the intracellular O_2 assay (10). The lack or the absence of optical response from the sample/cells is a rather common case. This can be due to low respiration rate so that the cells are unable to create measurable local O_2 gradients and changes in response to stimulation. Cell damage during sample preparation, phototoxicity, or probe photobleaching during the measurement can also occur. The following can be used to improve assay sensitivity: (i) increase cell confluence (overgrow cells); (ii) increase cell respiration by using glucose-free medium containing 10 mM of galactose and pyruvate (12); (iii) increase the height of medium layer in the cap; (iv) conduct imaging experiments under reduced pO_2 (hypoxia chamber set at 10% or 5% pO_2); (v) optimize cell culture and probe loading conditions.

18. If probe photobleaching is high, measured profiles should be corrected for it. The initial part of the profile, which corresponds to basal respiration, can be used to quantify the rate of photobleaching.
19. Try several different settings in steps 4 and 5, and select those which provide lowest photobleaching during image acquisition.
20. For our system, delay time is calculated from the ratio of the effective measurement time (between start and end of one cycle) and the number of measurement points less one point. Cycle time is the time between LED pulses. Gate is integration time of the fluorescence signal for each delay time.
21. To get approximately the same level of signal from the sample in the FLIM mode, exposure time on the camera needs to be increased approximately five times compared to the steady-state measurement mode.
22. If loading of the LCA with cells is poor, step 8 can be repeated.
23. The time of monitoring of metabolic responses of cells is determined by the mode of action of the effector. It is also limited by the changes in cell viability during the measurement (i.e., phototoxicity). Most of our imaging experiments were conducted over a period of 20–60 min. Since both MitoXpress and IC60 probes do not leak from the cells, longer experiments with reduced sampling frequency are possible.

Acknowledgments

Financial support of this work by the European Commission (FP7 project CP-FP214706-2) and by the Science Foundation of Ireland grant (07/IN.1/B1804) is gratefully acknowledged.

References

1. Aragonés J, Fraisl P, Baes M, Carmeliet P (2009). Oxygen sensors at the crossroad of metabolism. Cell Metabol **9**(1):11–22.
2. Duchen MR (2004). Mitochondria in health and disease: perspectives on a new mitochondrial biology. Mol Aspects Med **25**:365–451.
3. Dykens JA, Will Y (2007). The significance of mitochondrial toxicity testing in drug development. Drug Discov Today **12**:777–85.
4. Nicholls DG (2005). Mitochondria and calcium signaling. Cell Calcium **38**: 311–7.
5. Hynes J, Marroquin LD, Ogurtsov VI, Christiansen KN, Stevens GJ, Papkovsky DB Will Y (2006). Investigation of drug-induced mitochondrial toxicity using fluorescence-based oxygen-sensitive probes. Toxicol Sci **92**:186–200.

6. Matoba S, Kang JG, Patino WD, Wragg A, Boehm M, Gavrilova O, Hurley PJ, Bunz F, Hwang PM (2006). p53 regulates mitochondrial respiration. Science **312**:1650–3.
7. Dunphy I, Vinogradov SA and Wilson DF (2002) Oxyphor R2 and G2: phosphors for measuring oxygen by oxygen-dependent quenching of phosphorescence. Anal Biochem **310**:191–198.
8. Papkovsky DB (2004). Methods in optical oxygen sensing: protocols and critical analyses. Meth Enzymol **381**:715–35.
9. Rumsey WL, Vanderkooi JM, Wilson DF (1988). Imaging of phosphorescence: a novel method for measuring oxygen distribution in perfused tissue. Science **241**:1649–51.
10. Hynes J, O'Riordan TC, Papkovsky DB (2008). The use of oxygen sensitive fluorescent probes for the assessment of mitochondrial function. In JA Dykens and Y Will (Eds.), Mitochondrial Dysfunction in Drug Induced Toxicity. Wiley & Son, New York.
11. Will Y, Hynes J, Ogurtsov VI, Papkovsky DB (2006). Analysis of mitochondrial function using phosphorescent oxygen-sensitive probes. Nat Protoc **1**(6):2563–72.
12. Marroquin LD, Hynes J, Dykens JA, Jamieson JD, Will Y (2007). Circumventing the Crabtree effect: replacing media glucose with galactose increases susceptibility of HepG2 cells to mitochondrial toxicants. Toxicol Sci **97**:539–47.
13. Wu M, Neilson A, Swift AL, Moran R, Tamagnine J, Parslow D, Armistead S, Lemire K, Orrell J, Teich J, Chomicz S, Ferrick DA (2007). Multiparameter metabolic analysis reveals a close link between attenuated mitochondrial bioenergetic function and enhanced glycolysis dependency in human tumor cells. Am J Physiol Cell Physiol **292**(1):C125–36.
14. Hynes J, O'Riordan TC, Zhdanov AV, Uray G, Will Y, Papkovsky DB (2009). In vitro analysis of cell metabolism using a long-decay ph-sensitive probe and extracellular acidification assay. Anal Biochem **390**(1):21–8.
15. O'Hagan KA, Cocchiglia S, Zhdanov AV, Tambawala MM, Cummins EP, Monfared M, Agbor TA, Garvey JF, Papkovsky DB, Taylor CT, Allan BB. (2009). PGC-1alpha is coupled to HIF-1alpha-dependent gene expression by increasing mitochondrial oxygen consumption in skeletal muscle cells. Proc Natl Acad Sci USA **106**(7):2188–93.
16. O'Riordan TC, Fitzgerald K, Ponomarev GV, Mackrill J, Hynes J, Taylor C, Papkovsky DB. (2007). Sensing intracellular oxygen using near-infrared phosphorescent probes and live-cell fluorescence imaging. Am J Physiol Regul Integr Comp Physiol **292**:R1613–20.
17. Ji J, Rosenzweig N, Jones I and Rosenzweig Z (2001). Molecular oxygen-sensitive fluorescent lipobeads for intracellular oxygen measurements in murine macrophages. Anal Chem **73**:3521–3527.
18. Xu H, Aylott JW, Kopelman R, Miller TJ and Philbert MA (2001).A real-time ratiometric method for the determination of molecular oxygen inside living cells using sol-gel-based spherical optical nanosensors with applications to rat C6 glioma. Anal Chem **73**: 4124–4133.
19. Koo YE, Cao Y, Kopelman R, Koo SM, Brasuel M and Philbert MA (2004). Real-time measurements of dissolved oxygen inside live cells by organically modified silicate fluorescent nanosensors. Anal Chem **76**: 2498–2505.
20. Schmalzlin E, van Dongen JT, Klimant I, Marmodee B, Steup M, Fisahn J, Geigenberger P, Lohmannsroben HG (2005). An optical multifrequency phase-modulation method using microbeads for measuring intracellular oxygen concentrations in plants. Biophys J **89**:1339–1345.
21. O'Riordan TC, Zhdanov AV, Ponomarev GV and Papkovsky DB (2007). Analysis of intracellular oxygen and metabolic responses of mammalian cells by time-resolved fluorometry. Anal Chem **79**:9414–9.
22. Finikova OS, Lebedev AY, Aprelev A, Troxler T, Gao F, Garnacho C, Muro S, Hochstrasser RM, Vinogradov SA (2008). Oxygen microscopy by two-photon-excited phosphorescence. Chemphyschem **9**(12): 1673–9.
23. Zhdanov AV, Ward MW, Prehn JH and Papkovsky DB 2008. Dynamics of intracellular oxygen in PC12 Cells upon stimulation of neurotransmission. J Biol Chem **283**: 5650–61.
24. Deutsch M, Deutsch A, Shirihai O, Hurevich I, Afrimzon E, Shafran Y, Zurgil N (2006). A novel miniature cell retainer for correlative high-content analysis of individual untethered non-adherent cells. Lab Chip **Aug;6**(8):995–1000.

Chapter 17

Analysis of Mitochondrial pH and Ion Concentrations

Martin vandeVen, Corina Balut, Szilvia Baron, Ilse Smets, Paul Steels, and Marcel Ameloot

Abstract

Detailed practical information is provided with emphasis on mapping cytosolic and mitochondrial pH, mitochondrial Na^+, and briefly also aspects related to mitochondrial Ca^{2+} measurements in living cells, as grown on (un)coated glass coverslips. This chapter lists (laser scanning confocal) microscope instrumentation and setup requirements for proper imaging conditions, cell holders, and an easy-to-use incubator stage. For the daily routine of preparing buffer and calibration solutions, extensive annotated protocols are provided. In addition, detailed measurement and image analysis protocols are given to routinely obtain optimum results with confidence, while avoiding a number of typical pitfalls.

Key words: Renal epithelial cells, MDCK, mitochondrial pH, cytosolic pH, mitochondrial sodium, mitochondrial calcium, SNARF-1, CoroNa Red, rhod 2, confocal microscopy.

1. Introduction

1.1. Scope of this Chapter

It is known that the transport function and fate of renal tubular cells are determined by the degree of ATP depletion (1), which will lead to a gradual increase in free cytosolic Ca^{2+}, due to the disturbance of ATP-dependent ion channels and the loss of activity of Ca^{2+}- and Na^+/K^+-ATPases. This will induce an activation of proteases and phospholipases, and will result in a progressive loss of cytoplasmic membrane integrity and eventually collapse of the cell. Prevention of increased Ca^{2+} influx in the renal tubular

The authors Martin vandeVen and Corina Balut have equally contributed

D.B. Papkovsky (ed.), *Live Cell Imaging*, Methods in Molecular Biology 591,
DOI 10.1007/978-1-60761-404-3_17, © Humana Press, a part of Springer Science+Business Media, LLC 2010

cells or mitochondrial sequestration will delay the onset of cell injury (2). Therefore we were interested in studying how mitochondria could buffer elevations in cytosolic Ca^{2+} in the presence of a disturbed cytosolic sodium and pH homeostasis. We used a culture of Madin–Darby canine kidney (MDCK) cells, inflicted with metabolic inhibition (MI) and assessed cytosolic and mitochondrial pH, and mitochondrial sodium and calcium concentrations. The results of these investigations have been reported elsewhere (3–5).

In this chapter we treat subsequently the protocols employed by our group to measure in living MDCK cells: (1) mitochondrial and cytosolic pH by dual emission confocal microscopy of carboxy SNARF-1; (2) mitochondrial Na^+ concentrations using CoroNa Red; (3) mitochondrial Ca^{2+} concentrations using rhod 2.

Methods and protocols as presented, take cellular and mitochondrial motility into account and may be applicable with some optimization to other cell types that adhere to glass coverslips, provided that the dyes can be loaded predominantly into the proper target cellular compartments. Our protocols can most likely not be applied to, for example, heart muscle cells since the number and density of mitochondria precludes the location of cytosol volumes free of contamination with mitochondrial signal.

1.2. Rationale Behind Choice and Properties of Fluorophore Dyes

1.2.1. MitoTracker Green Mitochondrial Stain

MitoTracker Green (MTG) is used to visualize the mitochondrial network. This is a cell-permeant dye, well retained by mitochondria at any mitochondrial membrane potential ($\Delta\Psi_m$). In methanol, its absorption maximum is close to the 488 nm laser line with an emission peaking near 515 nm, allowing the use of fluorescein filter sets. The dye is bright and allows mitochondrial imaging for extended periods of time when exposure to laser light is not too excessive (typically 10 μW at 488 nm and at the sample position).

1.2.2. SNARF-1 pH Indicator Dye

The methodology employed by our laboratory to measure the pH changes in the cytosol (pH_i) and mitochondria (pH_m) of living MDCK cells relies on using the pH-sensitive fluorophore 5-(and 6-)carboxy SNARF-1 (SNARF-1).

SNARF-1 is a fluorescent dye with a pK_a of approximately 7.3 at 37°C. This dual emission probe exhibits a large emission spectral shift in response to pH changes, making it suitable for monitoring pH_i and pH_m in living cells in the range of 6.8–7.8.

SNARF-1 is loaded into the cells as an acetoxymethyl ester (AM) analog, which is readily permeable to cell membranes (6). The ultimate intracellular distribution of the dye is dependent on the activity of cytosolic and organelle esterases relative to the rate of uptake of the AM form of the dye into the cellular compartments. Post-incubation of SNARF-1 loaded cells for 2.5–3 h results in the accumulation of the dye into the mitochondrial compartment (7).

When excited at 568 nm, SNARF-1 demonstrates an isosbestic point at 585 nm and pH-dependent changes in emission intensity at 640 nm. Estimation of pH from the ratio of emission intensities at the two wavelengths has the advantage to correct for shortcomings such as photobleaching and leakage of the dye. Photobleaching appeared fairly strong for this dye loaded into MDCK cells. Limitation of the total exposure time to approximately 10 min is one of the aspects of the protocols listed.

1.2.3. CoroNa Red Na^+ Concentration Indicator Dye

Mitochondrial Na^+ concentrations ($[Na^+]_m$) can be monitored with CoroNa Red, a bright cationic dye that can be directly loaded into cells without using an AM ester precursor. Cellular uptake is rapid and produces predominantly mitochondrial localization (4). The dye has a good photostability as the fluorescence intensity decreases only about 1–3% after 30 times illumination. Since many physiological processes may involve the decrease of the $\Delta\Psi_m$ (e.g., ischemia), it is important to verify that even in extreme conditions the cationic CoroNa Red dye is retained in the mitochondrial matrix. Experiments using the mitochondrial uncoupler carbonyl-cyanide-4-(trifluoromethoxy)-phenylhydrazone (FCCP) prove that changes in $\Delta\Psi_m$ did not affect substantially the mitochondrial retention of CoroNa Red (4).

The spectral properties allow simultaneous imaging in the presence of MTG. Absorption and emission maxima occur at 554 and 578 nm. This allows for excitation with the 543 nm Green He-Ne laser line and detection with standard emission filters suitable for tetramethylrhodamine (TRITC).

The probe does not possess ratiometric properties preferred for easy correction and elimination of systematic errors. For MDCK cells the protocols were adjusted such that image series could be collected without an overwhelming influence of photobleaching.

1.2.4. Rhod 2 Ca^{2+} Concentration Indicator Dye

Mitochondrial Ca^{2+} concentrations ($[Ca^{2+}]_m$) can be monitored with rhod 2. The AM form is required for loading cells. Rhod 2-AM carries a delocalized positive charge and is readily taken up into polarized mitochondria. Upon hydrolysis of the ester moieties, the rhod 2-free acid remains trapped inside the mitochondria. An increase in $[Ca^{2+}]_m$ is reported by an increase in fluorescence intensity. Its spectral properties, excitation maximum at 540 nm and emission peak around 575 nm, allow imaging in the presence of the green MTG emitter dye. As rhod 2 is a strongly bleaching dye, we had to limit in our protocols the total exposure time to approximately 5 min. For more information about these dyes consult the website of the rhod 2 supplier Invitrogen: http://www.invitrogen.com/.

1.3. General Measurement Strategy

The basic protocol for measuring mitochondrial pH, Na^+, and Ca^{2+} concentrations involves the following general steps: cells are trypsinized and then plated onto glass coverslips and allowed to reach 100% confluency under cell-type required conditions. Next, cells are simultaneously loaded with the ion sensitive fluorophore of interest (e.g., SNARF-1, CoroNa Red, or rhod 2) and the ion insensitive mitochondrial stain MTG. This double labeling allows discrimination between cytosolic and mitochondrial pixels, and avoids challenges due to mitochondrial or cell motility. The glass coverslip containing the cell monolayer is mounted in a holder and visualized on the stage of an inverted confocal microscope. The images in the MTG and the ion sensitive indicator channels are recorded under conditions that ensure the minimization of dye bleaching and a good signal–to-noise ratio. The mitochondrial and cytosolic regions of interest are carefully selected using a "MTG masking" procedure. The fluorescence recordings are calibrated to reflect the ion concentrations of interest. Corrections for autofluorescence, emission detection channel cross talk, acquisition bleaching, and detector linearity are required to ensure the reliability of the data.

In **Sections 2** and **3**, we list required ingredients for cell and tissue culture growth and measurement media, cell support preparation, the various buffer solutions, and their preparation methods with caveats as to their practical use. Instrumentation and measurement protocol issues to consider for proper imaging are provided in detail to make reliable data collection feasible even for the novice user.

2. Materials

2.1. Cell Culture

2.1.1. Coverslip-Related Material

1. 24-mm no. 1.5 round glass coverslips (Menzel-Gläser, Braunschweig, Germany).
2. 100% analytic quality Normapur absolute ethanol (VWR Intl., Leuven, Belgium).
3. Forceps.
4. 6-well tissue culture plates.
5. Optionally, poly-L-lysine (Sigma), depending on cell type and support.

2.1.2. Cell Culture and Medium

1. Cells of interest, growing in cell culture. We have used MDCK cells intermediate between type I and type II (ohmic resistance of 400–500 Ωcm^2 (8)), low passage 23–27, kindly provided by Dr. H. De Smedt, Laboratory of Molecular and Cellular Signalling, KULeuven, Belgium.
2. Cell culture medium. For MDCK cells use a 1:1 mixture of DMEM and Ham's F-12 (Sigma or Gibco), supplemented with 10% fetal calf serum (Sigma-Aldrich), 14 mM L-glutamine, 25 mM $NaHCO_3$, 100 U/ml penicillin, and 100 μg/ml streptomycin (Sigma). Store in the dark at 4°C.
3. Cell-culture-grade trypsin 0.05% trypsin–EDTA solution (Sigma). Store as 2–3 ml aliquots in the freezer.
4. Sterile phosphate-buffered saline solution (PBS), pH 7.4.
5. Light microscope with 10× and 20× objectives for cell inspection and counting.
6. Humidified incubator at 37°C supplemented with 5% CO_2.

2.2. Fluorophores

1. MitoTracker Green (Invitrogen): 400 μM stock solution in dimethyl-sulfoxide (DMSO).
2. 5-(and 6-)carboxy SNARF-1 acetoxymethyl (AM) ester acetate (Invitrogen): 5 mM stock solution in DMSO.
3. CoroNa Red (Invitrogen): 1 mM stock solution in DMSO.
4. Rhod 2-AM (Invitrogen): 1 mM stock solution in DMSO.
5. For Na^+ and Ca^{2+} measurements Pluronic F-127 (Invitrogen) (0.025% (w/v) from 25% (w/v) stock solution in DMSO) was used to facilitate the solubilization of the dyes.
6. All fluorophores stored in darkness at –20°C are stable for months (*see* **Note 1**).
7. Use aliquot stock solutions to avoid repeated freeze–thaw cycles. Prepare the stock solutions and the aliquots under the hood, using sterile pipette tips and eppendorf tubes, to avoid any contamination of the samples. When preparing the aliquots, keep the stock solution tube wrapped in aluminium foil to prevent prolonged light exposure (*see* **Note 2** for preventing contamination of DMSO with water).

2.3. Solutions and Chemicals

2.3.1. Normal Saline Solution (NSS)

140 mM NaCl, 5 mM KCl, 1.5 mM $CaCl_2$, 1 mM $MgSO_4$, 10 mM HEPES, and 5.5 mM glucose (pH adjusted to 7.4 with Tris, at 37°C). HEPES and Tris buffer prevent pH alteration due to CO_2. Store solution at 4°C in the dark (*see* **Notes 3–5**).

2.3.2. pH Calibration Solution

1. High-K^+ buffer: 145 mM KCl, 1.5 mM $CaCl_2$, 1 mM $MgSO_4$, 10 mM HEPES, and 5.5 mM glucose.

2. Nigericin (Sigma) stock solution: 13 mM in ethanol (to equilibrate the pH gradient across the plasma membrane).
3. Carbonyl-cyanide-4-(trifluoromethoxy)-phenylhydrazone (FCCP, Sigma-Aldrich) stock solution: 10 mM in ethanol (to equilibrate pH gradients across the mitochondrial membrane).
4. Oligomycin (mixture of types A, B, C) (Sigma) stock solution: 5 mg/ml in ethanol (to inhibit the mitochondrial H^+-ATPase).

2.3.3. Na^+ Calibration Solutions

1. Solution 145 mM Na^+ (30 mM NaCl and 115 mM Na-gluconate) and no K^+.
2. Solution 145 mM K^+ (30 mM KCl and 115 mM K-gluconate) and no Na^+.
3. Both calibration solutions also contain 1.5 mM $CaCl_2$, 1 mM $MgSO_4$, 10 mM HEPES, and 5.5 mM glucose. pH is adjusted to 7.4 with Tris. Store all solutions at 4°C.
4. Gramicidin D (Sigma): 2 mM stock solution in ethanol (to equilibrate the Na^+ gradient across the plasma membrane).
5. Nigericin (Sigma): 13 mM stock solution in ethanol.
6. Monensin (Tocris): 10 mM stock in ethanol (to equilibrate Na^+ gradients across the mitochondrial membrane).
7. FCCP: 1 mM stock solution in DMSO.
8. Oligomycin (Tocris, Bristol, UK): 5 mg/ml stock in ethanol.
9. Store nigericin, FCCP, gramicidin, and monensin stock solutions at 4°C. The oligomycin stock solution should be stored as aliquots at –20°C.

3. Methods

3.1. Cell Culture

3.1.1. Coverslip Preparation

1. In a tissue culture hood, under sterile conditions, soak the coverslips in ethanol possessing a very low autofluorescence and briefly flame them, to remove any drops of ethanol. Place the sterilized coverslips in a 6-well plate.
2. For cells that do not adhere very well on glass, the coverslip surface should be treated with poly-L-lysine: cover each sterilized coverslip with several drops of 1× poly-L-lysine solution and place the dish in the incubator at 37°C for 15 min. Next, place the dish in the tissue culture hood, using the overhead ultraviolet lamps, for another 15 min. Wash twice with sterile PBS.

3.1.2. Cell Culture Preparation

1. MDCK cells are maintained in a humidified 5% CO_2 atmosphere at 37°C. The growth medium is renewed every 3–4 days.
2. Deplate and split apart the cells by trypsinolysis (or the splitting method appropriate for that cell type). Add 2–3 ml 0.05% trypsin–EDTA for 10 min typically.
3. Collect floating cells. Centrifuge and resuspend in medium.
4. Perform the cell counting in order to plate the cells on the coverslips at a density of 0.5–2 × 10^5 cells/coverslip for pH, Na^+, and Ca^{2+} imaging. The droplet containing the cells has to be spread on the coverslip during the seeding, to have an equal distribution of cells over the whole coverslip.
5. Place the 6-well plates containing the seeded coverslips in the incubator (37°C, 5% CO_2) and use them after 3–5 days of culturing as follows: on the third day the monolayers seeded at 2 × 10^5 cells/coverslip; on the fourth and fifth day the monolayers seeded at 1 × 10^5 cells and 0.5 × 10^5 cells/coverslip, respectively. The moment of harvesting is determined by regular visual inspection on a sterile microscope stage under low magnification and looking for monolayer confluency without any dome formation (**Fig. 17.1**). Cell growth rate depends on cell type and passage number (*see* **Notes 6** and 7 on seeding density and passage number and **Note 8** on checking for mycoplasma contamination).

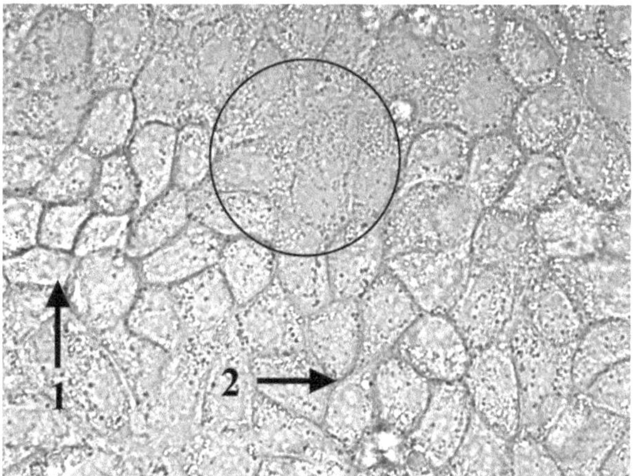

Fig. 17.1. Passage of MDCK cells after 5 days of growing (37°C, 5% CO_2). Changes in the monolayer morphology due to changes in the transport properties of the cells. Areas with cells forming domes (arrow 1) and patches of rounded cells (arrow 2) should be avoided during the measurement. Typically confluent areas, well attached to the glass are preferred for imaging (circled cells). Images collected with a Zeiss LSM 510 META confocal laser scanning microscope with 63×/oil objective, zoom 1, image size: 108.14 μm × 146.25 μm.

3.2. Cell Coverslip Holder and Imaging Hardware Material

1. Standard one-photon Zeiss LSM 510 META confocal laser scanning microscope (CLSM) with analog detection attached to an Axiovert 200 M frame (Zeiss, Jena, Germany), or equivalent.
2. Select 63×/1.4 Plan Apochromat oil-immersion objective or similar, with a suitable objective heater (PeCon GmbH, Erbach, Germany). This is meant to minimize the temperature differences between sample and objective, which otherwise impede image quality.
3. Install incubator adapted for the microscope stage, to maintain the cells at 37°C during the imaging. Our microscope is equipped with a model P type S (small) heated specimen holder with type S incubator (PeCon GmbH, Erbach-Bach, Germany) (*see* **Fig. 17.2c**).
4. Prepare holder and mount the glass coverslip covered with a nice confluent cell layer as checked with a standard transmission microscope image. We have used a homemade holder, but commercially available models are provided by most of the manufacturers of imaging systems. A layout of the in-house holder we have used is shown in **Fig. 17.2a and** an illustration of the complete assembly showing the plastic insert spacer with rubber seal and stainless-steel clamping cap is given in **Fig. 17.2b. Figure 17.2c** shows the assembly inside a small incubator mounted on top of the confocal stage.

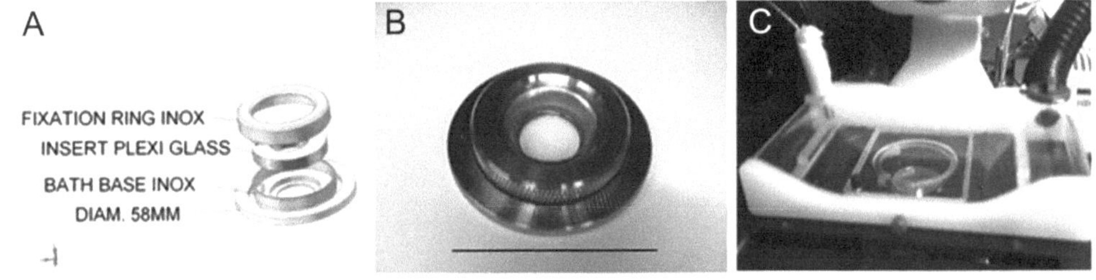

Fig. 17.2. (a) Exploded view of the homemade cell holder with plastic spacer clamp and screw-on cap. (b) Completely assembled unit showing the plastic insert spacer with rubber seal and stainless-steel clamping ring. (c) Cell holder unit mounted inside PeCon GmbH heated incubator stage with warmed air overflow for 37°C verified sample temperature. Rear left of small incubator: Pt temperature sensor feedback. 63×/1.4 oil objective is also heated with a heating coil to minimize temperature differences between cells and environment. Device base diameter (bar) 58 mm, assembled height 15 mm (b).

3.3. Loading Protocols

3.3.1. SNARF-1 and MTG Loading for pH_i and pH_m Measurements

1. This procedure is to be performed immediately prior to pH_i and pH_m measurements. Serum contains esterases that may cleave the AM groups on SNARF-1. Therefore, all steps of the loading procedure should be carried out in serum-free medium. We have chosen to perform the loading in NSS.

2. Monolayers should be simultaneously loaded with SNARF-1 and the mitochondrial stain MTG. This allows during the processing step to discriminate between cytosolic and mitochondrial SNARF-1 pixels, and to avoid challenges due to mitochondrial or cell motility. Two loading procedures were developed: *Protocol 1*, where mitochondria are predominantly stained and *Protocol 2*, where SNARF-1 is loaded into both mitochondria and cytosol.
3. *Protocol 1:* Pre-warm the NSS solution at 37°C. Gently wash cells two times with NSS, to remove any serum traces from the tissue. Add on top of the cells 1 ml of NSS containing 10 μM SNARF-1 and 200 nM MTG from the stock solutions. Allow 30 min incubation time for the dyes to be internalized. Next, wash the monolayers two times with 1 ml NSS, and incubate for 2.5 h in NSS, to allow for complete hydrolysis and preferential mitochondrial compartmentalization of the dye. At the end of the incubation time wash the cells two times with NSS. Mount the coverslip on the microscope holder (**Fig. 17.2**). Place 1 ml of fresh NSS on top of the cells. Cover the top of the holder with a transparent plastic dish to prevent untimely evaporation and proceed with imaging (*see* **Note 9** on cell washing protocol).
4. *Protocol 2:* Wash cells with NSS. Load the cells for 30 min with 1 ml of NSS containing 10 μM SNARF-1 and 200 nM MTG. Wash the monolayers with NSS and allow 1.5 h incubation period, then apply a second loading step with 5 μM SNARF-1 for 30 min. Wash off the dye and incubate the cells in fresh NSS for another 30 min to allow for hydrolysis of the dye into the cytosol. Perform a two times wash with NSS in between each loading step (*see* **Note 10**).
5. Perform all the loading and washing steps in the 6-well plate, under the tissue culture hood. For all the required incubation times, keep the cells at 37°C in 5% CO_2.
6. The result of the two loading protocols is illustrated in **Fig. 17.3**.

3.3.2. CoroNa Red and MTG Loading for Mitochondrial Na^+ Determination

Cells are washed with NSS, then incubated simultaneously with 200 nM MTG and 2 μM CoroNa Red in 1 ml of NSS that contained 0.025% wt/vol pluronic F-127 for 30 min at 37°C. Loading has to proceed in the dark or with a dim red roomlight in order to protect the fluorescence dyes from bleaching.

3.3.3. Rhod 2-AM and MTG Loading for Mitochondrial Ca^{2+} Measurements

1. Cells were washed gently with NSS twice. Loading of MDCK cells with MTG and rhod 2 occurred in two subsequent steps. At first, MTG (200 nM) was loaded into the cells for 30 min at 37°C. This was followed by incubation

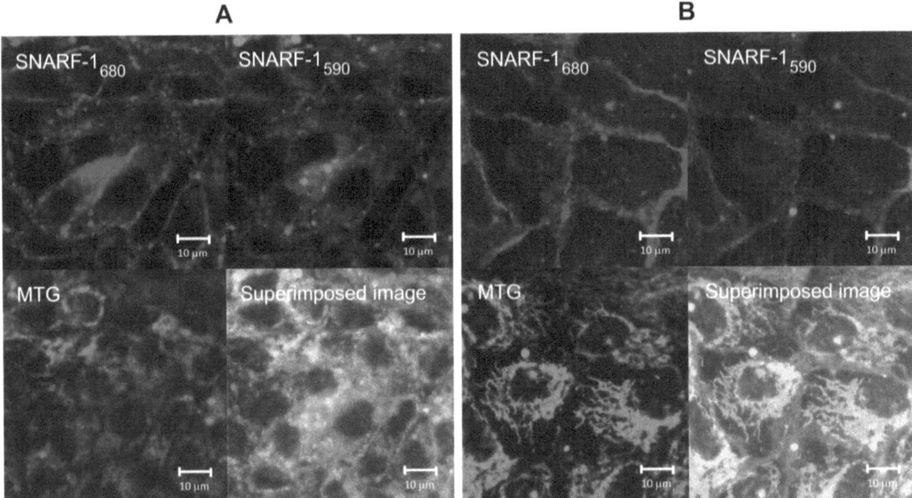

Fig. 17.3. Images of confluent monolayers of MDCK cells immediately after loading with both SNARF-1 and MitoTracker Green (MTG). Shown are emission detection channels for SNARF-1 (680 and 590 nm) and MTG (525 nm) according to two loading protocols: (**a**) Protocol 1; (**b**) Protocol 2 (for details see text). The superimposed images in each panel represent an overlay of the three others. The images in panels A and B were measured with identical settings of the confocal microscope. Reproduced with permission from *Kidney Int* (3).

of the cells for 30 min at 37°C in a loading solution containing rhod 2-AM (4 μM), MTG (200 nM), and pluronic acid (0.025% w/v) (5). Loading has to proceed in the dark or with a dim red roomlight.

2. Performing the loading procedure on the microscope stage, even in a standard atmosphere, offers the distinct advantage of imaging the background on exactly the same cells for later background subtraction. The buffer composition (HEPES and Tris) will maintain a constant pH independent of the presence of CO_2. For prolonged loading protocols like for SNARF-1 (3 h), it is better to maintain the cells in conditions that will ensure their normal physiological behavior.

3.4. Hints for Image Collection on Adherent Motile Cells

3.4.1. Optimization of Confocal Settings

1. The main tasks involve the following: (a) juggling the optimization of pinhole size, optical slice thickness for maximum detail and contrast; (b) minimizing fluorochrome bleaching and cellular stress. At the same time the cells and tissues have to stay in viable condition during imaging; (c) dealing with motile cells which may render automation of image processing protocols impractical and manual selection of regions of interest may become an arduous task; (d) obtaining workable signal and low background levels with a sustainable signal-to-noise ratio to numerically extract information from collected images.

2. Typical mitochondrial size is of the order of 200 nm. This is very close to the optical resolution of a microscope equipped with a 63×/1.4 oil objective. One may expect for 488 nm excitation and detection at 520 nm a radius for the Airy disk (influences the optical resolution) in the *xy* plane of 223 nm and in the Z-direction about three times worse: 669 nm. Reported values for the shape factor S (ratio of *z*-axis axial/*xy* plane radial Airy radius) may even be 5 or 7 and vary slightly from day to day. For 633-nm excitation with fluorescence emission observed at 650 nm these numbers relax about 25%. For maximum detail in living cells and tissues with excellent contrast, a small pinhole size is necessary. Even with minimal pinhole size (1 Airy), images of the mitochondrial network stay rather fuzzy even after deconvolution. Moreover, in the Z-direction the spatial nature of this network creates overlap. For studying mitochondria, an extended confluent layer of cells with a vast thin cytosol layer will make life easy.
3. When more than one detection channel is used, pinhole size has to be adjusted for each channel in order to have the same observed optical slice thickness for all emission channels. An emission channel consists of the collection of all optical elements, pinholes, lenses, optical filters, and detector. Using a confocal pinhole in front of the detector providing optical section capabilities to a microscope results in both stray-light and fluorescence reduction from layers not in focus, as well as an increase in image crispness. Look always for user-friendly features in CLSM software packages that will show the pinhole size and optical slice thickness directly in micrometers.
4. Photobleaching is always present and varies for each fluorochrome used. It is recommended to always perform pilot studies not only to optimize the loading protocol but also to evaluate the bleaching properties of the stains and changes in mitochondrial shape or levels of autofluorescence. For our measurements on MDCK cells, we had to decide to drastically minimize the number of exposures to laser light. Once the illumination protocol is optimized one can gain confidence in the validity of the collected images. With each new plating cycle the level of autofluorescence should be checked again for proper image correction. Damage caused by the illumination intensity will be somewhat relaxed when using a Nipkow type spinning disk which utilizes lamp illumination and parallel multiple pinholes of fixed diameter.
5. Both one- and two-photon illumination give rise to photochemical and photophysical effects, i.e., bleaching of fluorophores and photo-induced stress on cells. This stress will

show as a retraction and clustering of the mitochondrial network to the nuclear perimeter. In worse cases mitochondrial swelling may be observed with an increase in autofluorescence.

6. Practically the fluorescence signals may be very low (partially on purpose because of the minimal illumination to limit photobleaching) but should stay above the background signal level. Confocal microscope pinholes are usually NOT set to obtain diffraction-limited performance, but their settings should allow collection of images with a reasonable signal-to-noise ratio.
7. For our mitochondria images the "software derived" optical slice thickness was <2.4 μm. **Fig. 17.4c, d** show radial *xy* and axial *z*-direction point spread functions as measured on sub-resolution fixed 175-nm beads (Invitrogen, Merelbeke, Belgium) under experimental conditions as similar as possible with respect to the cell or tissue measurements. This apparently relatively poor resolution along the *z*-axis due to a rather extended detector pinhole size is a compromise in order to observe weak fluorescence with acceptable signal-to-noise ratio for extended periods of time.

3.4.2. Image Collection Settings

This section will present what settings one should use to image pH_i, pH_m, and mitochondrial ion concentrations in living MDCK cells on a standard one-photon Zeiss LSM 510 META CLSM or similar instrumentation. These settings, chosen to minimize bleaching and get a reasonable signal for the fluorophores used (listed in **Section 2.2**), should be tested and optimized if other cell types or confocal imaging systems are to be used.

1. Collect 512 × 512-pixel 8-bit (or 12 bit for better dynamic range) images, averaged twice (via software-selected repeated line scan mode) to improve the signal-to-noise ratio. A bit depth of 8 carries 256 gray values with black being zero and white being 255. In gray scale images pixel values below about 50 appear pitch black to the eye. It is very helpful during image collection and processing to switch to the Zeiss supplied Rainbow2 Look Up Table (LUT). With this LUT differences between a glass background and autofluorescence levels become easy to detect.
2. All Na^+ images were collected with a digital zoom factor of 1, while pH and Ca^{2+} images were collected with a digital zoom factor of 2. Pixel dwell time was always 25.6 μs.
3. MTG was excited by an Ar laser (488 nm line, 1% = 10 μW laser power at the sample position). SNARF-1, CoroNa Red and rhod 2 were excited with a Green HeNe laser (543 nm line, respectively 10 and 3.5% of 0.1 μW at the sample position).

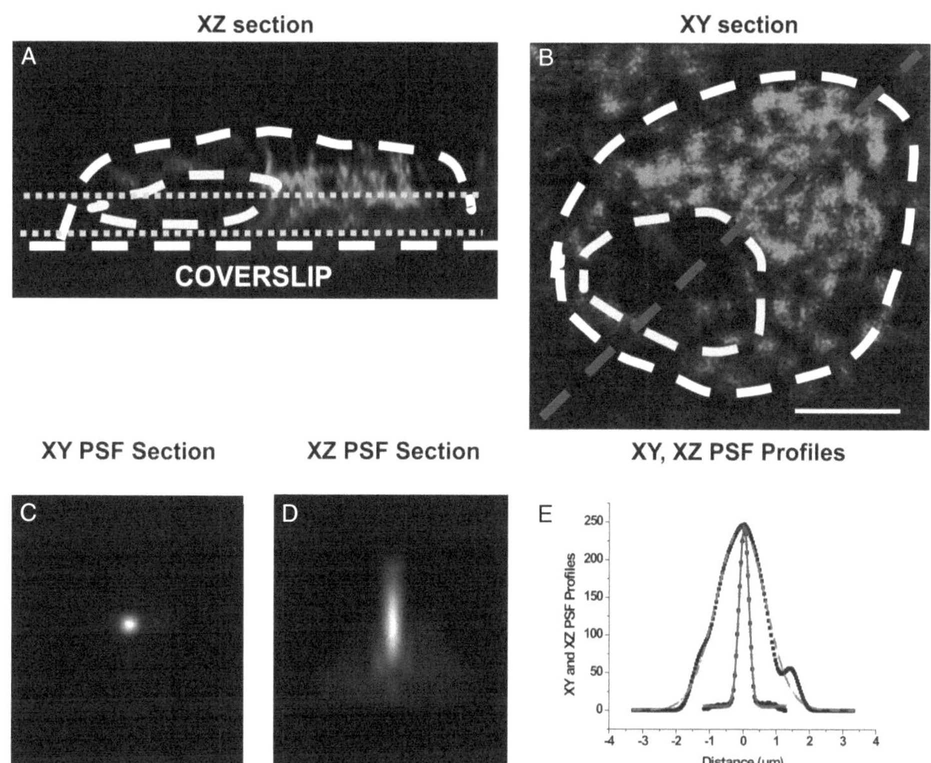

Fig. 17.4. Pixel selection in MDCK cells. Diagonal *xz* cross-section (**a**) and *xy* optical slice near support (**b**) as collected. Scale bar 10 μm. Image collection parameters: Zeiss 63×/1.4 oil plan-ApoChromat objective, zoom 2, pixel dwell time 6.4 μs, pinhole 325 μm for 488 nm excitation, 0.8% 488 nm excitation laser light with 10 μW on sample. Designation for the demarcation lines: *horizontal dotted lines* (**a**) – actual optical slice position with thickness to scale; external *circular dashed line* (**b**) – cell perimeter; inner *circular dash line* (**b**) – nucleus of MDCK cell; *diagonal grey dashed line* (**b**) – position of diagonal vertical cross-section. Cytosolic SNARF-1 signal retrieval in cells loaded according to Protocol 2: selection of pixels underneath the nucleus not having any discernable MTG contribution after digital signal enhancement. Mitochondrial SNARF-1 signal retrieval in cells loaded according to Protocol 1. Panels (**c**) and (**d**): *xy* and *xz* point spread function (PSF) cross-sections at 488 nm excitation for averaged unresolved microbead intensity images. Panel (**e**) horizontal (*xy*) $1/e^2$ width 0.56 μm, vertical (*xz*) $1/e^2$ full width 3.3 μm as obtained by Gaussian curve fit with $R^2 = 0.99$. Estimated error is 5%. *xz* for slice thickness (Zeiss software indicates slice thickness as < 2.4 μm). Theory predicts for the FWHM width of a diffraction-limited spot with 488 nm excitation 0.19 μm (*xy* plane) and 0.45 μm (*xz* plane). Reproduced with permission from *Kidney Int* (3).

4. SNARF-1 and rhod 2 are fast bleaching dyes, while CoroNa Red and MTG are very stable. Therefore, set the scanning sequence in such a way to capture first the fluorescence emission of the fastest bleaching dye, followed by the less bleaching one. For pH measurements: scan first SNARF-1 at 590 and 680 nm, through the 545 dichroic mirror with 565–615 and 655–705 nm bandpass (BP) filters, respectively. For $[Na^+]_m$ measurements: scan CoroNa Red via the 545 dichroic and 560 nm longpass (LP) filter. For $[Ca^{2+}]_m$ measurements: scan rhod 2 via the 545 dichroic and 590/25 nm BP filter. For all protocols, collect the MTG

signal the last, using a 490 dichroic and a 525/25 nm BP filter.

5. To minimize photobleaching and phototoxicity as much as possible while still maintaining discernable image detail toward the end of our experiments, laser power had to be kept as low as possible. At the same time, for pH imaging a 325-μm-diameter (~3 Airy units) pinhole setting had to be used for 488 nm and was adjusted accordingly for 543 nm excitation. This ensured an equal optical slice thickness of <2.4 μm (*see* **Note 11**). For Na^{+} and Ca^{2+} measurements, the thickness of the optical slices was <1.4 μm for all experiments.
6. The detector gain should be optimized to avoid saturation in any pixel while maintaining a proper signal-to-background noise ratio. When measuring pH_m (cells loaded as in Protocol 1) and also mitochondrial Na^{+} and Ca^{+}, the detector gain was always kept constant (e.g., 1100 for pH_m measurements). During the measurement of pH_i (cells loaded as in Protocol 2), the detector gain had to be adjusted between 800 and 1100 (Zeiss software settings) due to the rapid loss of dye from the cytosol. In this case, detector gain settings have to be separately calibrated (*see* **Section 3.4.3**).
7. For pH measurements: to facilitate the subsequent image analysis (see below), it is strongly recommended to choose the observation volume close to the basolateral membrane. In this way, the optical slice thickness enables part of the cytosol to be visualized underneath the nucleus region (**Fig. 17.4a**).

3.4.3. Required Image Intensity Corrections

Typical image intensity corrections include removal of the influence of autofluorescence, cross-talk between detection channels, effects of photobleaching, photomultiplier (PMT) detector linearity, and gain settings. SNARF-1 related pH image corrections serve as an example.

1. *Corrections for autofluorescence.* The autofluorescence levels should be measured on blank, unloaded cells, under similar experimental conditions and microscope settings as for the loaded cells. For blank cells, the pattern of mitochondrial areas and cytosolic areas underneath the nucleus appear clearly distinct in the SNARF-1 detection channels, allowing the background autofluorescence level to be measured for each compartment. Use Palette – Rainbow mode in the Zeiss LSM Image browser, to easily visualize the mitochondrial pattern of blank cells (**Fig. 17.5**). Use the image processing protocol as described in **Section 3.5.4** to calculate the background autofluorescence values related to the mitochondria and cytosol respectively. *See* **Note 12** on the importance of autofluorescence corrections, especially for

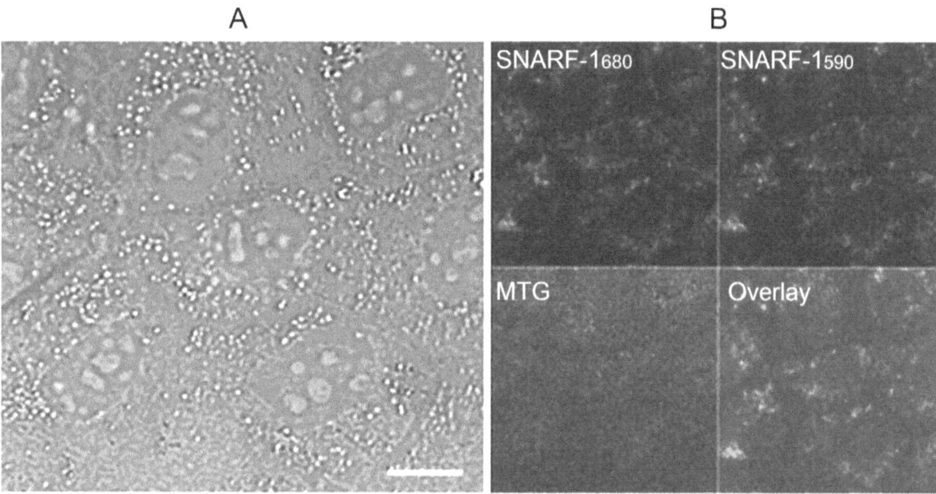

Fig. 17.5. Confluent monolayer of unlabeled MDCK cells, passage 25. (**a**) Transmission image; (**b**) background-signal on the three measured channels. All four panels display in gray-scale the Rainbow LUT provided by the Zeiss software. Cells were observed at detector gain 1100 (Zeiss software settings). Scale bar = 10 μm.

the pH protocol. In our measurements the mitochondrial background was consistently higher than the cytosolic background, although constant for the observed monolayers and individual passage numbers. Averaged values corresponding to mitochondria- and cytosol-related pixels were used to correct for the respective background signal. Glass coverslip background with buffer signal contributions can be estimated from areas free of any cells. Typically this background signal was 16 17 on an 8-bit (0–255) scale. Whenever the protocol requires the adjustment of the detector gain values during the measurement, the autofluorescence values corresponding to mitochondrial and cytosolic regions should be measured on blank cells at the respective detector gain values and used accordingly for background subtraction. **Figure 17.6** presents the changes in mitochondrial autofluorescence as a function of the detector gain, under our experimental conditions. Cytosolic autofluorescence was less sensitive to the changes in the detector gain, ranging between 17 and 20. Repeated exposure of cells to laser illumination may influence the level of autofluorescence of the cells and may prove to be phototoxic to them. In our experimental protocol, the autofluorescence proved to be constant over the duration of the experiment with identical illumination protocols, both for untreated control cells as well as for cells exposed to MI. Autofluorescence in CoroNa Red and rhod 2 images was always very small (*see* **Note 13** on published conditions to avoid phototoxicity).

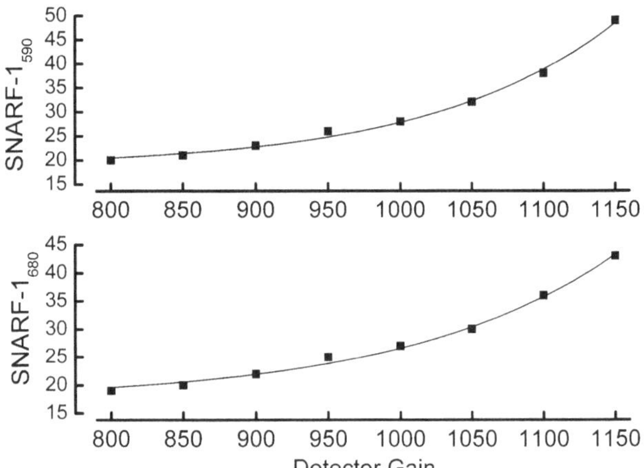

Fig. 17.6. Autofluorescence intensity (arbitrary units) of mitochondrial area as a function of detector gain values applied (Zeiss software settings) on each of the two SNARF-1 channels.

2. *Confocal laser scanning microscopy detection channel crosstalk.* To check for the crosstalk between the SNARF-1 and MTG channels, fluorescence should be measured in the corresponding channels for SNARF-1_{590}, SNARF-1_{680}, and MTG for monolayers loaded either only with SNARF-1 (following loading Protocol 1) or only with MTG. With the chosen settings, when SNARF-1 emission was measured in the MTG channel, only the autofluorescence signal was observed. Similarly, the MTG emission in the SNARF-1 channels only gave the basal background value, ensuring that there is no measurable crosstalk between the detection channels (**Fig. 17.7**). For the detection conditions as listed,

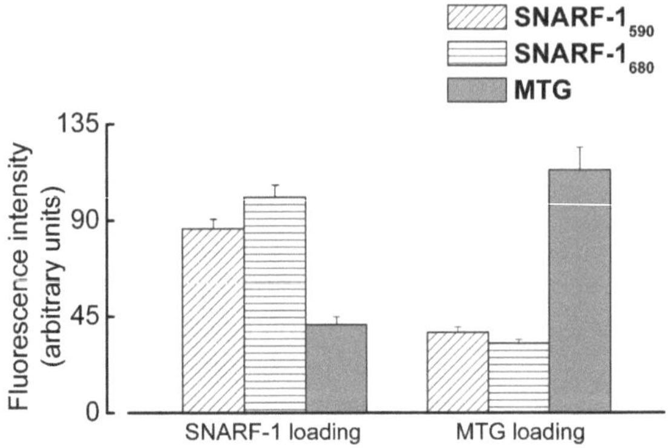

Fig. 17.7. Crosstalk check between the SNARF-1 and MTG channels. Fluorescence was measured in the corresponding channels for SNARF-1_{590}, SNARF-1_{680}, and MTG for monolayers loaded either only with SNARF-1 or only with MTG. Detector gain 1100 (Zeiss software settings). Values from four different monolayers are reported as mean $\pm$ SEM (standard error of the mean). Reproduced with permission from *Kidney Int* (3).

crosstalk was checked and found to be absent between MTG and CoroNa Red (Na^+) as well as rhod 2 (Ca^{2+}) and MTG detection channels.

3. *Photobleaching.* SNARF-1 and MTG bleaching should be checked on cells loaded separately with the individual dyes or when both were present. In our experimental protocol, MTG was never affected by bleaching. Data indicated pronounced bleaching of SNARF-1 during image sequence collection, but the SNARF-1 ratio (680 nm/590 nm) proved constant over time (**Fig. 17.8**). CoroNa Red was found to be very photostable, whereas rhod 2 rapidly bleached under our experimental conditions. Therefore the number of image collection events in rhod 2 loaded cells was restricted to four measurement points (*see* **Note 14**).
4. *Photomultiplier linearity.* To be able to see the weakened cytosolic SNARF-1 signals, PMT voltages should be adjusted and calibrated. PMT calibration was performed on

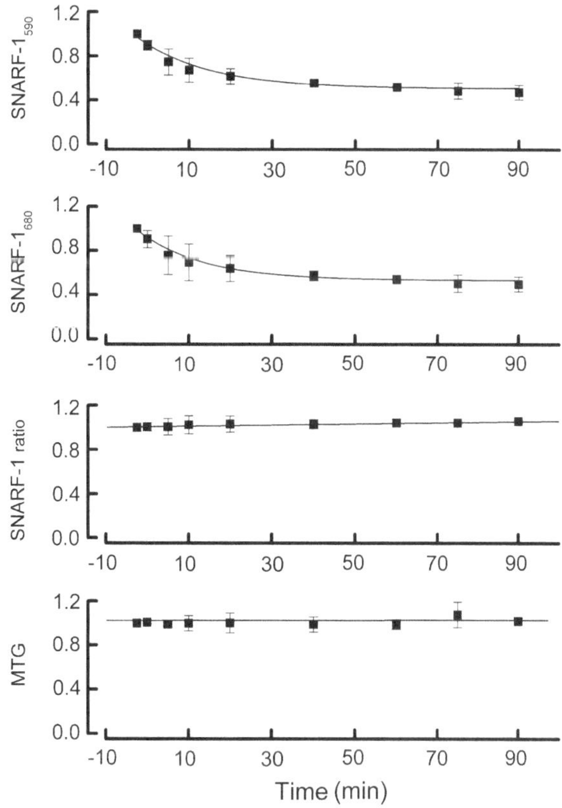

Fig. 17.8. Bleaching of SNARF-1 and MTG. Bleaching rates for SNARF-1_{590}, SNARF-1_{680}, Ratio SNARF-1_{680}/SNARF-1_{590}, and MTG measured under Protocol 1 loading conditions. Values are indicated as mean values ± SEM for two monolayers. Data are normalized in each case to the starting value to clearly show the trend over time. Reproduced with permission from *Kidney Int* (3).

cells loaded only with MTG, which showed no photobleaching for the conditions reported here. To obtain the calibration curve, the fluorescence in the MTG channel was recorded at different detector gain values in the range of 800–1100, while maintaining the same scanning parameters (laser intensities, pinhole, zoom, etc.) as presented in **Section 3.4.2**. An example of a typical calibration curve obtained is presented in **Fig. 17.9**. This PMT calibration was not important for CoroNa Red and rhod 2 experiments, since all images were detected with the same detector gain initially optimized to be in its linear response region.

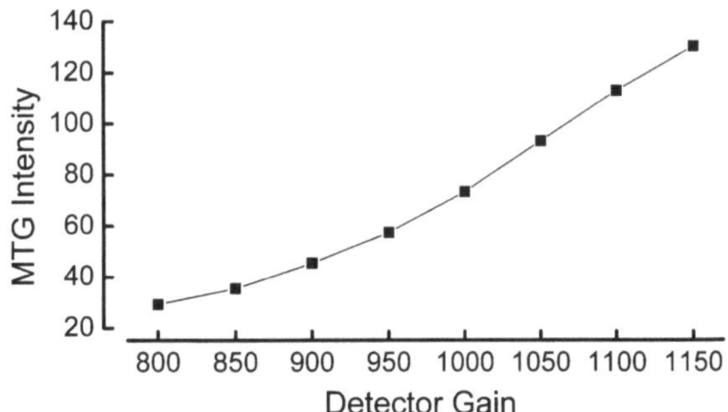

Fig. 17.9. Detector response and linearity as a function of detector gain applied (Zeiss software settings) as measured in the MTG channel.

3.5. Extracting Image Information

3.5.1. Software

For image transfer, conversion, and processing, the freeware Zeiss LSM Image Browser and ImageJ Java-based freeware (Research Services Branch, National Institute of Mental Health/National Institute of Neurological Disorders and Stroke, Bethesda, MD) plug-in routines were used. Excel (Microsoft Corp, Redmond, WA) and Origin (OriginLab Corp, Northhampton, MA) were used for statistical data analysis. Huygens Essential and the point distiller package (Scientific Volume Imaging, Hilversum, the Netherlands) was used for point spread function analysis.

3.5.2. Selection of Pixels Containing Mitochondrial pH Information Based on SNARF-1 and MTG Image Processing

1. Data collection. To avoid as much as possible contamination of the mitochondrial related pixels with cytosolic information, the mitochondrial SNARF-1 information was obtained from cells loaded with SNARF-1, according to Protocol 1. At each time point, a set of three sequential (hereafter called multi-track) images was collected under presented settings (**Section 3.4.2**).
2. Create stack images. Using the ImageJ LSM Toolbox plug-in (Dr. J. Mutterer et al., University of Strasbourg, France) or the Zeiss supplied export routine, convert raw confocal images to 8-bit TIFF files (TIFF = Tagged Image File Format, with extension .tif). Store the files corresponding

to each confocal channel, in a separate folder. Use number extensions to designate the time sequence of the files. In ImageJ use File→ Import to create a stack with the TIFF files measured on each confocal channel, for the entire experiment. Apply: Process → Noise → Despeckle. Save the despeckled stacks (e.g., SNARF590Stack.tif, SNARF680Stack.tif and MTGStack.tif) for subsequent analysis.

3. Eliminate saturated pixels by zeroing. Each of the stacks created for a certain experiment should be checked and corrected for any saturated pixels (occasionally present typically in the first image, since it was opted not to change standard detection settings once an experiment was under way). For this, open in ImageJ a Despeckled.tif stack. Go to Image → Adjust → Threshold. Set both low and high threshold sliders to the maximum pixel value: 255 (**Fig. 17.10a**). The saturated pixels will appear highlighted (arrows in **Fig. 17.10b**). Create the mask for saturated pixels: ImageJ→ Process → Binary → Make Binary. An image having all the saturated pixels 0 (black) and the remaining pixels 255 (white) will be created (**Fig. 17.10c**). Divide this image by the value 255, using Process → Math → Divide. This will create the mask image having all the saturated pixels zero (0) and the remaining desired ones with a value of one (1). Save this file (e.g., xxxxDespeckledSaturationMask.tif).

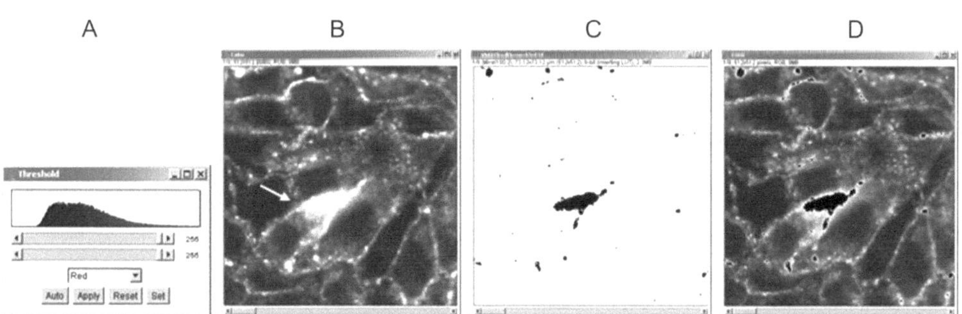

Fig. 17.10. The correction of saturated pixels exemplified on SNARF-1_{680} images. (**a**) Intensity histogram. (**b**) Highlighted saturated pixels (see *arrow*). (**c**) Mask image (255 = white and black = 0). (**d**) Saturation corrected image stack.

Multiply the original despeckled stack with the mask stack: Process → Image Calculator. The stack corrected for saturated pixels will be created in a new window (**Fig. 17.10d**). Save this file under a different name (e.g., xxxxDespeckled-SaturationCorrected.tif).

In the same way correct for any saturated pixels present in the other measured channels (SNARF-1_{590} and MTG).

4. Subtract background fluorescence. The background values corresponding to mitochondrial images should be

subtracted from SNARF-1_{680} and SNARF-1_{590} channels prior to any subsequent calculations. For this protocol, at the constant detector gain of 1100, the background values ranged between 37 and 38. The procedure for background values calculation is detailed in **Section 3.5.4**.

5. Create the MTG mask (*see* **Note 15**). MTG pixels related to a strong mitochondrial signal are selected by setting a high threshold. All pixel values in the MTG image between a variable lower limit (threshold value typically higher than 90–120) and the 255 maximum pixel value were used to create the mitochondrial mask. The careful pixel selection in the image analysis procedure allows obtaining mitochondrial SNARF-1 signals that are mainly of mitochondrial origin. The possibly few existing saturated pixels in the MTG image should be eliminated.

   After setting the threshold use Process → Binary → Make Binary. Then Process → Math → Divide by 255. This will create the MTG mask, having the mitochondria related pixels 1 and all the remaining ones 0. Save this file (e.g., MTGDespeckledMaskThres90.tif).

6. Mitochondrial SNARF-1_{680} and SNARF-1_{590} retrieval from MTG mask. Ratio calculation.

   Multiply the MTG mask with SNARF-1_{680} and SNARF-1_{590} stacks, respectively. Store the images in separate folders. To obtain the SNARF-1 ratio image, divide pixel by pixel the masked stacks SNARF-1_{680}/SNARF-1_{590}. Use: Process → Image Calculator → Divide (**Fig. 17.11**). Via Analyze → Measure, calculate the Ratio values corresponding to the entire stack. Save the result file as.txt for subsequent import in Excel or Origin and statistical analysis. pH_i and pH_m are expressed as the SNARF-1 (680 nm/590 nm) emission

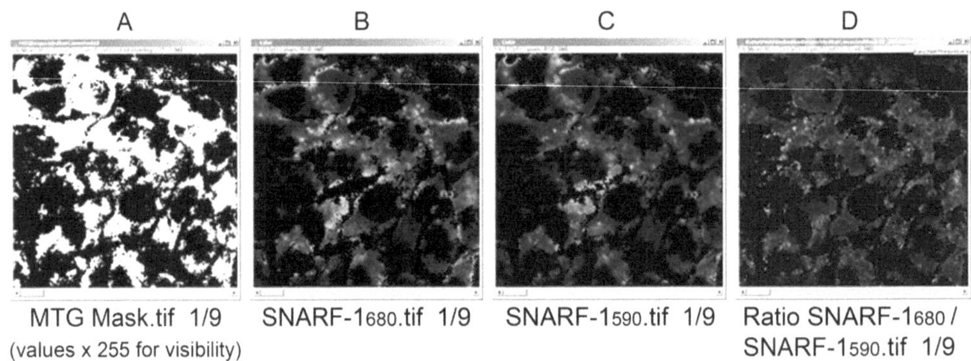

Fig. 17.11. MTG mask and retrieval of SNARF-1_{680}, SNARF-1_{590} mask images, and the ratio SNARF-1_{680}/SNARF-1_{590}.. First image from a stack of nine is shown.

ratio. A decrease in the SNARF-1 ratio represents a decrease in pH.

3.5.3. Selection of Pixels Containing Cytosolic pH Information Based on SNARF-1 Analysis in the Volume Beneath the Nucleus

1. Data collection. Based on the observation that SNARF-1 is not staining the nucleus at all, and given that the cytosol underneath the nuclear area contains very few mitochondria (**Figs. 17.3 and 17.4**), the cytosolic SNARF-1 intensity was measured by choosing only the SNARF-1 related pixels in the cytosol underneath the nucleus (**Fig. 17.4**), for cells loaded following Protocol 2. As mentioned, the observation volume was always chosen close to the basolateral membrane, the optical slice thickness enabling part of the cytosol to be visualized underneath the nucleus region. Collect at each time point a set of three multi-track images (SNARF-1_{680}, SNARF-1_{590}, and MTG, respectively).
2. Create the corresponding stacks for each channel, with data converted to 8-bit TIFF files, despeckled and corrected for any saturated pixels, as presented above in **Section 3.5.2**.
3. Subtract the corresponding background values from each SNARF-1 stack. For cytosolic pixels under the nucleus, this value was always low (~18–20) and was determined as described in **Section 3.5.4**.
4. Selection of the cytosolic pixels under the nucleus. To select the proper cytosol related region of interest (ROI) underneath the nucleus, the MTG signal should be set to maximum contrast via digital image enhancement and histogram stretching using the image analysis software (**Fig. 17.12a**). This prevents accidentally selecting pixels with a hardly discernable weak MTG signal. Store the enhanced MTG stack in a different file (e.g., xxxxEnhanced.tif). Next, per individual MTG image, delineate for each cell the darkest nucleus area (no trace of MTG) using ImageJ → Polygon symbol. Store the selected pixels in a separate folder using File\Save As\Selection\Polygon11.roi, Polygon12.roi, Polygon13.roi, etc. Typically, 5–15 small circular, elliptical, and hand-drawn ROI regions were selected. Check whether the chosen ROIs have any bright cell wall pixels in the corresponding SNARF-1_{680} and SNARF-1_{590} images (**Fig. 17.12b and c**). If necessary, adjust the ROIs to avoid wall pixels and store the newly defined ROI under the same name (*see* **Note 16**). Under Analyze\Tools\ROI Manager open all ROIs belonging to an image in a stack and Combine them. This binds the Polygon ROIs to a single composite one, in order to be able to open for each image in SNARF-1_{680} and SNARF-1_{590} all via MTG selected nucleus area pixels.

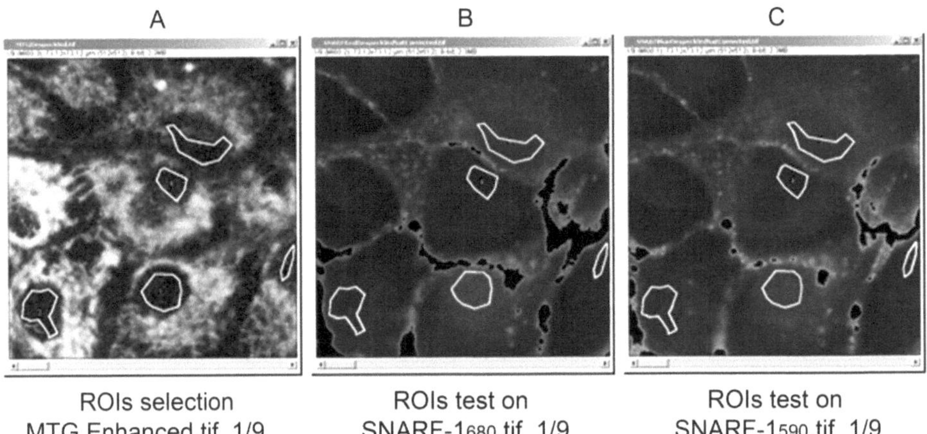

Fig. 17.12. Examples of the ROIs selection. First image of a stack of nine is shown. (**a**) MTG and (**b**) and (**c**) SNARF-1 images.

Repeat this procedure for all images in a MTG stack, starting again with the corresponding MTGEnhanced.tif file. It saves time to check whether the ROIs of the previous image (n–1) in the stack also fit the new MTG image, n, in the stack i.e., the cells did not move too much. When this is true, keeping the integrity of the Composite.roi, save Composite-n.roi. When, however, SNARF-1_{680} and SNARF-1_{590} show wall pixel overlap, the new Composite-n.roi has to be adjusted by changing the individual Polygon-(n–1).roi outlines.

5. Create the cytosolic pixels mask. Due to cell movement, it is necessary to create an individual cytosolic mask, based on the associated Composite.roi, for each image in the stack: in ImageJ create a new 8-bit TIFF file, then open the Composite-1.roi, for the first image in the stack. Invert and divide by 255. This will create an image having all the cytosol related pixels "1" (white) and the remaining ones "0" (black) (**Fig. 17.13**). Save as CytosolMask1.tif in a separate folder.

   Make in the same way, one by one, the CytosolMask-n.tif files, for the remainder of the images in the stack, using the associated Composite-n.roi.

   Bundle in a stack all the mask images, being careful to respect the correct sequence. Save in a separate file (e.g., CytosolMask.tif stack).

6. Cytosolic SNARF-1_{680} and SNARF-1_{590} retrieval from MTG mask and ratio calculation.

   Multiply the CytosolicMask.tif stack with SNARF-1_{680} and SNARF-1_{590} stacks, respectively. Store the images in separate folders. To obtain the SNARF-1 ratio image, divide

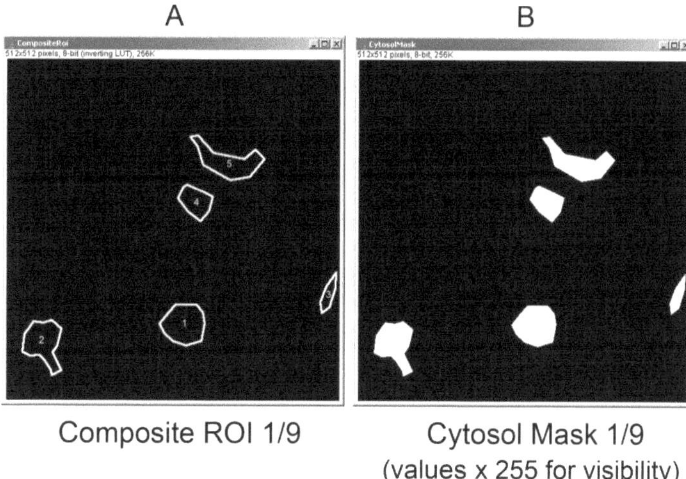

Fig. 17.13. (**a**) Example of a composite ROI for the first image out of a stack of nine. Each individual ROI in the composite image is numbered. This identification number allows retrieval of ROI statistics from the ROI export list feature in ImageJ; (**b**) cytosol areas free of any discernable MTG signal (*white* = 1) and area to be discarded (*black* = 0).

pixel by pixel the masked stacks SNARF-1_{680}/SNARF-1_{590}. Use: Process → Image Calculator → Divide. In Analyze → Measure, calculate the ratio values for each image in the stack. Save the result file as .txt for subsequent import in Excel or Origin and image statistics.

3.5.4. Estimation of Background Level for Mitochondrial and Cytosolic Areas on Blank Cells

This procedure is only required for the pH measurements protocol. It is not applied to the rhod 2 or the CoroNa Red experiments, which only use the brighter mitochondrial image regions outside the nucleus area.

After measuring the blank cells using the settings presented in **Section 3.4.2**, create the despeckled stacks from the images acquired for the three channels: SNARF-1_{680}, SNARF-1_{590}, and MTG.

1. Background values for mitochondrial pixels.
   Adjust via ImageJ the low threshold value, until the mitochondrial pattern becomes visible and well-defined dark nuclei areas are observed (**Fig. 17.14**). For our measurements, the low threshold value ranged between 20 and 30, depending on the value of detector gain during the measurement. Use Analyze → Measure and calculate the averaged values for the pixels related to mitochondria, at the selected threshold. Slide the image sequence cursor and calculate the corresponding values for all the images in a stack. Save the data as a .txt file. This will make it easier to import them into Excel or Origin for statistical analysis.

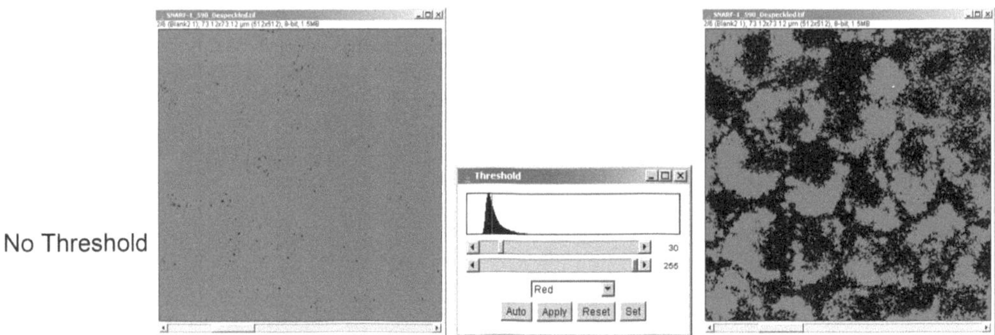

Fig. 17.14. Threshold selection to visualize the mitochondrial pattern on blank unstained cells.

2. Background for cytosolic pixels.
   To obtain the background values of the cytosol beneath the nucleus, apply the following sequence for both SNARF-1_{680} and SNARF-1_{590} stacks. Open stack, then enhance contrast, so all the bright areas related to mitochondria become clearly visible. Using the ROI manager in ImageJ, select small areas in the region of the nucleus, free of any discernable mitochondrial signal (**Fig. 17.15**). Combine all the ROIs selected for the first image in the stack in a composite image. Save it in a separate file, with the appropri-

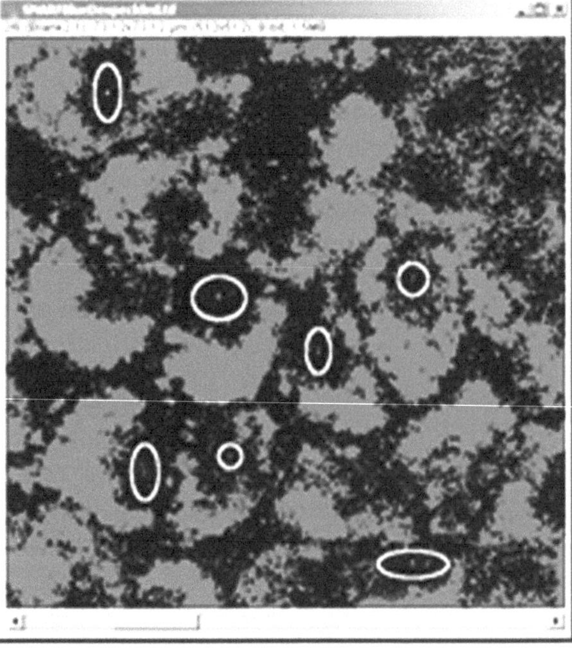

Fig. 17.15. ROIs selection for cytosol under nucleus areas on blank cells (SNARF-1_{590} channel). Each individual ROI in the composite image is numbered. This identification number allows retrieval of ROI statistics from the ROI export list feature in ImageJ.

ate extension number. Check if the Composite.roi applies to all the images in the stack. Due to cell movement, each image should be checked individually, and the position of the ROIs adjusted, to avoid any "contamination" with mitochondrial pixels. Save the adjusted ROI composite for each image in a separate file. Use Measure in RoiManager to calculate the averaged value for the pixels inside the selected ROIs for each image in the stack (i.e., cytosol beneath the nucleus).The background autofluorescence values obtained for our measurements, based on the above protocols, are presented in **Fig. 17.16**.

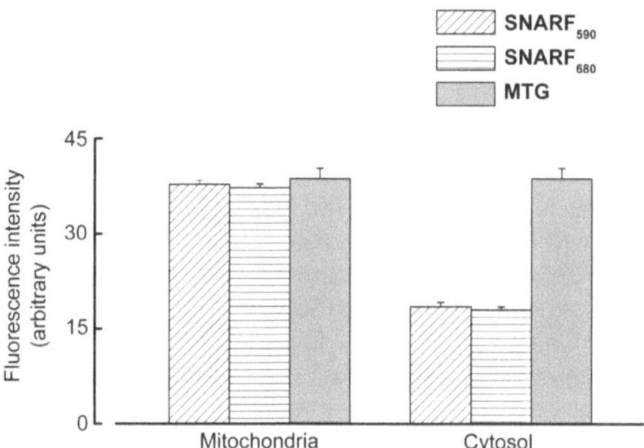

Fig. 17.16. Autofluorescence levels for mitochondria and cytosol. Autofluorescence was measured in the corresponding channels for SNARF-1_{590}, SNARF-1_{680}, and MTG based on the distinct pattern of mitochondrial and nuclear areas for unstained MDCK monolayers. Mean ± SE (*N*=6). Detector gain 1100 (Zeiss software settings). Reproduced with permission from *Kidney Int* (3).

Another way of calculating the background autofluorescence for mitochondrial area and cytosol beneath the nucleus, respectively, relies on the observation that there is no crosstalk between the MTG and SNARF-1 channels (**Fig. 17.7**). Therefore, by performing three multi-track measurements on cells loaded only with MTG, it is possible to create a very precise mask of the pixels related to mitochondria and the pixels beneath the nucleus, free of any mitochondrial contribution. Applying this mask on both SNARF-1_{680} and SNARF-1_{590} stacks will provide the background values for the mitochondria and cytosolic regions, respectively, during the entire measurement. This procedure is identical with the ones explained previously in **Sections 3.5.2** and **3.5.3**, respectively.

3.5.5. Selection of Pixels Containing Mitochondrial Na^+ Information Based on CoroNa Red and MTG Processing

The CoroNa Red fluorescence has to be corrected for non-mitochondrial CoroNa Red fluorescence, including the weak cytosolic CoroNa Red staining and the pronounced staining of

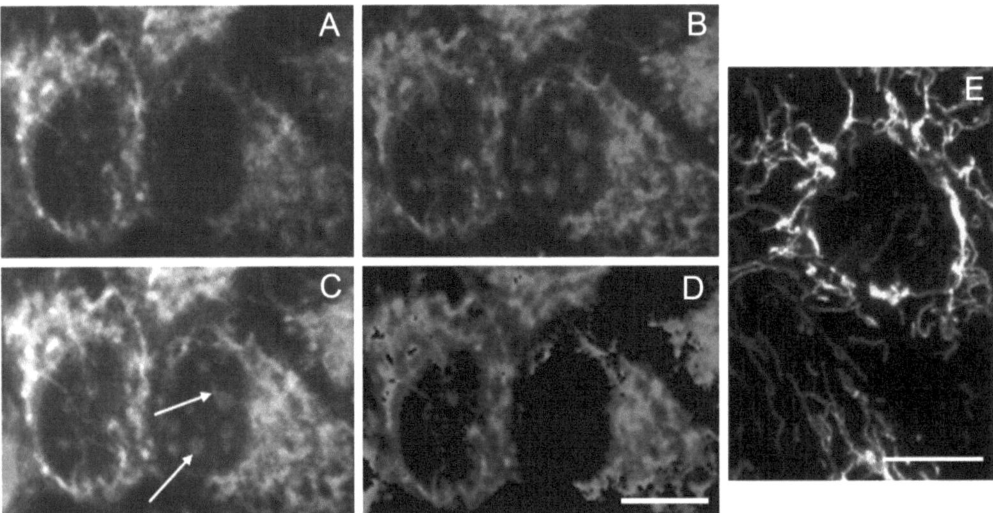

Fig. 17.17. Confocal images of MDCK cells loaded with MitoTracker Green (MTG) and CoroNa Red (**a–d**) and cells only loaded with MTG (**e**). MTG (**a**) and CoroNa Red (**b**) staining of mitochondria in MDCK cells. The staining with the Na^+-sensitive probe CoroNa Red was spatially correlated with the mitochondrial marker MTG (**c**), except for a weak CoroNa Red staining of the cytosol and a more pronounced staining of nuclear structures, presumably nucleoli (*arrows*). Application of the "mask" procedure (for details *see* **Section 3.5.5**) yielded an image consisting of only mitochondrial CoroNa Red fluorescence intensities (**d**) (scale bar=10 μm). (**e**) Mitochondrial staining in cells showing an extensive dynamic mitochondrial network under control conditions. Only weak rhod 2 staining of the cytosol (data shown in (5)) was observed and pronounced staining of nuclear structures similar to (**c**). Scale bar = 10 μm. (**a–d**) Reproduced with permission from *J Am Soc Nephrol* 2005 (4). (**e**) Reproduced with permission from the *Am J Physiol Renal Physiol* (5).

structures within the nuclei, presumably nucleoli (indicated by arrows in **Fig. 17.17c**). An MTG masking procedure was developed to retrieve only the mitochondrial CoroNa Red fluorescence intensity from the measured data (**Fig. 17.17d**).

1. Data collection in brief. Sets of multitrack images are collected: a CoroNa Red, followed by an MTG image under published conditions. Three multitrack images are collected for CoroNa Red at every time point to establish an average. An image set was collected every 3 min, for 1 h, since the dye proved to be very stable. Focus was readjusted when required.

2. Convert and export raw image data into uncompressed.tif. Split the multitrack.lsm confocal microscope images with Zeiss data collection software or use the LSM Toolbox in ImageJ. Put .tif converted CoroNa Red and MTG images with a sequence number in the filename in separate folders. ImageJ possesses convenient features, e.g., loading an entire stack upon selecting the first file in a folder. Using the File\Import\Image sequence feature in ImageJ image stacks are imported. Save as raw data MTGStack.tif and CoroNaRedStack.tif 8-bit (numerical values between 0 and

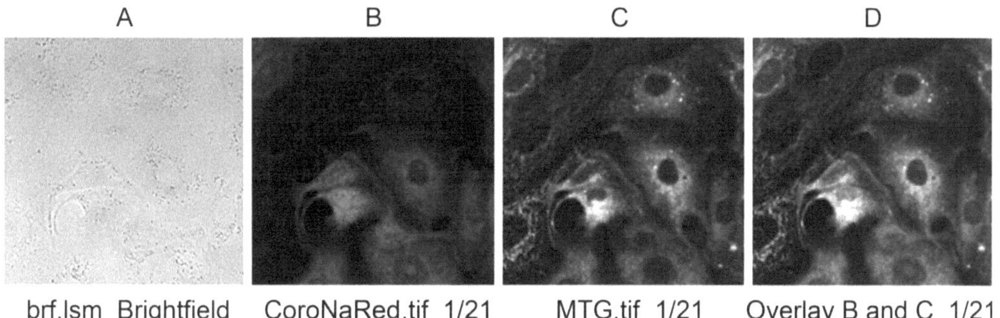

Fig. 17.18. Raw transmission and fluorescence images of MDCK cells loaded with CoroNa Red and MTG.

255) and as a merged StackRGB.tif for a color representation (**Fig. 17.18b–d**).

3. Despeckle raw image stacks.
4. Eliminate saturated pixels both for MTG (always required) and CoroNa Red (hardly ever). Set threshold to 255 numerical value. Process\Binary\Convert to Mask and Invert. Divide by 255 and multiply the mask with images.
5. Subtract autofluorescence of support microscope slide and buffer solution when required. Background autofluorescence intensity from mostly the glass support and maybe buffer solution contribution was always at most 16 or 17 as determined from the pixels with the lowest signal. These numbers were checked also on unstained and single-dye stained cells on a regular basis and after each new cell passage number. Save StackMaskSatBkgndCorrected and StackCoroNa RedSatBkgndCorrected image sets.
6. MTG "mask" creation. For StackMaskSatBkgndCorrected.tif select best rather high threshold 70à 85–200 with a strong mitochondrial contribution. Adjust threshold. Click Apply to convert to binary and divide by 255. Save as StackMask.tif (**Fig. 17.19a**) with numerical value one for mask pixels to be retained and zero otherwise (pixels not to be used). This includes pixels set to zero for accidental saturation, although the instrument settings used avoided this situation in general as much as possible.
7. CoroNa Red: retrieval of MTG mask related pixels. Multiply StackMask.tif with StackCoronaRedSatBkgndCorrected.tif. Save as Result.tif stack into AnalysisExp01 folder, for example (**Fig. 17.19b, c**). Note the differences in MTG area and the perinuclear location of saturated pixels regions in C as compared with B.
8. ROI selection. Open the Result.tif and StackMask.tif stacks. Select five (5) separate ROI areas on the first Result image (write down X, Y, W, H coordinates) (**Fig. 17.19d**).

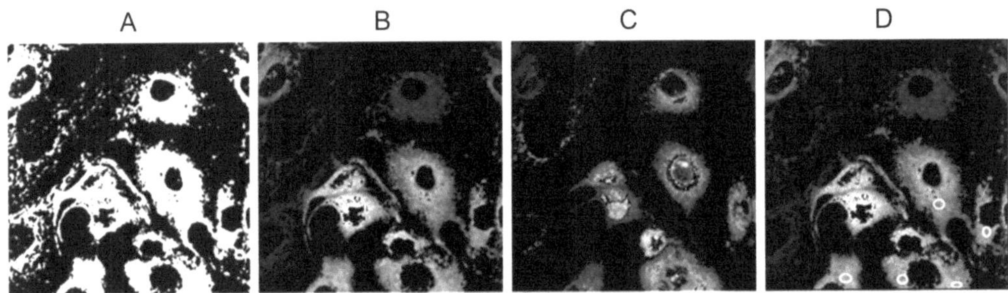

Fig. 17.19. MTG mask creation and retrieval of CoroNa Red mask images. (**a**) Binary MTG mask (×255 for visibility). A rather high threshold level ranging from 70 to 200 ensures selection of pixels with a strong mitochondrial contribution. (**b**, **c**) Masks applied to CoroNa Red images. Differences in the extend of saturated pixels in the perinuclear regions of 1/21 (**c**) as compared with 18/21 (**b**) due to contraction over time of the mitochondrial network. The notation *n/m* refers to the *n*th image in a stack comprising *m* images. (**d**) Selection of five separate ROI areas on the first result image (1/21). Bright areas are avoided altogether. Due to cellular motility and the requirement of validity over the whole stack, sizes of individual ROIs are small.

Use ROI manager. Avoid very bright areas. Check that each created ROI properly covers intended cellular areas over the whole image stack in order to avoid lacking or wrong area data due to cell motion.

For bleach correction and result normalization select StackMask.tif stack and Edit\Restore selection (ROI is transferred).

Create a *z*-axis intensity profile table via Image\Stacks\Plot Z-axis profile of StackMask.tif.

In the Result window: Edit\Copy All.

Paste the data from StackMask, Result.tif windows, and ROI summed intensity values as well as the ROI areas selected into Excel ∗.xls files.

9. Excel and Origin statistical analysis. Create three separate Excel sheets.

10. ROI normalisation procedure. Due to contraction of the mitochondrial network near the end of a MI experiment MTG image, areas may start to saturate. Saturated pixels are set to zero and are eliminated. The obtained CoroNa Red and rhod 2 fluorescence values for the ROIs were normalized with regard to the number of pixels with a mask value of one in the mask area to compensate for differences in mitochondrial density in each ROI of subsequent images in the stack. **Figure 17.20** shows an enlarged region of the ROI in the lower right corner of **Fig. 17.19d**.

    DataROI 1 (Raw *z*-axis intensity profile data). To be stored for both the MTG and CoroNa Red sets of, for example, five ROIs for each image in the stacks.

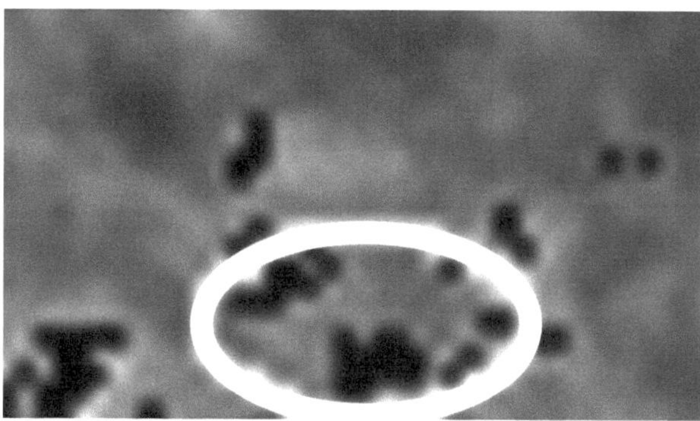

Fig. 17.20. Enlarged ROI showing MTG saturated pixels set to zero. Enlarged region around the ROI in the lower right corner of **Fig. 17.19d**.

SheetROI 1 (final table with averaged values + normalized percentages + error analysis).

3.5.6. Selection of Pixels Containing Mitochondrial Ca^{2+} Information Based on Rhod 2 and MTG Processing

Identical to the CoroNa Red protocol presented in **Section 3.5.5**. To minimize as much as possible the photobleaching of rhod 2, a multi-track series (consisting of one MTG image and one rhod 2 image) was collected every 20 min during 1 h. Focus was readjusted when required.

3.6. In Situ Dye Calibration Protocols

3.6.1. In Situ pH Calibration of SNARF-1

Given the fast bleaching of SNARF-1, it is not possible to perform the calibration at the end of each experiment, but this is done in separate measurements. Calibration of the ratio of the fluorescence signals from SNARF-1 is performed according to the method of Thomas et al. (9).

1. Load the cells with SNARF-1 in the cytosol and with SNARF-1 plus MTG in mitochondria, according to Protocol 2 (**Section 3.3.1**).
2. Prepare a solution of high-K^+ (**Section 2.3.2**) containing also 13 μM nigericin and 1 μM FCCP (to equilibrate the H^+ gradient across the plasma and mitochondria membrane, respectively) and 20 μg/ml oligomycin to inhibit the mitochondrial F_1F_0-ATPase.
3. Set the pH of this solution to four different values in the range of 6.8–7.8. The exact pH of each solution at 37°C should be recorded.
4. Place the holder containing the monolayer on the microscope stage. Remove the NSS medium from the cells and add the first pH calibrating solution. Allow 20 min for equilibration before measuring each pH point. When changing the pH calibration solutions, gently wash cells twice with the new pH solution to be measured.

5. To ensure that pH equilibration reached a steady state, each calibration point was measured after 20 and 30 min, respectively, of equilibration. At both time points the measured pH values were similar.
6. To avoid any possible systematic errors, perform the calibration procedure in both directions, from low to high pH and vice versa.
7. Following the mitochondrial "MTG mask" analysis, plot the calibration solution pH versus the SNARF-1 ratio in the cytosol and in the mitochondria, respectively, for all four pH points. Perform linear regression on the line obtained in each case to determine the slope (multiplier) and the *y*-intercept. Use these values to perform a linear scaling of the SNARF-1 fluorescence ratio.

The result of SNARF-1 calibration in living MDCK cells following the presented protocol are illustrated in **Fig. 17.21**.

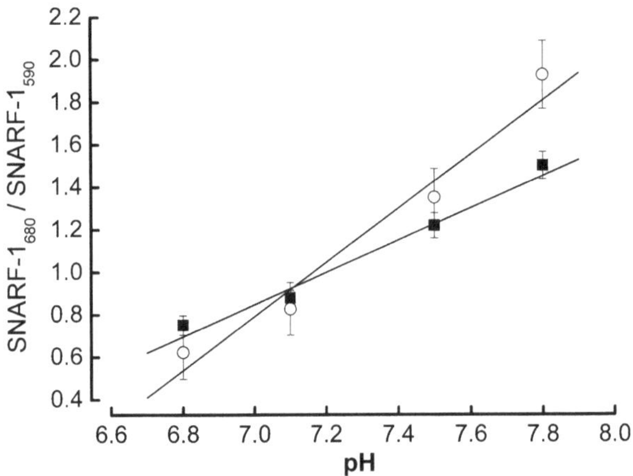

Fig. 17.21. pH calibration for the SNARF-1 signal. Averaged SNARF-1 ratios (680/590 nm) for cytosol (o) and mitochondria (■) were plotted against the pH of the bathing solution. *Error bars* indicate the SEM on the SNARF-1 ratio averaged over five experiments. Reproduced with permission from *Kidney Int* (3).

3.6.2. In Situ [Na⁺]m Calibration Protocol for CoroNa Red

1. Expose the cells to various extracellular Na^+ concentrations in the presence of 10 μM gramicidin D, 10 μM nigericin, 20 μM monensin, 1 μM FCCP, and additionally 20 μg/ml oligomycin (*see* **Note 17**). Allow 15 min of equilibration before measuring each [Na^+] point.
2. The solutions with various sodium concentrations were prepared by mixing in different proportions the two calibration solutions (**Section 2.3.3**) of equal ionic strength and osmolality. The sodium concentration of the solution was set to five different values in the range of 145 to 0 mM (145, 120, 72.5, 60, and 0). The plot of the normalized fluorescence

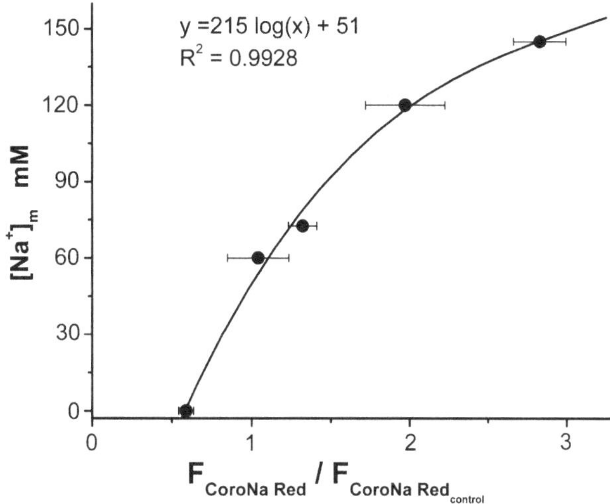

Fig. 17.22. In vivo calibration curve of the mitochondrial CoroNa Red signal in MDCK cells (4). Mitochondrial Na^+ concentration ($[Na^+]_m$) as a function of the fluorescence intensity of the normalized CoroNa Red signal (N=7) ± SE (standard error). The control CoroNa Red fluorescence intensity is defined as the fluorescence intensity in normal, untreated MDCK cells after loading with CoroNa Red. Reproduced with permission from *J Am Soc Nephrol* 2005 (4).

intensity of CoroNa Red against $[Na^+]_m$ resulted in a logarithmic curve (**Fig. 17.22**).

3. Since CoroNa Red is not a ratiometric probe, non-Na^+-dependent changes in the mitochondrial CoroNa Red fluorescence, e.g., changes induced by shrinkage or swelling of mitochondria, might lead to misinterpretation of the results. For assessing the contribution of non-specific changes in mitochondrial CoroNa Red fluorescence intensities, the CoroNa Red_{mito}/MTG fluorescence ratio was used to calculate $[Na^+]_m$. Because changes in the mitochondrial volume induce equal relative changes in the fluorescence of both mitochondrial CoroNa Red and MTG, the CoroNa Red_{mito}/MTG ratio is expected to be volume independent. The $[Na^+]_m$ values obtained by the "CoroNa Red_{mito}/MTG fluorescence ratio" method or calculated directly from the mitochondrial CoroNa Red fluorescence intensities are not significantly different. Therefore, the latter method is applied to calculate $[Na^+]_m$.

3.6.3. No Calibration Protocol for Rhod 2

Calibration was not performed in rhod 2 experiments. Only *relative* changes in mitochondrial rhod 2 fluorescence as induced by treatment of the cells with metabolic inhibitors were assessed. The observed relative increase in mitochondrial rhod 2 fluorescence during MI is underestimated because signal is lost throughout the 60 min protocol. This loss of signal might be ascribed to either photobleaching of the probe or leakage of the rhod 2 dye out of the mitochondrial matrix. Observed changes were not related to focus drift of the instrument since the focus was re-optimized before every image collection.

4. Notes

1. Fluorescence dye solutions have to be protected from light.
2. When removed from the freezer, allow the aliquots to equilibrate to room temperature in the dark (aluminum foil wrapped). This will prevent the contamination of DMSO with water (as is the case when a bottle is opened before the content has warmed sufficiently to room temperature).
3. Unless stated otherwise all solutions should be prepared in water that has a resistance of 18.2 MΩ; this standard is referred to as "water" in this text.
4. All chemicals are analytical grade.
5. All solutions can be stored at 4°C for about 1 month (check regularly that fungi do not appear in them).
6. The recommended seeding density allows tight confluent monolayers of MDCK cells to be obtained after 3 days of culturing, as well as an increased stability of the cells on the glass surface. For other cell types the seeding density should be adjusted accordingly.
7. At higher passage numbers we observed a delay in cell growth, the confluency being reached only after 5 days in culture. Moreover, the monolayer became less uniform, the morphology of the cells changing from flat, "polygonal" shape, to a more rounded shape. This change in morphology observed for older passages is illustrated in **Fig. 17.1**. We noticed that when the passage number was high, growth to confluency was slow and patchy with in general fuzzy small rounded cells. Switching to new low passage numbers remedied all issues.
8. On a regular monthly basis potential mycoplasma contamination should be checked with, e.g., the mycoplasma PCR ELISA kit from Roche (cat no 11 663 925 910) (Vilvoorde, Belgium).
9. Washing gently means avoiding direct contact with any cell covered areas, not hitting attached cells head-on with a syringe flow but breaking the flow by very gently squishing against the walls of the cell holder. Each wash is performed with 2 ml buffer, letting it sit for 30 s. Use very

gentle re-suction, keeping the needle tip as far away from the cell covered surfaces and maintaining a low liquid flow.

10. The second loading protocol was developed based on the observation that during the experimental time course the cytosolic fluorescence intensity is decreasing continuously due to a gradual loading of the mitochondria. Therefore, to be able to work with a sufficiently high signal-to-noise ratio of SNARF-1 in the cytosol during longer time periods, the mitochondrial "sink" was loaded first, followed by a second loading of the cytosolic pool with SNARF-1.
11. When dual labeling is used, the pinhole size has to be adjusted for each channel in such a way that the same observed optical slice thickness is observed for all emission channels.
12. This step is very important for the reliability of the method, as using the averaged autofluorescence intensity over the whole cell can induce large errors in the calculated values for pH_i and pH_m. This will affect mainly the data from the final part of the experiments, when SNARF-1 intensity is lowered due to photobleaching (see below) and dye leakage, and therefore the contribution of the background to the calculations becomes substantial.
13. As discussed in **Section 3.4.1**, one concern raised by repeated laser illumination on living cells is related to the phototoxicity which might occur. Previous phototoxicity studies (10, 11) have shown that cellular stress induced by laser excitation in the visible range (as applied in our experimental protocol) can be detected via an increase in the cellular autofluorescence. Monitoring changes in autofluorescence during the experimental protocol can therefore be a useful method to study phototoxicity.
14. These settings allowed us to perform the following protocol with minimal light exposure and photobleaching in order to monitor the pH_i and pH_m changes in MDCK cells exposed to MI: we first acquired two pre-MI control images, followed by SNARF-1 ratio imaging at 5, 10, 20, 40, and 60 min during MI. After replacing the MI solution by fresh NSS, the SNARF-1 ratio was again measured after 15 and 30 min into the recovery period of the cells.
15. Mitochondrial and cytosolic regions of interest should be carefully selected since at the given pinhole opening (**Section 3.4.2**) the optical slice volume contained both cytoplasmic and mitochondrial localized fluorescence information. Just taking the average of all pixels (related to both mitochondrial and cytosolic signals) is completely

erroneous and incorrect since the background signal levels differ between the two. To properly create the MTG mask for either the pixels under the nucleus (lack of any MTG signal) or the mitochondria-related pixels (maximum MTG signal), the MTG image stack was enhanced and stretched in order to observe even the weakest observable signal.

16. It is to be understood that this part is not easy to automate. Due to cellular motion usually none of the selected ROI areas carries over to the next image and the whole process has to be repeated separately for each image in the stack.
17. Calibration is done starting from high concentration to zero.

Acknowledgments

The authors acknowledge Mr. J. Janssen for his excellent technical assistance with the cell culture. We also thank Mrs. R. Beenaerts, Mr. W. Leyssens, Mr. P. Pirotte, and Mr. R. Van Werde for their technical help and Mrs. M. Ieven for artwork. We are also grateful to various groups for the free software for confocal image analysis: LSM Toolbox plug-in (University of Strasbourg, Strasbourg, France; Dr. J. Mutterer, Dr. Y. Krempp, and Dr. P. Pirrotte), ImageJ (NIH, Bethesda, MD; Dr. W. Rasband; available online: http://rsb.info.nih.gov/ij/plugins/lsm-reader.html), and VolumeJ (Utrecht University, Utrecht, the Netherlands; Dr. M. Abramoff).

This work was supported by a bilateral research collaboration program between Flanders and Hungary (BIL01/18 HON), between Flanders and Romania (BOF 05 B03), and by the Research Foundation Flanders (FWO, GO270.07). Support by IAP P6/27 *Functional Supramolecular Systems* (BELSPO) and by the FWO-onderzoeksgemeenschap *Scanning and Wide Field Microscopy of (Bio)-organic Systems* are also thankfully acknowledged.

References

1. Sekine, T., Miyazaki, H., Endou, H. (2008) Solute transport, energy consumption and production in the kidney, in Seldin and Giebisch's, in *The Kidney, Physiology and Pathophysiology, vol. I*, (Alpern, R.J., Hebert, S.C., eds.), Academic Press, Elsevier Inc: Burlington, MA. pp. 185–210.
2. Schrier, R.W. (2003) Acute renal failure: pathogenesis, diagnosis and management, in *Renal and Electrolyte Disorders* (Schrier, R.W., ed.), 6th Edition, Lippincott Williams & Wilkins: Philadelphia, pp. 401–455.
3. Balut, C., vandeVen, M., Despa, S., Lambrichts, I., Ameloot, M., Steels, P., Smets, I. (2008) Measurement of cytosolic and mitochondrial pH in living cells during reversible

metabolic inhibition. *Kidney Int*, **73**, 226–232.
4. Baron, S., Caplanusi, A., vandeVen, M., Radu, M., Despa, S., Lambrichts, I., Ameloot, M., Steels, P., Smets, I. (2005) Role of mitochondrial Na^+ concentration, measured by CoroNa red, in the protection of metabolically inhibited MDCK cells. *J Am Soc Nephrol*, **16**, 3490–3497.
5. Smets, I., Caplanusi, A., Despa, S., Molnar, Z., Radu, M., vandeVen, M., Ameloot, M., Steels, P. (2004) Ca^{2+} uptake in mitochondria occurs via the reverse action of the Na^+/Ca^{2+} exchanger in metabolically inhibited MDCK cells. *Am J Physiol Renal Physiol*, **286**, F784–F794.
6. Tsien, R.Y., (1981) A non-disruptive technique for loading calcium buffers and indicators into cells. *Nature*, **290**, 527–528.
7. Takahashi, A., Zhang, Y., Centonze, E., Herman, B. (2001) Measurement of mitochondrial pH in situ. *Biotechniques*, **30**, 804-808, 810, 812 passim.
8. Despa, S., (2000) *Microfluorimetry of Epithelial Cells: Lifetime-Based Sensing of* Na^+ *Concentration and the Effects of Chemical Hypoxia*, Physiology Department, Hasselt University: Diepenbeek (Royal Library Albert I, Brussels, D-2000-2451-6), p. 18.
9. Thomas, J.A., Buchsbaum, R.N, Zimniak, A., Racker, E. (1979) Intracellular pH measurements in Ehrlich ascites tumor cells utilizing spectroscopic probes generated in situ. *Biochemistry*, **18**, 2210–2218.
10. Dailey, M.E., Manders, E., Soll, D., Terasaki, M., (2006) Confocal microscopy of living cells, in *Handbook of Biological Confocal Microscopy*, (Pawley, J.B., ed.), 3rd Edition, Springer Science + Business Media, LLC: New York, pp. 381–403.
11. König, K., (2001) Cellular response to laser radiation in fluorescence microscopes, in *Methods in Cellular Imaging*, (Periasamy, A., ed.), Oxford University Press: Oxford, pp. 236–251.

Chapter 18

Live Cell Imaging Analysis of Receptor Function

Daniel C. Worth and Maddy Parsons

Abstract

Cell surface receptors are crucial in the regulation of a wide variety of signalling responses to extracellular stimuli such as soluble growth factors or matrix proteins. To respond effectively to rapidly changing environmental cues, many receptors are rapidly endo- or exo-cytosed to either subcellular or membrane compartments or they recruit specific intracellular binding partners. Recent advances in microscopy techniques have made it possible to study receptor behaviour in live cells to gain a better understanding of dynamics, binding partners and sub-cellular localisation. Here we describe several common currently used techniques to study receptor behaviour in living cells.

Key words: Microscopy, receptor, fluorescence, FRAP, FRET, FLIM, TIRF, Green fluorescent protein variants, endocytosis.

1. Introduction

Receptors are key to many fundamental processes in cell biology. Analysis of receptor function using biochemical assays has provided a great deal of information regarding posttranslational modification and binding partners, but does not allow precise determination of subcellular localisation or dynamics. Recent advances in microscopy techniques have enabled direct visualisation of receptor function during processes such as internalisation, dimerisation and intracellular trafficking in response to ligand binding. Such experiments have been made possible by both advancements in imaging technology and protein labelling techniques. In this paper we outline several imaging techniques and strategies that are currently used to analyse receptor function

D.B. Papkovsky (ed.), *Live Cell Imaging*, Methods in Molecular Biology 591,
DOI 10.1007/978-1-60761-404-3_18, © Humana Press, a part of Springer Science+Business Media, LLC 2010

in live cells. The advantages, disadvantages and troubleshooting suggestions are discussed for each technique.

1.1. Fluorescent Labelling of Receptors

The most popular method for receptor labelling for live imaging is by generating a genetically encoded fluorescent tag on either the N or C terminus of the protein (**Fig. 18.1a**). Typically this is the green fluorescent protein (GFP) or a variant of it, which can be excited at specific wavelengths to produce a fluorescent signal. Many variants of the original GFP have been produced within the past 15 years and when these are exposed to light of a specific wavelength, they produce characteristic emission (1, 2). Photo-activatable (PA) or photo-convertible variants of the GFP tag have also been used to study localised protein dynamics (3). These can only be viewed when exposed to a burst of light, for example, PA-GFP requires activation with light of a wavelength of 405 nm, only then will the activated PA-GFP be viewable using excitation at a wavelength of 488 nm. These PA-constructs allow activation and tracking of the behaviour of a specific population of molecules (4). Since the generation of the first PA-GFP (5), variants have been developed with similar photo-activation or conversion properties but at different wavelengths of fluorescence (6–8). These fluorophores have been used successfully to study protein behaviour (9) and trafficking of integrin receptors (10). Fluorescent protein tags are relatively large (around 27 kDa) and as such may affect protein folding, function or disrupt interactions with potential binding partners. It is important therefore to rigorously test localisation and biochemical behaviour of the expressed tagged receptor prior to imaging to ensure function is not compromised.

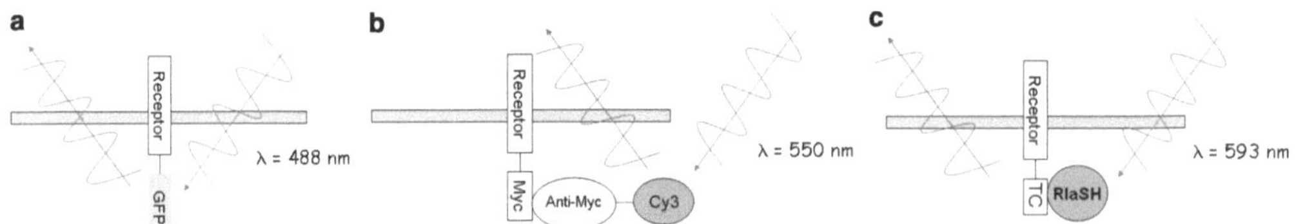

Fig. 18.1. Three most commonly used methods for labelling receptors in live cell imaging studies: (A) receptors labelled with genetically encoded fluorescent tags (e.g. GFP); (B) small molecule tags (e.g. myc) followed by anti-tag antibody detection; or (C) tetracycteine motifs for ReASH detection.

An alternative strategy is to use antibody labelling for live cell imaging (**Fig. 18.1b**). Kits are available which allow direct labelling of an antibody to the receptor of interest with a fluorescent reporter such as the Cy™ (GE Life Sciences), Alexa Fluor® (Molecular Probes) or DyLight™ Fluor (Pierce) dyes. This technique has been used to image a number of receptors including the $p75^{NTR}$ (11). In some cases this technique may prove more valuable as it permits the use of antibodies engineered to recognise different regions of the receptor or different conformations and post-translational modifications. The antibodies being used may also have a blocking or inhibitory effect on receptor function; this needs to be considered when choosing an antibody. An additional way to use labelled antibodies is to genetically encode a small peptide tag on to the receptor of interest and then use a fluorescently labelled anti-tag antibody to study receptor behaviour. Examples of this include the use of both HA and myc tags to study nicotinic acetylcholine receptors (nAChR) (12) and flag tags to look at Class 1 metabotropic glutamate receptors (mGluRs) (13). These labelled anti-tag antibodies can be introduced into live cells either by direct microinjection into the cytoplasm or by brief treatment of cells with a mild permeabilising agent (such as saponin) to allow entry of antibody added to the medium. The same principle also applies to some fluorescent dyes such FIAsH-EDT_2 and RIAsH-EDT_2 dyes (Invitrogen; **Fig. 18.1c**). These particular dyes are fluorescent upon binding to the tetracysteine tag, consisting of a Cys-Cys-Pro-Gly-Cys-Cys consensus sequence (14). This has the benefit of being relatively small compared to fluorescent protein tagging and has been used successfully in the study of AMPA receptors (15). In some cases direct labelling of a receptor may not be possible and in these cases, direct labelling of a ligand may be more appropriate. The same labelling strategies discussed for receptors can also be considered for the ligand. The advantage of labelling ligand is that in many cases this can then be produced as a recombinant protein and labelled directly with dyes prior to treating cells. Labelled epidermal growth factor (EGF) is a good example of this with successful imaging being performed with Cy3- or Cy5-tagged EGF (16) or EGF tagged with quantum dots (17). Quantum dots are highly photostable semiconductor crystals that emit light when exposed to UV radiation or visible light depending on the crystal (18). One of the common studies to perform with labelled ligands is pulse-chase experiments. Here, cells are exposed to a dose of labelled ligand (pulse) and then at a later time point, unlabelled ligand (chase). In cases where both ligand and receptor are labelled, it is possible to investigate interactions between the two.

1.2. Overview of Strategies for Imaging Receptors in Living Cells

A number of different imaging techniques can be used to study fluorescently labelled receptors (**Fig. 18.2**). Wide-field (or epifluorescence) microscopy allows cells to be imaged with fluorescence and phase contrast, thus the position of a fluorescently tagged receptor can be seen in relation to cell movement. Imaging can be performed on a fluorescent microscope equipped with a charge-coupled device (CCD) camera and the appropriate excitation and emission filter sets to distinguish spectral regions of interest. When using sensitive CCD cameras, this method of imaging is relatively non-toxic to cells due to the low illumination/exposure times that are required to achieve high-quality images (19).

A: Widefield time-lapse

0 mins 3 mins 6 mins 9 mins

β3 integrin-GFP

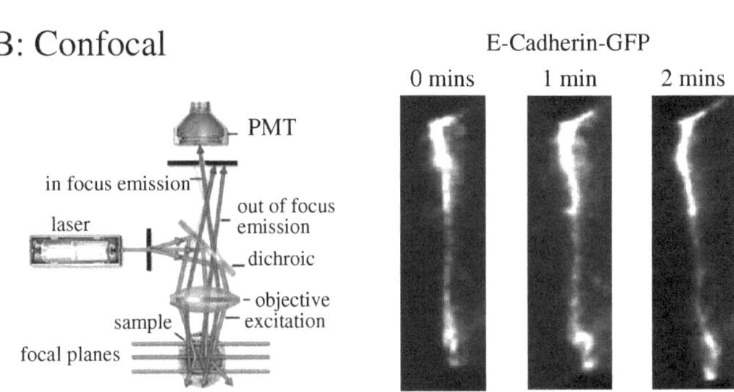

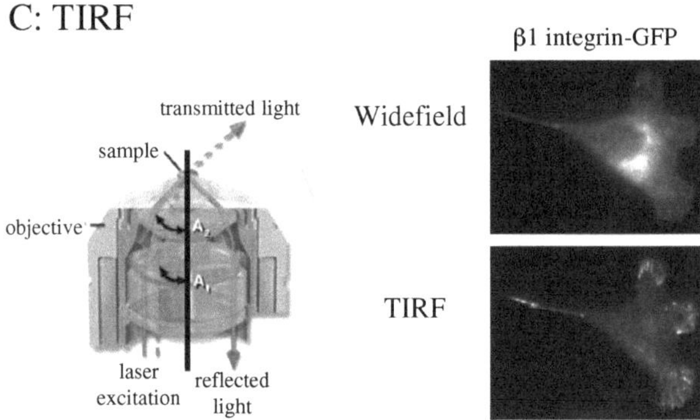

Fig. 18.2. Overview of methods to image receptors in live cells. (A) Examples of images acquired using wide-field fluorescence time-lapse microscopy to image dynamics of β3 integrin-GFP in fibroblasts. (B) Cartoon schematic of basic set-up of a confocal microscope (left panel) and (right panels) example images of E-Cadherin expressed in human epithelial cells imaged over 2 min. (C) Cartoon schematic of the principles of TIF microscopy (left panels) and example comparative images of β1 integrin-GFP in a mouse fibroblast by wide-field microscopy (top panel) and the same cell imaged by TIRF (bottom panel).

An alternative to wide-field microscopy is the highly sensitive confocal microscope. Here, instead of polychromatic light passing through filter sets, lasers are used to excite at specific pre-defined wavelengths. This then excites the tagged receptor and the emitted light from the sample is received by a photo-multiplier tube detector. Using this form of microscopy allows for greater resolution in the *z*-axis than wide-field microscopy as the emitted light is from a narrow focal plane. Depending on the speed of the system, images can be taken anywhere from 4 to 30 frames/s using a fast resonant scan head. Faster scan speeds are more suited for live imaging and some systems will also allow Z-stack images to be acquired over time. A drawback of using this form of microscopy for live imaging is that the exposure of cells to lasers can result in tag photobleaching and phototoxicity, which can affect cell behaviour thus leading to artefacts.

Another method for viewing cell surface receptors in contact with the substrate (such as integrins) is total internal reflection (TIRF) microscopy. This technique works through production of an evanescent wave, which is achieved when light is totally reflected when passing from a solid phase to a liquid phase (**Fig. 18.2**). The resulting wave decays exponentially and so only penetrates a short distance into the cell, approx 100nm (20). Thus, this makes TIRF an ideal method for viewing events near to the plasma membrane, such as viewing protein recruitment to a receptor or dynamics of a receptor to, from or within the membrane.

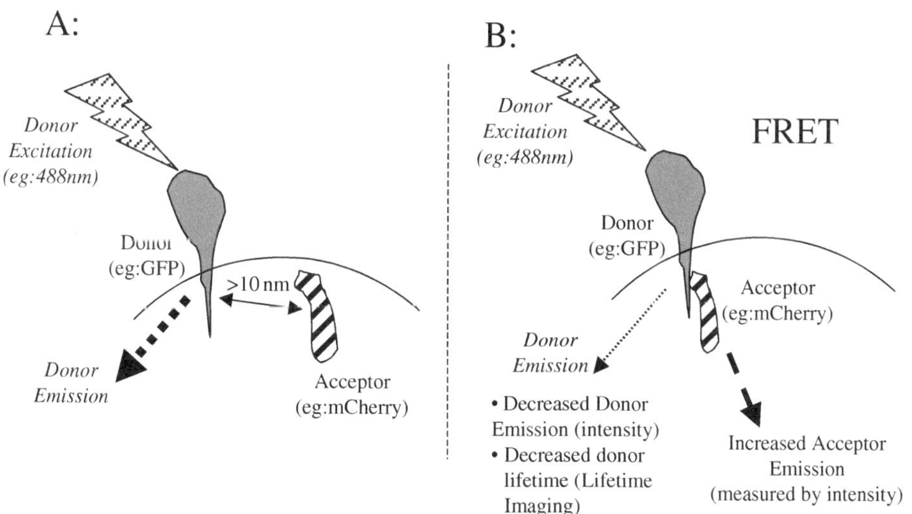

Fig. 18.3. Using FRET to analyse receptor function. (A) Donor and acceptor molecules are more than 10 nm apart, therefore excitation of the donor does not result in transfer of energy to acceptor. (B) When acceptor fluorophore is less than 10 nm from donor fluorophore, excitation of donor can result in both increased acceptor emission and decreased donor emission. FRET efficiencies can be calculated by measuring these parameters.

Fluorescence resonance energy transfer (FRET) is a technique used to measure the interaction between two fluorescently tagged proteins, when the donor (a protein tagged with a shorter wavelength tag) comes in close proximity with an acceptor located on the binding partner (**Fig. 18.3**). Non-radiative resonance energy transfer takes place from the donor to the acceptor, resulting in increased fluorescence of the acceptor, decreased fluorescence of the donor and a reduction in fluorescence lifetime of the donor. Any of these three occurrences can be measured to ascertain the levels of FRET between the two interacting proteins. FRET is a powerful technique to study both receptor homo- and hetero-dimerisation and receptor association with intracellular binding partners in live cells (21).

Fluorescence recovery after photobleaching (FRAP) is an excellent method to study the dynamics of receptors and to quantify the speed of mobilisation within certain compartments of a cell (**Fig. 18.4**). Using a confocal microscope, regions of interest are defined and bleached until a total loss of fluorescent signal in that area is achieved. The rate of recovery of fluorescent signal within the region of interest is then recorded over time. The resulting data can be used to calculate the speed of recovery of intensity: the faster the time the faster the dynamics of your receptor. It can also be used to calculate the mobile and immobile fractions of the receptor, with the percentage recovery seen being the mobile fraction and the percentage recovery not seen being the immobile fraction (22). Fluorescence loss in photobleaching (FLIP) uses the same principle as FRAP apart from instead of measuring recovery, loss of intensity is recorded. The chosen region of interest is bleached repeatedly while images of the whole cell are taken. If fluorescence is lost outside the bleached region, then that indicates that your tagged protein has moved from, or through, this area (23).

A: FRAP of E-Cadherin-GFP

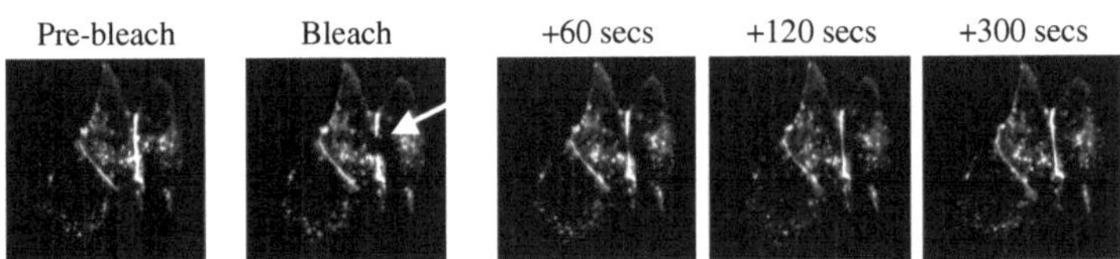

B: Example of FRAP recovery curve for E-Cadherin-GFP

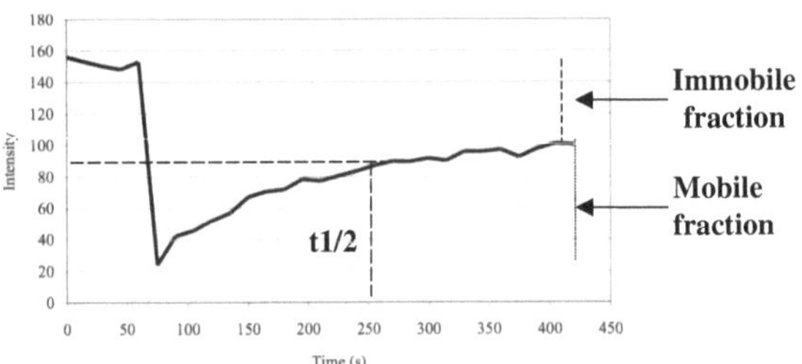

Fig. 18.4. FRAP analysis of receptor dynamics. (A) Sample images of a time course of images acquired on a confocal microscope to analyse dynamics of E-Cadherin-GFP by FRAP. Panels show a single pre-bleach image followed by an image acquired immediately post-bleach (bleached region of interest box highlighted in white) and subsequent recovery images. (B) An example of a recovery curve obtained from the FRAP data set.

2. Materials

1. Substrates designed for live cell imaging, such as glass coverslips (e.g. Menzel-Gläser) glass-bottom minidishes (MatTek, Ashland, MA) or LabTek® 2- or 8-well chamber slides (Nunc, Rochester, NY).
2. Extracellular matrix protein, e.g. fibronectin or vitronectin (*see* **Note 1**).
3. Phenol-red free medium (*see* **Note 2**).
4. Buffering solution such as HEPES (4-(2-hydroxyethyl)-1-piperazineethanesulfonic acid) (*see* **Note 3**).
5. Cells to be analysed – normal or transfected with tagged receptor.
6. Labelled antibodies or fluorescent ligands. These must be prepared in advance of imaging and diluted in an appropriate buffer. Labelled proteins should be filtered using a 0.22-mm syringe filter (Millipore, Bedford, MA) prior to use to avoid interference from aggregates of unbound fluorescent dye.
7. Proper filter sets (GFP, filters labels).
8. Immersion oil (e.g. from Cargille Laboratories, NJ, or Leica Microsystems, Wetzlar, Germany).

3. Methods

3.1. Wide-field Imaging of Fluorescently Tagged Receptors

1. Plate cells of interest into imaging dishes (*see* **Note 1**). Time of plating prior to imaging depends upon the experimental aims.
2. Prepare microscope. If an environmental chamber is used (surrounding microscope) ensure the chamber is pre-heated to 37°C and temperature is stable before use to avoid fluctuations in focus during imaging. Switch on fluorescence illuminator 20 min before use to allow stabilisation.
3. For pulse-chase experiments using labelled ligand, add ligand to cells on ice, allow binding to occur, wash off excess unbound ligand and replace with fresh media.
4. Place the dish onto the stage ensuring both the stage insert and dish are securely held in place. Allow around 10 min for stabilisation of temperature (this will reduce focal drift) and settling of the dish/plate.

5. If fluorescently labelled ligand or exogenous proteins are to be used, add these to the media at this stage to allow binding prior to imaging.
6. Identify a labelled/transfected cell of interest. This can be done either using illumination down the eyepieces or using direct visualisation onto the camera (*see* **Note 4**).
7. Set up acquisition software (this will vary – refer to manufacturers guidelines) and acquire an image of the cell in both GFP and phase-contrast channels.
8. Once the desired cell is in focus and stable, begin time-lapse acquisition (*see* **Note 5**).
9. Once acquisition is complete, the acquired images can be saved as a movie file (e.g. avi or quicktime) or saved as individual files (e.g. tiffs) and be presented as a montage as in **Fig. 18.2a**, where adhesions containing GFP-tagged β3-integrin can be seen. The change in localisation of these adhesions can be seen in relation to the cell's movement. Also, the assembly and disassembly of these sites can be quantified over the movie time period.

3.2. Confocal Microscopy Analysis of Receptor Localisation

1. Prepare cells and confocal microscope as in **Section 3.1**.
2. Once dishes containing cells are secure on the microscope stage, use the wide-field illuminator to find cells of interest down the eyepieces (*see* **Note 6**).
3. Once cells are located, switch to confocal PMT detector mode. Select the correct laser and filter set configuration.
4. Acquire a single-scan image of cells of interest and ensure focal plane/laser gain levels are correct and the resulting image is not saturated.
5. Set the time-lapse protocol window to acquire a single-scan image every 10–15 s over 5 min. Modify as necessary as in **Section 3.1**.
6. Z-stacks can also be acquired throughout the entire cell at every time point to enable collection of three-dimensional data sets over time. This enables a 3D reconstruction of the cell to be generated post-acquisition to analyse receptor behaviour within the entire cell rather than at a single focal plane. An example of this is shown in **Fig. 18.2b** where Z-stacks of cells expressing E-Cadherin-GFP have been taken (*see* **Note 7**).
7. In the analysis of dynamics of photo-activatable variants of GFP (PA-GFP) or photo-switchable fluorescent proteins, avoid exposure of cells to UV light prior to experimental photo-activation, as this can lead to low-level GFP activation.

Image and focus cells initially using brightfield illumination and a single scan using phase contrast/differential interference contrast (DIC) settings on the confocal scan head acquired.

8. Set up the time-lapse acquisition protocol to undergo a “bleach/activation” cycle (at 405 nm for PA-GFP) followed by acquisition of the desired number of images every 10–15 s using the 488-nm laser (to detect the activated GFP).
9. Highlight the area of interest to be photo-activated (e.g. plasma membrane) with a region of interest box using the bleaching function protocol. Activate the 405-nm laser at 100% power and illuminate the cell for a set number of iterations within the defined region of interest.
10. Once activation is complete, visualise and track the GFP over time within the cell.

3.3. TIRF Analysis of Receptor Behaviour

TIRF is only suitable for imaging receptors that are presented at the interface between the cell and coverslip. Correct alignment and calibration of TIRF illumination angle (both for wide-field and laser-based systems) must be checked prior to imaging to ensure a good TIRF signal is achievable.

1. Cell preparation as in **Section 3.1** (*see* **Note 8**).
2. Place imaging chamber securely onto microscope stage ensuring that the sample is completely flat on the stage.
3. Acquire starting images of cells under wide-field and TIRF illumination. Optimise focal plane and adjust TIRF laser angle if necessary. If an automatic focussing device is available on the microscope, activate this to ensure correct focal plane is maintained throughout the acquisition period.
4. Set up time-lapse software to acquire an image every × seconds over × number of repeats. Start experiment.
5. The images obtained again can be saved as in **Section 3.1**, step 9. Sample images can be seen in **Fig. 18.2c** where a fibroblast expressing GFP-tagged β1 integrin is seen in wide-field and in TIRF. The TIRF image shows β1 integrin containing adhesions which are in contact with the coverslip.

3.4. FRET Analysis of Receptor Dimerisation or Binding Partners

The common method used to measure FRET in live samples is using confocal acquisition of intensity-based measurements. Other techniques such as fluorescence lifetime imaging (FLIM) are used but tend to be less common, and due to lengthy imaging periods required for photon counting are often not best placed for live cell imaging studies. The following is an example of how to set up a FRET experiment for analysis by fluorescence intensity ratiometric method.

1. Co-transfect cells with donor- and acceptor-labelled molecules of choice. Common fluorophore pairs used are CFP/YFP and GFP/mRFP (or mCherry) (*see* **Note 9**).
2. Plate and mount cells on microscope stage as in **Section 3.1**.
3. Locate expressing cells and acquire a single-confocal scan to optimise focal plane and laser settings. Use a pseudocolour look up table (LUT) to ensure images are not saturated at any point.
4. Set up the time-lapse and bleach control windows to acquire a single-scan pre-bleach, followed by a bleach at a defined region of interest, followed by a single-scan post-bleach with no delay between acquisition scans. Set the bleaching laser to 100% output (see **Note 10**).
5. Set up region(s) of interest for bleaching. Run protocol and save data. If bleaching is incomplete, repeat on a different sample using increased bleach time.
6. Many software packages provide FRET analysis tools for post-acquisition purposes. Use these plug-ins where possible. If software is not installed on imaging system, the programme ImageJ (available from: http://rsb.info.nih.gov/ij/) can be used. Numerous plug-ins are available to download that allow you to analyse FRET data.
7. When analysing data consider the following: (a) corrections for fading during image acquisition (changes in intensity across the whole cell before and after bleaching); (b) corrections for differences in starting intensity of donor and acceptor fluorophores (this is essential when analysing ratiometric FRET data); (c) crosstalk detected between different fluorophores (i.e. wavelength "bleedthrough"), which can be tested by acquiring each channel as a single line scan individually with the other laser line inactivated.
8. Images should be smoothed using a 3×3 box mean filter (this can be done in ImageJ or photoshop), background subtracted and post-bleach images fade compensated. A FRET efficiency ratio map over the whole cell is calculated using the following formula: $(\text{donor}_{\text{postbleach}} - \text{donor}_{\text{prebleach}})/\text{donor}_{\text{postbleach}}$. Ratio values are extracted from pixels falling inside the bleach region as well as an equally sized region outside of the bleach region and the mean ratio determined for each region and plotted on a histogram. The non-bleach ratio is then subtracted from the bleach region ratio to give a final value for the FRET efficiency ratio. Data from images must be used only if acceptor bleaching efficiency is greater than 70%.

3.5. FRAP/FLIP Analysis of Receptor Kinetics and Recruitment

1. Set cells up as in **Section 3.1**.
2. Set up cells/imaging chambers on microscope stage securely.
3. Find cell(s) of interest and focal plane as in **Section 3.2**. Acquire a single scan at each of the required wavelengths to ensure good signal in each channel.
4. Set up time-lapse and bleach acquisition control panels. Define the region(s) of interest for photobleaching as in **Section 3.4**, and the laser for bleaching (e.g. for bleaching GFP, use 488 nm) set to 100% for bleach cycle (*see* **Note 11**).
5. Set up time-lapse acquisition software to acquire a minimum of four scans before bleaching (to ascertain pre-bleach intensity and fluorophore fading) followed by a single bleach scan, followed by a single scan every x seconds over x cycles (depending upon kinetics of receptor to be imaged).
6. Run imaging experiment.
7. Once the time-lapse protocol has finished, analyse data for recovery (or loss of signal for FLIP) kinetics. Plot intensity over time for the region of interest.
8. The final data set must be corrected for fading of the fluorophore over time (due to illumination). To achieve this, apply the slope of the curve of the pre-bleach image intensities plotted as a function of time, or a background region of interest for the entire time-lapse sequence.
9. Re-plot corrected recovery values as percentage recovery over time.

4. Notes

1. If cells do not adhere, coverslips or imaging chambers can be pre-coated with an extracellular matrix protein such as fibronectin or vitronectin.
2. Phenol red dye is auto-fluorescent and can result in high background levels during imaging. Media such as OptiMEM® can be used. If cells are unhealthy, the media can be supplemented with 1–10% fetal bovine serum (FBS).
3. Buffering solutions are only required if cells are to be imaged in an environment where the CO_2 levels are not regulated.

4. Avoid excessive exposure of cells to fluorescent light to preserve fluorescent signal and protect cells.
5. The interval period and number of images acquired will depend upon the dynamics of the receptor being studied. A good starting point is to acquire an image every 10–15 s for 5 min in both fluorescence and phase contrast channels. This will provide an initial idea of speed of movement of the receptor/ligand and the acquisition protocol can be subsequently modified if necessary.
6. Using Wide-field as opposed to confocal will avoid unnecessary exposure to lasers and help prevent photobleaching and phototoxicity.
7. Two potential caveats exist with acquiring Z-stacks over time. First, depending upon the final resolution of the image obtained and the optical depth of the samples, the acquisition of a full Z-stack can be rather time-consuming. In this case, it may be necessary to reduce either the resolution (from 1024×1024 pixels to 512 or 256 pixels) or the number of Z-stacks to speed up acquisition times. Second, the increased exposure to laser illumination during acquisition of Z-stacks may cause considerable fading of the fluorophore and/or prove toxic to some cells.
8. For TIRF analysis, cells must be plated on glass coverslips and mounted in aqueous medium, e.g. growth media or PBS (phosphate-buffered saline).
9. Ideally, each fluorophore-tagged protein should be expressed at similar levels in each cell to be analysed.
10. The number of iterations (or bleach scan passes) should be optimised for each fluorophore depending upon the intensity of the signal to be bleached and the output power of the laser line used.
11. For FRAP, the bleach region will be a relatively small, defined area of interest such as a membrane compartment or vesicle. For FLIP analysis, the protocol set-up is essentially the same, but the bleach region is much larger to allow calculation of loss of fluorescence in the remaining un-bleached region of interest.

Acknowledgements

The authors would like to thank the Biotechnology and Biological Sciences Research Council (BBSRC) and the Royal Society for funding.

References

1. Shaner, N.C., Steinbach, P.A. & Tsien, R.Y. A guide to choosing fluorescent proteins. *Nat Methods* **2**, 905–909 (2005).
2. Giepmans, B.N., Adams, S.R., Ellisman, M.H. & Tsien, R.Y. The fluorescent toolbox for assessing protein location and function. *Science* **312**, 217–224 (2006).
3. Patterson, G.H. & Lippincott-Schwartz, J. Selective photolabeling of proteins using photoactivatable GFP. *Methods* **32**, 445–450 (2004).
4. Lippincott-Schwartz, J. & Patterson, G.H. Fluorescent proteins for photoactivation experiments. *Methods Cell Biol* **85**, 45–61 (2008).
5. Patterson, G.H. & Lippincott-Schwartz, J. A photoactivatable GFP for selective photolabeling of proteins and cells. *Science* **297**, 1873–1877 (2002).
6. Stiel, A.C. et al. 1.8 A bright-state structure of the reversibly switchable fluorescent protein dronpa guides the generation of fast switching variants. *Biochem J* **402**, 35–42 (2007).
7. Dedecker, P. et al. Fast and reversible photoswitching of the fluorescent protein dronpa as evidenced by fluorescence correlation spectroscopy. *Biophys J* **91**, L45–47 (2006).
8. Chudakov, D.M. et al. Photoswitchable cyan fluorescent protein for protein tracking. *Nat Biotechnol* **22**, 1435–1439 (2004).
9. Betzig, E. et al. Imaging intracellular fluorescent proteins at nanometer resolution. *Science* **313**, 1642–1645 (2006).
10. Caswell, P.T. et al. Rab25 associates with alpha5beta1 integrin to promote invasive migration in 3D microenvironments. *Dev Cell* **13**, 496–510 (2007).
11. Dusan Matusica, M.P.F., Rogers, M.L., Rush, R.A. Characterization and use of the NSC-34 cell line for study of neurotrophin receptor trafficking. *J Neurosci Res* **86**, 553–565 (2008).
12. Gottschalk, A. & Schafer, W.R. Visualization of integral and peripheral cell surface proteins in live *Caenorhabditis elegans*. *J Neurosci Meth* **154**, 68–79 (2006).
13. Bhattacharya, M. et al. Ral and phospholipase D2-dependent pathway for constitutive metabotropic glutamate receptor endocytosis. *J Neuropsychiatry* **24**, 8752–8761 (2004).
14. Adams, S.R. et al. New biarsenical ligands and tetracysteine motifs for protein labeling in vitro and in vivo: synthesis and biological applications. *J Am Chem Soc* **124**, 6063–6076 (2002).
15. Ju, W. et al. Activity-dependent regulation of dendritic synthesis and trafficking of AMPA receptors. *Nat Neurosci* **7**, 244–253 (2004).
16. Sako, Y., Minoghchi, S. & Yanagida, T. Single-molecule imaging of EGFR signalling on the surface of living cells. *Nat Cell Biol* **2**, 168–172 (2000).
17. Lidke, D.S. et al. Quantum dot ligands provide new insights into erbB/HER receptor-mediated signal transduction. *Nat Biotechnol* **22**, 198–203 (2004).
18. Groc, L. et al. Surface trafficking of neurotransmitter receptor: comparison between single-molecule/quantum dot strategies. *J Neurosci* **27**, 12433–12437 (2007).
19. Coates, C.G., Denvir, D.J., McHale, N.G., Thornbury, K.D. & Hollywood, M.A. Optimizing low-light microscopy with back-illuminated electron multiplying charge-coupled device: enhanced sensitivity, speed, and resolution. *J Biomed Opt* **9**, 1244–1252 (2004).
20. Groves, J.T., Parthasarathy, R. & Forstner, M.B. Fluorescence Imaging of Membrane Dynamics. *Annu Rev Biomed Eng* **10**, 311–338 (2008).
21. Piston, D.W. & Kremers, G.-J. Fluorescent protein FRET: the good, the bad and the ugly. *Trends Biochem Sci* **32**, 407–414 (2007).
22. Reits, E.A.J. & Neefjes, J.J. From fixed to FRAP: measuring protein mobility and activity in living cells. *Nat Cell Biol* **3**, E145–E147 (2001).
23. Lippincott-Schwartz, J., Snapp, E. & Kenworthy, A.. Studying protein dynamics in living cells. *Nat Rev Mol Cell Biol* **2**, 444–456 (2001).

Chapter 19

Subcellular Dynamic Imaging of Protein–Protein Interactions in Live Cells by Bioluminescence Resonance Energy Transfer

Julie Perroy

Abstract

Protein functions rely on their ability to engage in specific protein–protein interactions and form complexes that are dynamically regulated by stimuli. Bioluminescence resonance energy transfer (BRET) is a highly sensitive technique, which allows monitoring of interaction between two proteins: one tagged with the luminescent donor *Renilla* luciferase, the other with a fluorescent acceptor such as YFP. We adapted this method to single-cell imaging. To this aim, we tag proteins of interest, transfect cells with these fusions, and use the high-sensitivity microscopy, combined with electron multiplying cooled charge-coupled device (EMCCD) cameras and improved bioluminescence probes. We thus achieve rapid acquisition of high-resolution BRET images and study the localization and dynamics of protein–protein interactions in individual live cells.

Key words: Bioluminescence, live cell BRET imaging, subcellular localization, protein–protein interactions, bioluminescence-dedicated microscope, electron multiplying cooled charge-coupled device camera.

1. Introduction

Bioluminescence resonance energy transfer (BRET) allows real-time analysis of interactions between proteins, which are expressed in their natural location in living systems. BRET is a proximity-based assay that relies on the non-radiative transfer of energy between donor and acceptor molecules according to the Förster mechanism. The efficiency of energy transfer depends primarily on (1) the overlap between the emission

D.B. Papkovsky (ed.), *Live Cell Imaging*, Methods in Molecular Biology 591,
DOI 10.1007/978-1-60761-404-3_19, © Humana Press, a part of Springer Science+Business Media, LLC 2010

spectrum of the donor and excitation spectrum of acceptor and (2) the distance (<10 nm) and orientation of the donor and acceptor entities (1, 2). The steep dependence of BRET on the distance between the donor and acceptor ($1/r^6$) allows monitoring of protein–protein interactions. This is achieved by attaching BRET-compatible donor and acceptor entities to the studied proteins. The energy donor is a bioluminescent enzyme such as *Renilla* luciferase (*R*luc) that generates light emission upon the addition of its substrate, whereas the acceptor is a fluorescent protein such as yellow fluorescent protein (YFP). To fully benefit from the excellent signal to background ratio provided by BRET (3), we developed BRET-based microscopy imaging, which provided the ability to study spatio-temporal dynamics of protein–protein interactions in live cells. Since it does not require sample illumination, bioluminescence imaging circumvents the problems of phototoxicity common in fluorescence microscopy, thus improving imaging in live objects (4, 5) and photosensitive tissue (6). More importantly, this methodology enables to visualize and quantify dynamics of protein–protein interactions at subcellular level in individual mammalian cells (7).

2. Materials

2.1. Cell Culture

1. Human Embryonic Kidney 293T cell line (HEK293T, Gentaur, Paris, France) (*see* **Note 1**).
2. Dulbecco's modified Eagle's medium (DMEM) (Gibco/BRL, Bethesda, MD) supplemented with 10% fetal bovine serum (FBS, HyClone, Ogden, UT).
3. Glass-bottom culture minidishes, type P35GC-0-14-C (MatTek Corporation, Ashland, MA).

2.2. Transfection

1. Plasmids: pRluc-N1 and pEYFP-N1 (PerkinElmer, Boston, MA). pDsRed-N1 (DsRed, Clontech, Saint-Germain-en-Laye, France).
2. Transfection reagent: HEPES-buffered saline (HBS, 280 mM NaCl, 50 mM HEPES [pH 7], 1.5 mM Na_2HPO_4) and 2.5 M $CaCl_2$.
3. Phosphate-buffered saline (PBS): 137 mM NaCl, 2.7 mM KCl, 1.4 mM NaH_2PO_4, 4.3 mM Na_2HPO_4, pH 7.4.

2.3. BRET

1. 2 mM Coelenterazine H (CoelH, Interchim, Montluçon, France) dissolved in pure ethanol, kept at –20°C.

2. Bioluminescence-dedicated microscope: Axiovert 200 M (Zeiss, Le Pecq, France). Plan-Apochromat 63×/1.40 Oil M27 objective (Zeiss, Le Pecq, France). Filters: Exciter HQ480/40 #44001 – emitter HQ525/50 #42017 – exciter HQ540/40 #59313 – emitter HQ600/50 #65886 – emitter D480/60 #61274 – emitter HQ535/50 nm #63944 (all from Chroma, Rockingham, VT).
3. Camera: cascade 512B (Photometrics, Evry, France).
4. Acquisition and analyses software: Metamorph software package (Molecular Devices, Downingtown, PA).

3. Methods

The catalytic oxidation of Coelenterazine H (CoelH) by the bioluminescent enzyme *R*luc results in the emission of light with maximum at 480 nm (Em480). When an appropriate energy acceptor such as the YFP is present within BRET-permissive distance from *R*luc, part of the energy can be transferred to the acceptor non-radiatively. This process leads to the excitation of the YFP and subsequent emission of light with a characteristic spectrum having maximum at 535 nm (Em535) (8) (*see* **Fig. 19.1**). One of the difficulties in establishing BRET imaging is to reliably discriminate the signal originating from the transfer of energy from that resulting from an overflow of the energy donor output into the energy acceptor detection channel. To control for the basal signal of the donor, the *R*luc fused protein (P1-*R*luc) is expressed in the absence of YFP-tagged protein. In this case, DsRed is used as a reporter of transfection to identify the cells that do not display BRET. In parallel, another pool of cells is transfected with the two proteins of interest: P1-*R*luc and P2-YFP. Twenty-four hours after transfection, these cells are pooled and cultivated for a further 24 h in the same culture dishes. After this, BRET experiments are carried out.

Once transfected cells (expressing DsRed or YFP protein and the donor protein) are identified by fluorescent imaging of DsREd or YFP, excitation light is switched off and Coelenterazine H solution is added to cells. Then emission of Rluc (at 450–510 nm, maximum at 480 nm) and YFP (510–560 nm, maximum at 535 nm) are recorded, using Em480 and Em535 filters, respectively. The ratiometric image 535 nm/480 nm reveals the subcellular localization of the interaction between P1-Rluc and P2-YFP. The cells used as BRET-negative control, i.e., expressing P1-Rluc alone, produce negligible signal.

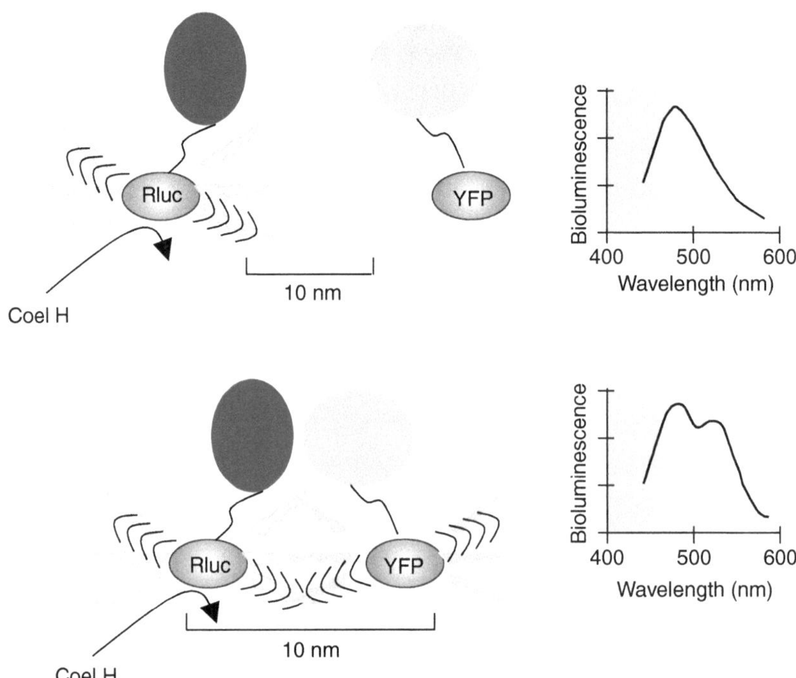

Fig. 19.1. Principle of bioluminescent resonance energy transfer (BRET). Upon binding of the two proteins, the donor and acceptor tags are brought in close proximity so that the luminescence energy resulting from the catalytic degradation of CoelH by *R*Luc is transferred to the YFP. YFP then emits fluorescence producing characteristic changes in emission spectrum. The typical effective distance between the donor and acceptor is 10–100 Å. This range correlates well with most of the biological interactions, thus making BRET an excellent tool for real-time monitoring of these interactions in living cells.

3.1. Cell Culture and Transfection

1. pRluc-N1 and pEYFP-N1 are used to fuse the DNA coding sequence of Rluc and YFP in frame with the DNA coding sequence of partner1 (P1-*R*luc) and partner 2 (P2-YFP), respectively (*see* **Note 2**) (9).
   Rluc protein:

   MTSKVYDPEQRKRMITGPQWWARCKQMNVLDSFI-NYYDSEKHAENAVIFLH

   GNAASSYLWRHVVPHIEPVARCIIPDLIGMGKSGKSG-NGSYRLLDHYKYLTA

   WFELLNLPKKIIFVGHDWGACLAFHYSYEHQDKIKA-IVHAESVVDVIESWDE

   WPDIEEDIALIKSEEGEKMVLENNFFVETMLPSKIM-RKLEPEEFAAYLEPFKEK

   GEVRRPTLSWPREIPLVKGGKPDVVQIVRNYNAYLR-ASDDLPKMFIESDPGFF

   SNAIVEGAKKFPNTEFVKVKGLHFSQEDAPDEMG-KYIKSFVERVLKNEQ –

YFP protein:

MVSKGEELFTGVVPILVELDGDVNGHKFSVSGEGE-GDATYGKLTLKFICTTG

KLPVPWPTLVTTFGYGLQCFARYPDHMKQHDFFK-SAMPEGYVQERTIFFKDD

GNYKTRAEVKFEGDTLVNRIELKGIDFKEDGNILG-HKLEYNYNSHNVYIMAD

KQKNGIKVNFKIRHNIEDGSVQLADHYQQNTPIGD-GPVLLPDNHYLSYQSALS

KDPNEKRDHMVLLEFVTAAGITLGMDELYK

2. For HEK293T transfection experiments, cells are seeded at a density of 2×10^6 cells per 100 mm dish and cultured for 24 h. Transient transfections are then performed using the calcium phosphate precipitation method (10). Distinct pools of cells are transfected with plasmids coding for P1-*R*luc and P2-YFP, or plasmids coding for P1-*R*luc and DsREd (transfection reporter) (*see* **Note 3**). Twenty-four hours after transfection, the two populations of transfected cells are pooled and cultured for an additional 24 h in glass-bottom culture dishes.

3.2. Imaging Set-Up

Standard inverted fluorescence microscope Axiovert 200 M is modified such that all light-emitting diodes are taken out and the light path is blocked with a 1.5-m optical fiber to limit optical interferences. The microscope is placed in a black box which protects from ambient light (*see* **Fig. 19.2**). Images are recorded at room temperature with a 63× objective. Identification of transfected cells necessitates the use of two-color detection. Exciter HQ480/40 #44001 and emitter HQ525/50 #42017 are used for YFP, whereas exciter HQ540/40 #59313 and emitter HQ600/50 #65886 are used for DsRed. BRET experiment, i.e., without photoexcitation but in the presence of luciferase substrate, requires the selection of 480 nm (filter D480/60 nm) and 535 nm (filter HQ535/50 nm) emission wavelengths. Images are collected with a cascade 512B camera equipped with an EMCCD detector, back-illumination and On-chip Multiplication Gain, which is mounted on the camera baseport of the microscope.

3.3. Identification of Cells Expressing P1-R luc Alone or P1-R luc and P2-YFP

1. Wash once the sample in glass-bottom dish containing both types of transfected cells with 200 μl of PBS and then add 200 μl of PBS.
2. Place the sample on the microscope. To identify positively transfected cells, excite the sample with appropriated filters: 540 nm excitation and 600 nm emission for DsRED;

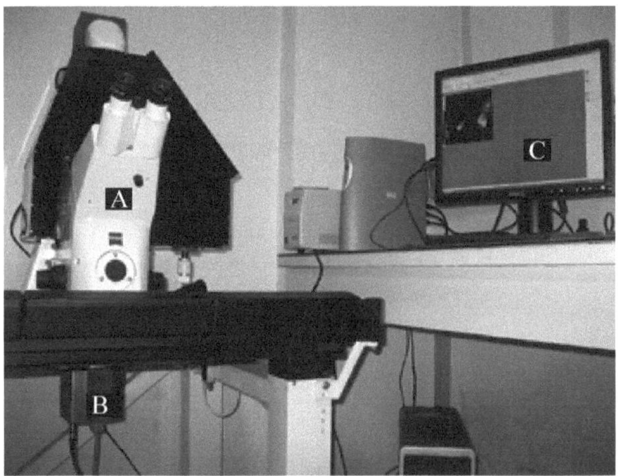

Fig. 19.2. Bioluminescence imaging set-up. (**a**) The Axiovert 200 M Microscope is an inverted fluorescence microscope, which is dedicated to bioluminescence. Thus to get rid of the ambient light pollution and collect the weak luminescent signal, the luminescent diodes were taken off, the light source deviated with an optical fiber (1.5 m long), and the microscope was installed into a *black box*. (**b**) The Cascade 512B camera is mounted on the base port of the microscope. (**c**) We use the acquisition software Metamorph.

480 nm excitation and 525 nm emission for YFP. Identify the cells expressing DsRED and P1-*R*luc alone and cells co-expressing P1-*R*luc and P2-YFP (*see* **Fig. 19.3a**). At the end of this experiment, switch off the excitation light source.

3.4. BRET Images Acquisition

1. Dilute the 2 mM CoelH stock solution in PBS to obtain 50 μl of a 100-μM CoelH solution.
2. Apply 50 μl of CoelH 100 μM in the culture dish, to get a final concentration of 20 μM CoelH, and wait for 5 min to allow diffusion of CoelH into the cell and initiation of *R*luc-catalyzed oxidation of CoelH by oxygen which produces light (*see* **Note 4**).
3. Perform sequential 30 s acquisitions (*see* **Note 5**), using the following parameters: Gain – 3950 at 5 MHz, binning – 1. Using emission filters D480/60 nm and HQ535/50 nm, obtain sets of Em480 and Em535 images, respectively. Record these raw images (*see* **Fig. 19.3b**). To follow the dynamic of the interaction between P1-Rluc and P2-YFP, subsequent acquisitions could be repeated (*see* **Note 4**).

3.5. BRET Images Analyze

1. Analyze recorded images in Methamorph software. Open the Em480 and Em535 images. In each image, define a square region (10 × 10 pixels) in the area without cells ("background region") and subtract the averaged signal

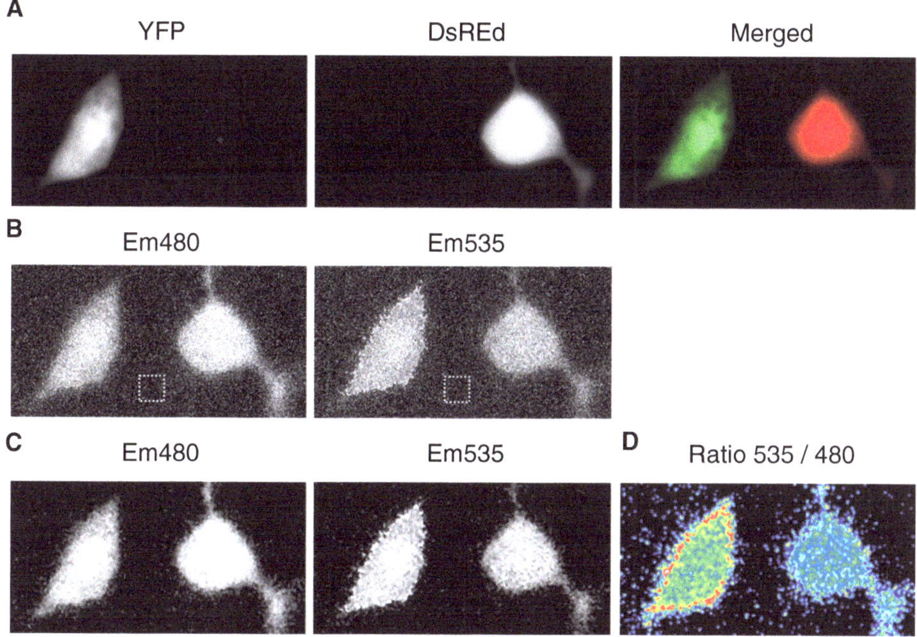

Fig. 19.3. Representative BRET images. (**a**) Identification of transfected cells. Cells are excited with a light at the appropriated wavelength for YFP or DsRED. In the same microscopic field cells express either the donor P1-*R*Luc and acceptor P2-YFP or only the donor P1-*R*Luc (+ DsRED). (**b**) Em480 and Em535 acquired 5 min after incubation with 20 μM CoelH. Em480 (*left*) corresponds to the light emitted by the P1-*R*Luc at 480 nm. Em535 (*right*) depicts light emitted at 535 nm coming from (1) the overflow of the energy donor output into the energy acceptor detection channel and (2) originating from the transfer of energy, if any. (**c**) Em480 and Em535 images obtained after subtraction of the background region (*dotted square* on **Fig. 19.3b**) and reduction of the noise with a median filter. (**d**) Ratio 535/480 image obtained by pixel-by-pixel division of the intensity emitted at 535 over 480 nm (illustrated on **Fig. 19.3c**), expressed in 255 pseudo-colors from *black* (minimum) to *red* (maximum). Note that the cell that expresses P1-*R*Luc and P2-YFP displays a much higher ratio signal than the control cell expressing only the donor. Intense BRET signals between the two partners are preferentially localized at the plasma membrane of the cell.

from this region from the raw image. Then apply a median filter to produce the Em480 and Em535 images (*see* **Fig. 19.3c**) (*see*, important, **Note 6**).

2. Generate ratiometric Em535/Em480 images by dividing the absolute intensities of each pixel of the yellow and blue images obtained at 535 and 480 nm, respectively.
3. Analyze the ratiometric images. Cells that express P1-*R*luc only donor display homogenous and relatively low ratio values resulting from donor emission leaking into the acceptor detection channel. In contrast, the cells transfected with both plasmids may display heterogenous ratio values. High values correspond to efficient energy transfer and interaction between the two proteins (*see* **Note 7**). These numerical ratios (from min to max) are processed further and visualized with a continuous 256 pseudo-color Look-Up Table, LUT (*see* **Fig. 19.3d**).

4. Notes

1. There is no restriction concerning the cell line type, provided that cells could be transiently or stably transfected. We successfully performed BRET imaging on COS cells or primary culture of hippocampal neurons.
2. The insertion of the tag should not damage the expression and function of the protein. The tag location therefore depends on the nature of the protein. The expression and function of the resulting fused protein should be compared with those of the native protein, prior to the initiation of protein–protein interaction experiments. It is worthwhile to construct several different tagged fusion proteins and choose the most efficient BRET couple. Since the BRET signal depends on the relative orientation of the donor and acceptor entities, testing several tagged couples may improve method performance.
3. DsRED has been shown to have no effect the BRET between *R*Luc and YFP (7). However, other fluorescent reporters of transfection can also be used, provided that their expression does not impair the BRET signal recorded in the presence of CoelH.
4. BRET signal usually reaches its maximum 5 min after the addition of CoelH and then remains constant for at least 25 min (7). Acquiring images within this time frame allows constant BRET values.
5. Acquisition times between 1 and 120 s can be used; however, each system has its optimum. Longer acquisition times usually improve image resolution; however, when they are too long signal-to-noise ratio can be impaired (7).
6. Median filter reduces the noise in the active image by replacing each pixel with the median of four neighboring pixel values. Therefore, median filter masks isolated pixels with ratio values in the areas having near-background signals. These areas tend to be highly variable or even undefined (division by zero).
7. To determine the minimal and maximal ratio values, identical square regions are drawn on Em535 and Em480 images on the areas of interest. The intensity of each pixel is determined and exported into Excel table. Calculate the ratio for each pixel by dividing Em535 by Em480, and find the minimal and maximal values of the ratio between which the 256-color scale will be set.

References

1. Tsien, R. Y., Bacskai, B. J. and Adams, S. R. (1993) FRET for studying intracellular signalling. *Trends Cell Biol* **3**, 242–245
2. Wu, P. and Brand, L. (1994) Resonance energy transfer: methods and applications. *Anal Biochem* **218**, 1–13
3. Boute, N., Jockers, R. and Issad, T. (2002) The use of resonance energy transfer in high-throughput screening: BRET versus FRET. *Trends Pharmacol Sci* **23**, 351–354
4. Loening, A. M., Wu, A. M. and Gambhir, S. S. (2007) Red-shifted Renilla reniformis luciferase variants for imaging in living subjects. *Nat Methods* **4**, 641–643
5. De, A., Loening, A. M. and Gambhir, S. S. (2007) An improved bioluminescence resonance energy transfer strategy for imaging intracellular events in single cells and living subjects. *Cancer Res* **67**, 7175– 7183
6. Xu, X., Soutto, M., Xie, Q., Servick, S., Subramanian, C., von Arnim, A. G. and Johnson, C. H. (2007) Imaging protein interactions with bioluminescence resonance energy transfer (BRET) in plant and mammalian cells and tissues. *Proc Natl Acad Sci U S A* **104**, 10264–10269
7. Coulon, V., Audet, M., Homburger, V., Bockaert, J., Fagni, L., Bouvier, M. and Perroy, J. (2008) Subcellular imaging of dynamic protein interactions by bioluminescence resonance energy transfer. *Biophys J* **94**(3), 1001–1009
8. Pfleger, K. D. and Eidne, K. A. (2006) Illuminating insights into protein–protein interactions using bioluminescence resonance energy transfer (BRET). *Nat Methods* **3**, 165–174
9. Perroy, J., Raynaud, F., Homburger, V., Rousset, M. C., Telley, L., Bockaert, J. and Fagni, L. (2008) Direct interaction enables cross-talk between ionotropic and group I metabotropic glutamate receptors. *J Biol Chem* **14**(11), 283
10. Mellon, P., Parker, V., Gluzman, Y. and Maniatis, T. (1981) Identification of DNA sequences required for transcription of the human alpha 1-globin gene in a new SV40 host-vector system. *Cell* **27**(2 Pt 1), 279–288

Chapter 20

Quantitative Analysis of Membrane Potentials

Manus W. Ward

Abstract

The changes that occur in electrochemical gradients across biological membranes provide us with invaluable information on physiological responses, pathophysiological processes and drug actions/toxicity. This chapter aims to provide researchers with sufficient information to carry out a quantitative assessment of mitochondrial energetics at a single-cell level thereby providing output on changes in the mitochondrial membrane potential ($\Delta\psi_m$) through the utilization of potentiometric fluorescent probes (TMRM, TMRE, Rhodamine 123). As these cationic probes behave in a Nernstian fashion, changes at the plasma membrane potential ($\Delta\psi_p$) need also to be accounted for in order to validate the responses obtained with $\Delta\psi_m$-sensitive fluorescent probes. To this end techniques that utilize $\Delta\psi_p$-sensitive anionic fluorescent probes to monitor changes in the plasma membrane potential will also be discussed. In many biological systems multiple changes occur at both a $\Delta\psi_m$ and $\Delta\psi_p$ level that often makes the interpretation of the cationic fluorescent responses much more difficult. This problem has driven the development of computational modelling techniques that utilize the redistribution properties of the cationic and anionic fluorescent probes within the cell to provide output on changes in $\Delta\psi_m$ and $\Delta\psi_p$.

Key words: Plasma, mitochondrial, membrane potentials, fluorescence microscopy, confocal fluorescence, TMRM, computational modelling.

1. Introduction

The desire to accurately monitor changes in mitochondrial membrane potential ($\Delta\psi_m$) at a single-cell level has increased dramatically as mounting evidence implicates mitochondria in many roles in both the physiology and pathology of many cell types (1, 2). Indeed, much of my research has focused on the characterization of mitochondrial function during the glutamate-induced neuronal injury associated with ischemic, traumatic, and

D.B. Papkovsky (ed.), *Live Cell Imaging*, Methods in Molecular Biology 591,
DOI 10.1007/978-1-60761-404-3_20, © Humana Press, a part of Springer Science+Business Media, LLC 2010

seizure-induced brain injury (3). Glutamate receptor overactivation is known to result in an excessive Ca^{2+} uptake leading to a rapid loss of mitochondrial bioenergetics and necrosis (4–10), a delayed apoptotic neuronal injury with a collapse of mitochondrial bioenergetics hours after the initial excitation (10–14) and neurons with a hyperpolarized $\Delta\psi_m$ that are tolerant to the stimulus (15, 16). In this injury paradigm it is also evident that rapid changes occur at the plasma membrane potential ($\Delta\psi_p$) during excitation with Ca^{2+} and Na^+ loading of the neurons that is followed by a recovery of $\Delta\psi_p$ when the glutamate stimulus is removed (16). Therefore, a specific need has been established that requires the accurate determination of changes in mitochondrial function (monitoring $\Delta\psi_m$) in situ relative to changes in the electrochemical gradient across the plasma membrane (monitoring $\Delta\psi_p$) at a single-cell level.

Fluorescent membrane-permeant cationic probes such as tetramethylrhodamine methyl and ethyl esters (TMRM, TMRE) and Rhodamine-123 have become some of the most frequently used probes for the analysis of $\Delta\psi_m$ in intact cells due to their minimal photo-toxicity, low photo-bleaching, and the ability to use them in either a quenched (aggregated probe) or non-quenched (no aggregation) mode (17, 18). Previously we have successfully utilized TMRM to monitor changes in $\Delta\psi_m$ in cerebellar granule neurons (10, 16). During these studies we also established that TMRM whole-cell fluorescence is highly sensitive to changes in $\Delta\psi_p$ (10, 18, 19). Indeed, this sensitivity of cationic fluorescent probes to changes in both $\Delta\psi_m$ and $\Delta\psi_p$ have often resulted in a misinterpretation of the fluorescent responses obtained, with changes in $\Delta\psi_m$ frequently overestimated (7, 8, 11, 18, 20–23). This anomaly has in turn driven the development of computational models that utilize the Nernstian behaviour of the cationic fluorescent probes to provide quantitative output on $\Delta\psi_m$ and $\Delta\psi_p$ from the fluorescent traces obtained (10, 16, 18, 24, 25).

Here, I outline microscopy techniques and modelling systems developed for the quantitative analysis of both $\Delta\psi_m$ and $\Delta\psi_p$ at a single-cell level using both anionic and cationic fluorescent probes. Through this series of protocols and control experiments, I aim to provide sufficient information to allow even the most novice researcher to acquire quantitative output on changes in $\Delta\psi_m$ and $\Delta\psi_p$ within any cellular system.

2. Materials

2.1. Chemicals

Poly-L-lysine, poly-D-lysine, collagen IV, soybean trypsin inhibitor, trypsin, bovine serum albumin (BSA), penicillin/

streptomycin, trypsin/EDTA solution, d-glucose, L-glutamine, nerve growth factor (NGF), *N*-methyl-D-aspartic acid (NMDA), kainic acid, glycine, tunicamycin, thapsigargin, epoxomicin, H_2O_2, menadione, carbonyl cyanide *p*-(trifluoromethoxy) phenylhydrazone (FCCP), phosphate buffer saline (PBS) tablets, NaCl, KCl, KH_2PO_4, $NaHCO_3$, Na_2SO_4, $MgCl_2$, NaOH (all from Sigma); $CaCl_2$ (Fluka), L-glutamic acid monosodium salt (Fluka), HEPES (Fluka), bortezomib (LCLabs). Distilled water (Barnstead, Milli-Q).

2.2. Media

RPMI 1640 (Lonza), MEM (Gibco/Invitrogen), horse serum (Gibco/Invitrogen), fetal bovine serum (FBS) (Sigma).

2.3. Tissue

Wistar rat pups, 7 days old (Harlan; in-house breeding), PC-12 cells (ATCC).

2.4. Instruments

Coverslips 13 mm, #1 thickness (VWR International), Willco dishes (Willco BV, the Netherlands), Haemocytometer (Hawksley, UK), Microlance 23G needles (Becton Dickinson), pH meter (Thermo Electro Corporation), perfusion chamber (Warner Instruments).

2.5. Software

MATLAB® software (Mathworks, UK). *ImageJ* free online image analysis software http://rsb.info.nih.gov/ij/features.html. *MetaMorph* Software (Molecular Devices, Berkshire, UK). *CellTrack* free online cell tracking software http://db.cse.ohio-state.edu/CellTrack.

2.6. Experimental Buffer for Microscopy Experiments

120 mM NaCl, 3.5 mM KCl, 0.4 mM KH_2PO_4, 20 mM HEPES, 5 mM $NaHCO_3$, 1.2 mM Na_2SO_4, 1.2 mM $CaCl_2$, 1.2 mM $MgCl_2$ and 5 mM glucose, pH 7.4, with NaOH (*see* **Note 1**).

2.7. Plating Cells onto Glass for Microscopy

Below are some examples of the culturing conditions required for the plating of neurons/cells on glass surface in order to carry out real-time analysis of $\Delta\psi_m$ and $\Delta\psi_p$ with fluorescence or confocal microscopy. Choice of glass coverslips or glass dishes used is dictated by the type of perfusion/cell chamber you have.

2.7.1. Plating Primary Cerebellar Granule Neurons onto Glass

1. All work carried out on animals must have ethical approval and a valid government licence prior to commencing.
2. Pre-coat glass coverslips and glass Willco dishes with poly-L-lysine or poly-D-lysine at a concentration of 5 μg/mL for at least 2 h. Then wash three times with autoclaved/filtered distilled water and leave to dry prior to the plating of the neurons (*see* **Note 2**).
3. Dissect cerebella from 7-day-old Wistar rats of both sexes and pool in ice-cold preparation buffer (500 mg glucose,

600 mg BSA, one tablet of PBS and make up to 200 mL of distilled water). Place the tissue in 0.25 mg/mL trypsin and incubate at 37°C for 20 min, with trypsinization terminated by the addition of 0.05 mg/mL soybean trypsin inhibitor.

4. Triturate the neurons in order to break up any remaining clumps of cells and resuspended in 20 mL of supplemented culture medium: 450 mL MEM, 50 mL FBS, 100 U/mL penicillin and 100 μg/mL streptomycin, 3 g d-glucose, 700 mg KCl 150 mg glutamine.
5. Place 100 μL of the cell suspension on a haemocytometer and count the cells. Live cells appear as phase bright spheres; dead cells are dark and of an irregular shape.
6. Dilute the cell suspension to a final concentration of 1×10^6 cells/mL and plate in the density required: 1 mL for Willco dishes (1×10^6 cells), 200 μL for 13-mm coverslips (2×10^5 cells).
7. Add extra media 1 h after plating (10).

2.7.2. Preparation and Plating of PC12 Cells

1. To differentiate PC12 cells, the cells are gently washed with sterile PBS and incubated for 2 min at 37°C with 0.25% trypsin/2 mM EDTA solution.
2. The PC12 cells are then resuspended in supplemented culture: RPMI 1640 (425 mL), 2 mM L-glutamine, 10% (50 mL) horse serum, 5% (25 mL) fetal bovine serum, 100 U/mL penicillin and 100 μg/mL streptomycin) and passed 6–8 times through the 23G needle to break down any clumps.
3. The cells are then counted using a haemocytometer and diluted in the supplemented culture media and seeded at the density required (Willco dish, 1 mL, 1×10^4 cells, 13 mm glass coverslips, 200 μL, 5×10^2 cells) on glass pre-coated with collagen IV (2 μg/mL) (26).
4. Plate cells to allow for 50–70% confluency on the day of experimentation.
5. The cells are allowed to settle and stick to the collagen coated glass for 8 h prior to switching on to differentiating media (RPMI 1640 (495 mL), 1% horse serum (5 mL), 100 U/mL penicillin, 100 μg/mL streptomycin and 100 ng/mL NGF) (27).

2.7.3. Preparation and Plating of Most Other Cell Lines

1. Cultured cells are washed twice with sterile PBS and incubated for 5 min at 37°C with 0.25% trypsin/2 mM EDTA solution.
2. The cells are then resuspended in the required supplemented culture media (RPMI 1640 (450 mL), 2 mM L-glutamine,

10% (50 mL) fetal bovine serum, 100 U/mL penicillin and 100 μg/mL streptomycin) with rigorous pipetting that is sufficient to break up any remaining clumps of cells.

3. The cells are then counted using a haemocytometer and diluted in the supplemented culture media and seeded to allow for 50–70% confluency (Willco dish, 1 mL, 1×10^4 cells, 13 mm glass coverslips, 200 μL, 5×10^3 cells) on the day of experimentation.
4. For most cell lines glass surfaces do not needed to be pretreated.
5. Extra media should be added 1 h after plating.

2.8. Fluorescent Probes

1. TMRE (Invitrogen) and TMRM (Invitrogen) are made up in DMSO at a concentration of 1 mM, dispensed into 50 μL and stored at –20°C.
2. Rhodamine 123 (Invitrogen) is made up in DMSO at a concentration of 10 mM, dispensed into 50 μL and stored at –20°C.
3. DiSBAC$_2$ (3) (Invitrogen) is made up in DMSO to a stock concentration of 10 mM, dispensed into 50-μL aliquots and frozen.
4. A single vial of "Membrane Potential Assay Kit" (Molecular Devices) containing the plasma membrane potential indicator (PMPI) is reconstituted in 1 mL of distilled water, and dispensed into 50-μL aliquots, and frozen (PMPI stock).

3. Methods

3.1. Examples of Image Acquisition

Fine details are dependent on the requirements of the user needs and the equipment available.

3.1.1. Confocal Fluorescence Microscopy

Carl Zeiss (Jena, Germany) 510, 5-live, 710 confocal microscopes. When using 40× and 60× oil-immersion objectives, the pinhole should be between 1.0 and 1.6 Airy units to allow for an optical slice of 1 μm. TMRM is excited at 543 nm with an HeNe laser at 2/3% of total power output and the emission collected above 560 nm with a 560 nm longpass filter. This will allow the user to resolve individual mitochondria. A larger optical slice will significantly increase fluorescence; however, individual mitochondria will not be readily resolved.

3.1.2. Wide-Field Fluorescent Microscopy

Fluorescence is observed using an Axiovert 200 M inverted microscope equipped with a 40× numerical aperture 1.3 oil-immersion objective (Carl Zeiss, Jena, Germany), polychroic mirror, and filter wheels in the excitation and emission light path containing the appropriate filter sets (TMRM, excitation 530 ± 25 nm, emission 592.5 ± 22.5 nm; dichroic mirrors for TMRM; Semrock (Rochester, NY)). Images are recorded using a back-illuminated, cooled electron multiplying CCD camera Andor Ixon BV 887-DCS (Andor Technologies, Belfast, Northern Ireland), and the imaging setup is controlled by MetaMorph 7.1 software (Molecular Devices Ltd., Wokingham, UK).

3.2. Determination of $\Delta\psi_m$ in Living Cells

TMRM and TMRE are preferred membrane potential sensors for the quantitative assessment of $\Delta\psi_m$ in living cells. They pass through the plasma membrane better than the related Rhodamine 123 and accumulate in mitochondria (17). The transmembrane distribution of these probes is directly related to the membrane potential, thereby allowing for the accurate determination of $\Delta\psi_m$. This section describes in brief a number of techniques that are commonly used to measure $\Delta\psi_m$ and $\Delta\psi_p$; however, most of it focuses on the utilization of TMRM to provide a quantitative assessment of $\Delta\psi_m$ at a single-cell level over time.

3.2.1. Measuring Changes in $\Delta\psi_m$ with TMRE

1. Make a 1 in 10 dilution (10 μL of stock TMRE added to 90 μL of experimental buffer) of your 1 mM stock solution of TMRE in experimental buffer.
2. From the 100 μM diluted solution add 1 μL to every 1 mL of experimental buffer to give a final concentration of 100 nM TMRE.
3. Incubate cells for 15 min at 37°C, then mount the glass coverslips/dishes in the perfusion/cell chamber in experimental buffer containing 100 nM TMRE.
4. Allow at least 20 min for the temperature of all components of the imaging setup to come to a steady state.
5. Measure TMRE fluorescence using wide-field or confocal fluorescence microscopy, with optimal excitation at 540 nm and emission collected above 560 nm.
6. Ensure that TMRE is equilibrated within the cytosol and mitochondria by monitoring the TMRE fluorescence for 10–15 min and that the whole-cell fluorescence remains stable (*see* **Note 3**).
7. When the fluorescence is stable the desired stimulation (*see* **Note 4**) is applied and the fluorescence monitored.

3.2.2. Measuring Changes in $\Delta\psi_m$ with Rhodamine 123

1. Make a 1 in 10 dilution (10 μL of stock Rhodamine 123 added to 90 μL of experimental buffer) of the 10 mM Rhodamine 123 stock solution.

2. From this 1 mM diluted stock add 2.6 μL to 1 mL of loading buffer to give a loading buffer with a final concentration of 2.6 μM.
3. Cells are incubated with 2.6 μM Rhodamine 123 for 15 min in experimental buffer at 37°C prior to mounting the cells in the perfusion chamber. Alternative loading paradigms with Rhodamine 123 (26 μM) may sensitize the neurons to photo-induced damage.
4. Unlike TMRM and TMRE, Rhodamine 123 TMRE is not present in the experimental buffer during the course of the experiment.
5. Following the placement of the cells on the microscope stage allow at least 20 min for the temperature of the imaging setup to come to a steady state.
6. Rhodamine 123 fluorescence is then measured with fluorescent/confocal microscopy; excitation at 485 nm and emission collected above 520 nm.
7. A baseline Rhodamine 123 fluorescence should be measured for at least 10–15 min (**Note 3**) prior to addition of any stimulation (**Note 4**) (9, 10).

3.2.3. Measuring Changes in $\Delta\psi_m$ with TMRM

1. Make a 1 in 100 dilution (10 μL of stock TMRM added to 990 μL of experimental buffer) of your stock (1 mM) TMRM solution in experimental buffer.
2. From this 10 μM solution add the desired volume to every 1 mL of experimental buffer to give a final concentration For example, 2 μL of 10 μM TMRM added to 1 mL of experimental buffer gives a final concentration of 20 nM.
3. Typically to work in the quenched condition, use a concentration of TMRM between 50 and 200 nM (*see* **Fig. 20.1a**) and for non-quenched conditions a concentration of less than 20 nM (*see* **Fig. 20.1b**) is desirable (*see also* **Section 3.2.5**).
4. Cells on coverslips/glass dishes have the media removed and replaced with experimental buffer containing TMRM for 30 min.
5. Replace with fresh buffer containing the loading concentration of TMRM when the coverslips/glass dish are mounted in the perfusion/cell chamber and maintain at 37°C. When using a non-perfusion chamber check to make sure that evaporation is not a major issue or seal with mineral oil.
6. Give the system a further 20–30 min for all components of the imaging setup to reach a constant temperature. Small fluctuations in temperature lead to small drifts in focus and

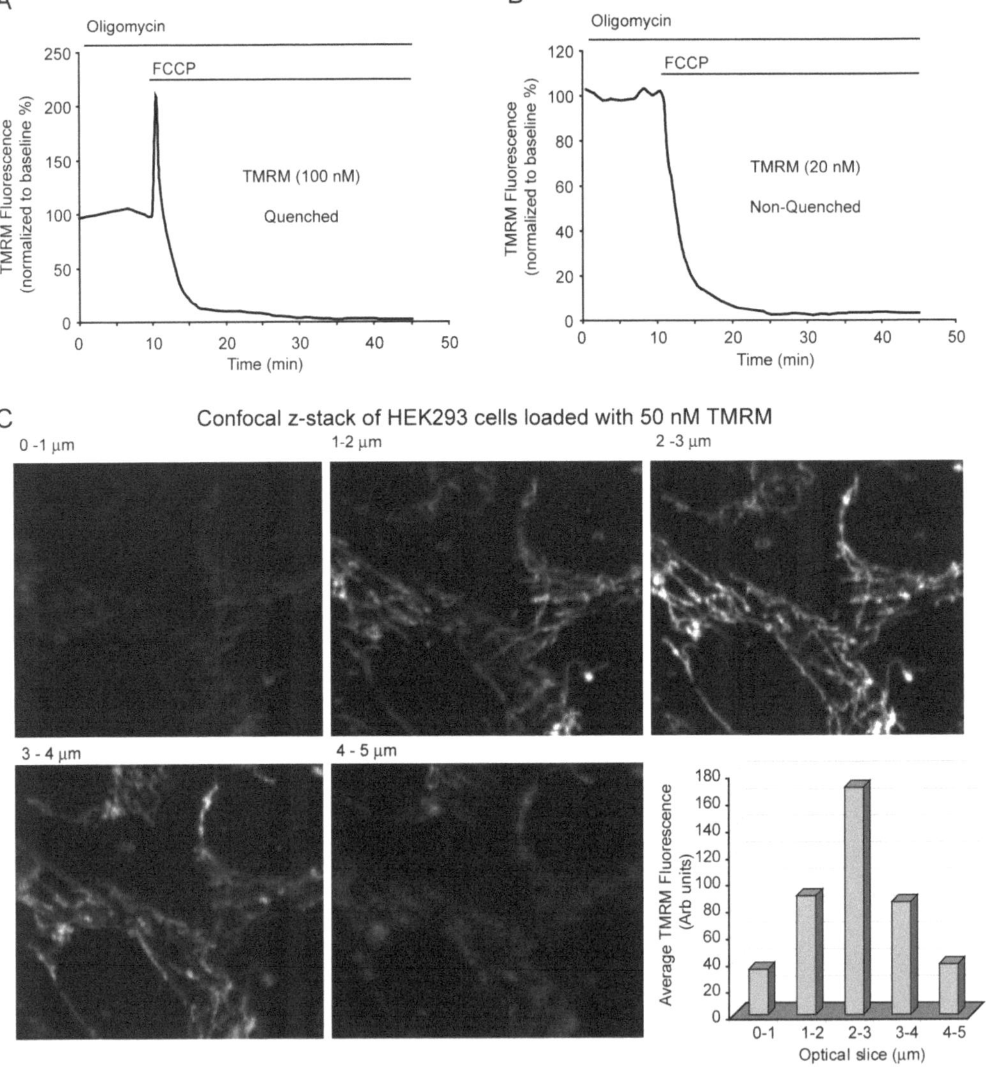

Fig. 20.1. Characterization of TMRM fluorescence in cells with high-resolution single-cell confocal microscopy. (**a, b**) Cerebellar granule neurons loaded with 100 nM TMRM (**a**) and 30 nM TMRM (**b**) were exposed to 2 μM FCCP in the presence of 2 μg/mL oligomycin and the changes in the whole-cell fluorescence monitored. (**c**) HEK293 cells were loaded with 100 nM TMRM and a high-resolution *z*-stack was acquired of the mitochondria within the cells. *Graph inset* indicates the average TMRM fluorescence within mitochondrial rich regions for each plane of the *z*-stack.

this can have major implications on the whole-cell fluorescence obtained (*see* **Fig. 20.1c, Notes 5 and 6**).

7. TMRM has a broad excitation and emission spectra with excitation of TMRM between 530 and 550 nm preferable; however, TMRM can be excited at 488 nm if required. The emission spectra for TMRM are also broad so a 560-nm longpass filter is preferred.

8. A baseline TMRM fluorescence should be measured for at least 10–15 min (**Note 3**) prior to any stimulation (*see* **Note 4**)

3.2.4. Image Analysis

The type of analysis carried out will be highly dependent on the software kinetics package associated with the imaging setup. If the associated software does not provide sufficient flexibility two other options are recommended: ***ImageJ*** (http://rsb.info.nih.gov/ij/features.html) and its Java source code are freely available and in the public domain. This software is continually updated with over 500 plug-ins available. ***MetaMorph*** Software (Molecular Devices) is a very powerful tool for image acquisition, processing, and analysis. The software package is very versatile and offers solutions for a wide range of applications. Both software packages are routinely used within our group and are highly recommended.

1. An associated kinetic package with the software will allow for the data (e.g. average intensity, integrated fluorescence and standard deviation) to be collected for each region of interest (ROI) at each time point.
2. The data is typically stored in a format that will allow it to be easily exported or opened in an Excel spreadsheet.
3. For an accurate comparison of changes in the average fluorescence intensity for multiple cells it is suggested to normalize the baseline fluorescence to 100% due to mitochondria within different cells not all being on the same *z*-plane.
4. To ensure that the traces for each region are representative of what is happening in the cell, carefully check to make sure that the cell remains within the ROI and that there are no major drifts in focus (*see* **Note 5**).
5. If cells move or change shape significantly on a regular basis, analysis should be carried out with cell tracking software (***CellTrack*** http://db.cse.ohio-state.edu/CellTrack (28) (*see* **Note 7**).

3.2.5. Determination of the Quench Limit for TMRM Within the Mitochondrial Matrix

1. The current modelling techniques for changes in $\Delta\psi_m$ are valid when working in both the quenched (1A) and non-quench (*see* **Fig. 20.1b**) modes. However, it is important to establish the concentration of TMRM in the matrix that initiates aggregation and quenching (10, 16, 24, 29).
2. Cells are equilibrated with TMRM in a concentration range from 100 to 10 nM.
3. Cells are pre-treated with oligomycin (2 μg/mL) for 10 min prior to collapsing $\Delta\psi_m$ with a protonophore (FCCP) in

order to block ATP synthase reversal and cellular ATP depletion.

4. The protonophore FCCP is added at a concentration of 2 μM, which should be sufficient to rapidly collapse $\Delta\psi_m$. If the concentration of FCCP is not sufficient then increase it to 5–10 μM. However, at these higher concentrations $\Delta\psi_p$ may also be affected (*see* **Note 8**).
5. Validation that the fluorescent changes are due to alterations of $\Delta\psi_m$ need to be carried out using the $\Delta\psi_p$-sensitive indicator $DiSBAC_2$ (3) or PMPI (**Section 3.2**).
6. If significant changes are occurring at a $\Delta\psi_p$ level, then the concentration of FCCP needs to be reduced or other channel inhibitors added (e.g. for neurons the glutamate receptor antagonists MK-801and NBQX are required (16)).

3.2.6. Estimation of the Rate Constant for TMRM Re-equilibration Across the Plasma Membrane

The relaxation constant seen in Eqs. 20.1–20.3 is due to the delay of movement of the TMRM through the plasma membrane. It is obtained in control experiments where either $\Delta\psi_p$ or $\Delta\psi_m$ are depolarized (high KCl, FCCP) and the redistribution constant for TMRM is calculated (10, 16, 24, 29). Nernstian equilibration of a generic membrane-permeant monovalent cation C^+ among the extracellular, cytoplasmic, and mitochondrial matrix phases at 37°C results in the following relationships for the concentrations of the free cations at equilibrium:

$$c^+_{[\text{matrix}]} = c^+_{[\text{cytoplasm}]} * 10^{-\Delta\Psi m/61.5} \quad (1)$$

$$c^+_{[\text{cytoplasm}]} = c^+_{[\text{medium}]} * 10^{-\Delta\Psi p/61.5} \quad (2)$$

$$c^+_{[\text{matrix}]} = c^+_{[\text{medium}]} * 10^{-(\Delta\Psi m+\Psi p)/61.5} \quad (3)$$

where $\Delta\psi_m$ and $\Delta\psi_p$ are, respectively, the plasma and mitochondrial membrane potentials (using accepted sign conventions), and the divisor 61.5 is the value (in mV) for RT/F at 37°C. Re-equilibration of the cytosolic and extracellular compartments due to changes in the potentials of $\Delta\psi_p$ and $\Delta\psi_m$ are taken into account by assuming a first order flux between the extracellular medium and the cytosol:

$$J_{[\text{cytoplasm}]} = \left(c^+_{[\text{cytoplasm,actual}]} - c^+_{[cytoplasm,equilibrium]}\right) k_{\text{kinetic}} \quad (4)$$

where $[k_{\text{kinetic}}]$ is a constant proportional to the permeability of the C^+ probe across the plasma membrane.

3.2.7. The Fraction of the Cell Volume Occupied by the Mitochondrial Matrices

Cellular TMRM fluorescence is not only highly dependent on the changes in both $\Delta\psi_m$ and $\Delta\psi_p$ but is also directly related to the total mitochondrial volume within the cell (10, 16, 24, 29).

Therefore, a stringent determination of the mitochondrial volume is required.

1. Cells are loaded with 50 nm TMRM.
2. High-resolution *z*-stacks with 1 μm optical slices and 0.2 μm steps are taken of 3–4 fields of cells (70–80 images per field).
3. Deconvolution software is used to create a three-dimensional image of the cells and the mitochondria. Individual mitochondria are counted and the volume of each estimated for at least three fields of cells.
4. The volume of the respective cells are calculated and the mitochondrial volume is then calculated as a percentage of the total cellular volume.

3.3. Protocols for Determination of $\Delta\psi_p$ in Living Cells

Patch clamping has long been considered the standard way to measure changes in $\Delta\psi_p$ in single cells; however, it can be a slow, labour-intensive process with relatively low throughput. To provide much faster methods that have applications within both research and industry a series of voltage-sensitive dyes have been devised that provide reliable fast responses to changes in $\Delta\psi_p$.

3.3.1. Measuring Changes in $\Delta\psi_p$ with DiSBAC$_2$ (3)

1. DiSBAC$_2$ (3) is a bis-barbituric acid oxonol compound that partitions into the membrane as a function of membrane potential. Hyperpolarization causes extrusion of the dye and decreased fluorescence, whereas depolarization causes enhanced fluorescence (30).
2. Make a 1 in 10 dilution (10 μL of stock DiSBAC$_2$ (3) added to 990 μL of experimental buffer) of your stock (10 mM) DiSBAC$_2$ (3) solution in experimental buffer.
3. From this 1 mM solution add 1 μL to every 1 mL of experimental buffer to give the desired working concentration.
4. Cells are incubated with experimental buffer containing 1 μM DiBAC$_2$ (3) for 30 min at 37°C. One micromolar DiSBAC$_2$ (3) is also present in the experimental buffer when cells are mounted on the microscope stage.
5. Optimal excitation of DiSBAC$_2$ (3) probe is at 543 nm with and the emission collected above 560 nm.
6. DiSBAC$_2$ (3) fluorescence is calibrated as a function of $\Delta\psi_p$ depolarization and is determined by quantifying the fluorescent enhancement obtained when $\Delta\psi_p$ is depolarized by increasing KCl concentrations.
7. A series of experiments can be performed in which step increases in K^+ concentrations are made by removing defined volumes of low-K medium and replacing this with an equal volume of high-K medium to give final K^+ concentrations from 3.5 to 75 mM (24).

3.3.2. Measuring Changes in $\Delta\psi_p$ with PMPI

1. The Molecular Devices "Membrane potential assay kit" has been utilized in a number of studies and has been termed PMPI, for "plasma membrane potential indicator" by David Nicholls (24).
2. Cell are washed with experimental buffer and then incubated with 0.5 μL/mL PMPI stock in experimental buffer (37°C) for 45 min prior to imaging. 0.5 μL/mL PMPI is also maintained in the experimental buffer during the course of the experiment.
3. PMPI is excited in single track mode with the 514 nm band of an argon laser and the emission collected above 570 nm with the laser intensity maintained at 1% to reduce phototoxicity.
4. PMPI fluorescence can be calibrated as above (**Section 3.2.1**, step 5).

3.4. Computational Modelling

Since fluorescently charged dyes such as TMRM respond to changes in both $\Delta\psi_m$ and $\Delta\psi_p$, the interpretation of changes in whole-cell fluorescence is often difficult. For this reason, mathematical models that exploit the Nernstian behaviour of TMRM have been used to create computer simulations and models that provide output on changes in $\Delta\psi_m$ and $\Delta\psi_p$ over time.

3.4.1. Variables

Certain variables are required prior to Data input for each cellular system. These can be obtained by carrying out the control experiments described above, namely, *establishing the mitochondrial volume fraction* of the cell (**Section 3.2.7**); inputting the *external TMRM probe concentration* (quenched and non-quench (**Section 3.2.3**)), the *TMRM rate constant k* (**Section 3.2.6**), the *quench limit* for TMRM (**Section 3.2.5**), the *PMPI rate constant* (**Section 3.2.2**) calculated from curve-fitting an experiment in which $\Delta\psi_p$ is rapidly depolarized by elevating external KCl concentration. The initial resting values for $\Delta\psi_m$ are in the region of –150 mV (24). The resting value for $\Delta\psi_p$ is typically between –60 and –80 mV (24).

3.4.2. Data Input to Computer Simulation

1. The computer simulation and accompanying notes are provided as supplemental material in http://www.jbc.org/cgi/content/full/M510916200/DC1 (24).
2. When the whole fluorescent traces for both the TMRM and PMPI have been obtained, open a copy of the master template in the supplemental data.
3. Enter the variables in the upper left hand corner (A).
4. Paste in a copy of the TMRM and PMPI traces into panel B.

5. Manually alter the potential profiles for $\Delta\psi_m$ and $\Delta\psi_p$ (H) in order that the simulation traces (C) can fit the raw data traces. This may require some time as multiple changes in both $\Delta\psi_m$ and $\Delta\psi_p$ may be required to obtain an optimal fit for the raw data.
6. When you are satisfied with the fitting a graphical form of the output for changes in $\Delta\psi_m$ and $\Delta\psi_p$ can be found in panel (E) and numerical values obtained in column (H).
7. When changing the external TMRM concentration in (A), significant changes will occur to the total simulated signal (T). To get good fits multiply the raw data by a factor so that the experimental trace and fitted trace can overlap in panel (D). A similar approach should also be adopted for fitting the PMPI experimental traces.

3.4.3. TOXI-SIM

The Nernstian properties of TMRM and how it responds to changes in both $\Delta\psi_m$ and $\Delta\psi_p$ allow for the mathematical analysis of the distribution of the probe to provide quantitative information on changes of both $\Delta\psi_m$ and $\Delta\psi_p$ over time (10, 16, 24, 25). Huber et al. (29), have built on the initial studies to develop an automated computational model TOXI-SIM (http://systemsbiology.rcsi.ie/tmrm/index.html) that is based on Newton-fitting algorithms implemented into MATLAB® software (Mathworks, UK).

Data Input for TOXI-SIM

1. In work carried in neurons stimulated with glutamate a number of clearly defined TMRM responses representing distinct injury paradigms (rapid necrosis, delayed (biphasic) necrosis, apoptosis and tolerance) were established (10, 16, 29).
2. For each injury model a mathematical model was created that can be opened by clicking on the appropriate header option (rapid necrosis, biphasic necrosis, apoptosis and tolerance) in TOXI-SIM main menu.
3. For each model the fitting variables (**Section 3.4.1**) are entered into the upper right hand panel on the web service.
4. Each variable is entered and a fitting range of either 10%, 100%, free fitting or no change is entered. The fitting range will depend on the range within which a certain variable may fall.
5. The experimental protocol parameters and cell specific initial values for $\Delta\psi_m$ and $\Delta\psi_p$ are entered in the lower right table.
6. Fluorescence traces for up to eight cells, normalized to an initial value of 100%, are entered either by copying them to the dedicated textbox in the left bottom or by directly accessing a locally stored Excel file via a browse facility.

7. The first column of the input data is dedicated to measurements time points (s, min, h) and the appropriate timescale is selected in the checkbox above the trace entry field.
8. The calculate options is then selected and the resultant fluorescent traces, fitted traces and output for $\Delta\psi_m$ and $\Delta\psi_p$ are rendered graphically and provided in an Excel format for each cell.
9. To achieve optimal fits for your data sets, the traces for the cells should be broken down into subsets that are similar to each of the models described and this option selected, i.e. do not try to fit traces similar to the apoptotic response with the necrotic model.

Data Base for Storage

1. TOXI-SIM is designed as a flat file based database for experimental storage and retrieval allowing for multiple users to access the same data sets.
2. By entering the "Submit experimental Model data" link on the main page, a form is automatically loaded that permits the submission of data sets for each of the four models.
3. Each data set allows for the entry of data time series for up to eight cells, modification of parameter and their fitting ranges as well as technical data from the experimental design.
4. The data set is given a unique name thus allowing for easy retrieval by multiple users.

Overall, the ability of computer models to simultaneously evaluate changes in $\Delta\psi_m$ and $\Delta\psi_p$ for multiple cells significantly enhances our understanding of cellular function and mitochondrial energetics in response to different stimuli, offering novel insights into how their cellular system respond. In addition the throughput capacity and reduced manual input provided by TOX-SIM allows for statistical analysis to be carried out on large populations of single cells and has the potential to be utilized in high-content screening and high-throughput screening platforms within research and industry thereby allowing for the rapid screening and characterization of drugs actions/toxicity.

The most accurate computational fittings and therefore the most accurate quantitative assessment off changes in $\Delta\psi_m$ and $\Delta\psi_p$ are obtained when the fluorescent response for the cationic and anionic fluorescent responses are representative of what is occurring within the cell. Therefore, it is essential that users ensure that the quality of the images obtained and the processing of the data is of the highest standard.

4. Notes

1. $MgCl_2$ is not added to the experimental buffer under conditions where an excitation of NMDA glutamate receptors is required.
2. A number of different coating techniques use mixtures of laminin, collagen, poly-L-lysine and poly-D-lysine for different types of neuronal populations. Please check which technique is most widely used as this can have a major influence on neuronal survival following plating, their development in culture, and behaviour on stimulation.
3. A baseline TMRM/TMRE/Rhodamine 123 fluorescence should be measured for at least 10–15 min prior to the stimulation of the cells to ensure that the loading of probe within the mitochondria and cells has come to a resting equilibrium.
4. Some example of stimuli routinely used are as follows: neuronal excitation; *NMDA (100 μM), kainate (300 μM), glutamate (100 μM)*. Endoplasmic reticulum stress; *tunicamycin (1 μM), thapsigargin (10 μM)*. Proteasomal stress; *epoxomicin (50 nM), bortezomib (100 nM)*. Oxidative stress; *H_2O_2 (50 μM), menadione (100 nM)*.
5. Even small drifts of focus can dramatically alter the whole-cell fluorescence. In **Fig. 20.1c**, a *z*-stack containing five optical slices was taken through a HEK293. It is readily apparent that only a 1-μm shift up or down from the optimal optical (2–3 μm) slice results in a marked alteration in the whole -cell TMRM fluorescence. To reduce focus drifts, allow all components of the system to come to a steady temperature prior to beginning an experiment. Room temperature should be maintained within a 1–2°C window. Use a temperature controller for the objective if possible. Have your microscope on an air table or in a stable environment to reduce even the smallest of vibrations. If the microscope system allows, carry out small *z*-stacks (3–4 images, 3–4 μm) at each time point, thereby allowing for some minor drifts in focus. If your cells have drifted out of focus, stop the experiment, save data, refocus and start acquisitions again. A great deal of data can still be obtained from the experiment.
6. Choosing the best field of focus is often difficult. Some cells spread out and are flat (HEK293, *see* **Fig. 20.1c**), making it relatively easy to resolve individual mitochondria even with a standard fluorescent microscope. However, most cells are not flat; therefore, it is advisable to focus on a *z*-plane that

cuts through mitochondria-rich areas within the majority of cells.

7. Many cells move and change shape on a regular basis throughout the course of an experiment. This is particularly relevant when monitoring most cell lines over long periods of time. For this reason many groups are currently working on the development of cell tracking software that can segment and track individual cells, thereby providing an accurate assessment of changes in whole-cell fluorescence over time. Free tracking software is now available online (***CellTrack*** http://db.cse.ohio-state.edu/CellTrack (28)) and this area is a major focus of development within many imaging groups with further advancements in cell tracking technology soon available.

8. In neurons, $\Delta\psi_p$ is altered following a rapid depolarization of synaptic mitochondria with the protonophore FCCP. This rapid depolarization is believed to allow for the release of glutamate from the pre-synaptic terminal into the synaptic cleft and the activation of post-synaptic NMDA receptors.

References

1. Duchen MR. (2004) Mitochondria in health and disease: perspectives on a new mitochondrial biology. Mol Aspects Med, 25, 365–451.
2. Nicholls DG. (2004) Mitochondrial dysfunction and glutamate excitotoxicity studied in primary neuronal cultures. Curr Mol Med, 4, 149–77.
3. Choi DW. (1994) Glutamate receptors and the induction of excitotoxic neuronal death. Prog Brain Res, 100, 47–51.
4. Choi DW. (1987) Ionic dependence of glutamate neurotoxicity. J Neurosci, 7, 369–79.
5. Tymianski M, Charlton MP, Carlen PL, Tator CH. (1993) Secondary Ca^{2+} overload indicates early neuronal injury which precedes staining with viability indicators. Brain Res, 607, 319–23.
6. Budd SL, Nicholls DG. (1996) Mitochondria, calcium regulation, and acute glutamate excitotoxicity in cultured cerebellar granule cells. J Neurochem, 67, 2282–91.
7. White RJ, Reynolds IJ. (1996) Mitochondrial depolarization in glutamate-stimulated neurons: an early signal specific to excitotoxin exposure. J Neurosci, 16, 5688–97.
8. Stout AK, Raphael HM, Kanterewicz BI, Klann E, Reynolds IJ. (1998) Glutamate-induced neuron death requires mitochondrial calcium uptake. Nat Neurosci, 1, 366–73.
9. Vergun O, Keelan J, Khodorov BI, Duchen MR. (1999) Glutamate-induced mitochondrial depolarisation and perturbation of calcium homeostasis in cultured rat hippocampal neurones. J Physiol, 519 Pt 2, 451–66.
10. Ward MW, Rego AC, Frenguelli BG, Nicholls DG. (2000) Mitochondrial membrane potential and glutamate excitotoxicity in cultured cerebellar granule cells. J Neurosci, 20, 7208–19.
11. Ankarcrona M, Dypbukt JM, Bonfoco E, Zhivotovsky B, Orrenius S, Lipton SA, Nicotera P. (1995) Glutamate-induced neuronal death: a succession of necrosis or apoptosis depending on mitochondrial function. Neuron, 15, 961–73.
12. Ward MW, Rehm M, Duessmann H, Kacmar S, Concannon CG, Prehn JH. (2006) Real time single cell analysis of bid cleavage and bid translocation during caspase-dependent and neuronal caspase-independent apoptosis. J Biol Chem, 281, 5837–44.
13. Budd SL, Tenneti L, Lishnak T, Lipton SA. (2000) Mitochondrial and extramitochondrial apoptotic signaling pathways in cerebrocortical neurons. Proc Natl Acad Sci USA, 97, 6161–6.
14. Luetjens CM, Bui NT, Sengpiel B, Munstermann G, Poppe M, Krohn AJ, Bauerbach E, Krieglstein J, Prehn JH. (2000) Delayed mitochondrial dysfunction in

excitotoxic neuron death: cytochrome c release and a secondary increase in superoxide production. J Neurosci, 20, 5715–23.
15. Iijima T, Mishima T, Akagawa K, Iwao Y. (2003) Mitochondrial hyperpolarization after transient oxygen-glucose deprivation and subsequent apoptosis in cultured rat hippocampal neurons. Brain Res, 993, 140–5.
16. Ward MW, Huber HJ, Weisova P, Dussmann H, Nicholls DG, Prehn JH. (2007) Mitochondrial and plasma membrane potential of cultured cerebellar neurons during glutamate-induced necrosis, apoptosis, and tolerance. J Neurosci, 27, 8238–49.
17. Ehrenberg B, Montana V, Wei MD, Wuskell JP, Loew LM. (1988) Membrane potential can be determined in individual cells from the nernstian distribution of cationic dyes. Biophys J, 53, 785–94.
18. Nicholls DG, Ward MW. (2000) Mitochondrial membrane potential and neuronal glutamate excitotoxicity: mortality and millivolts. Trends Neurosci, 23, 166–74.
19. Farkas DL, Wei MD, Febbroriello P, Carson JH, Loew LM. (1989) Simultaneous imaging of cell and mitochondrial membrane potentials. Biophys J, 56, 1053–69.
20. Khodorov B, Pinelis V, Vergun O, Storozhevykh T, Vinskaya N. (1996) Mitochondrial deenergization underlies neuronal calcium overload following a prolonged glutamate challenge. FEBS Lett, 397, 230–4.
21. Keelan J, Vergun O, Duchen MR. (1999) Excitotoxic mitochondrial depolarisation requires both calcium and nitric oxide in rat hippocampal neurons. J Physiol, 520 Pt 3, 797–813.
22. Kiedrowski L. (1998) The difference between mechanisms of kainate and glutamate excitotoxicity in vitro: osmotic lesion versus mitochondrial depolarization. Restor Neurol Neurosci, 12, 71–9.
23. Prehn JH. (1998) Mitochondrial transmembrane potential and free radical production in excitotoxic neurodegeneration. Naunyn Schmiedebergs Arch Pharmacol, 357, 316–22.
24. Nicholls DG. (2006) Simultaneous monitoring of ionophore- and inhibitor-mediated plasma and mitochondrial membrane potential changes in cultured neurons. J Biol Chem, 281, 14864–74.
25. Dussmann H, Rehm M, Kogel D, Prehn JH. (2003) Outer mitochondrial membrane permeabilization during apoptosis triggers caspase-independent mitochondrial and caspase-dependent plasma membrane potential depolarization: a single-cell analysis. J Cell Sci, 116, 525–36.
26. Tomaselli KJ, Damsky CH, Reichardt LF. (1987) Interactions of a neuronal cell line (PC12) with laminin, collagen IV, and fibronectin: identification of integrin-related glycoproteins involved in attachment and process outgrowth. J Cell Biol, 105, 2347–58.
27. Zhdanov AV, Ward MW, Prehn JH, Papkovsky DB. (2008) Dynamics of intracellular oxygen in PC12 cells upon stimulation of neurotransmission. J Biol Chem, 283, 5650–61.
28. Sacan A, Ferhatosmanoglu H, Coskun H. (2008) CellTrack: an open-source software for cell tracking and motility analysis. Bioinformatics, 24, 1647–9.
29. Huber HJ, Plchut M, Weisova P, Dussmann H, Wenus J, Rehm M, Ward MW, Prehn JH. (2009) TOXI-SIM-A simulation tool for the analysis of mitochondrial and plasma membrane potentials. J Neurosci Methods, 176, 270–5.
30. Freedman JC, Novak TS. (1989) Optical measurement of membrane potential in cells, organelles, and vesicles. Methods Enzymol, 172, 102–22.

Chapter 21

Image Correlation Spectroscopy to Define Membrane Dynamics

Jeremy Bonor and Anja Nohe

Abstract

Fluorescent imaging techniques are powerful tools that aid in studying protein dynamics and membrane domains and allow for the visualization and data collection of such structures as caveolae and clathrin-coated pits, key players in the regulation of cell communication and signaling. The family of image correlation spectroscopy (FICS) provides a unique way to determine details about aggregation, clustering, and dynamics of proteins on the plasma membrane. FICS consists of many imaging techniques which we will focus on including image correlation spectroscopy, image cross-correlation spectroscopy and dynamic image correlation spectroscopy. Image correlation spectroscopy is a tool used to calculate the cluster density, which is the average number of clusters per unit area along with data to determine the degree of aggregation of plasma membrane proteins. Image cross-correlation spectroscopy measures the colocalization of proteins of interest. Dynamic image correlation spectroscopy can be used to analyze protein aggregate dynamics on the cell surface during live-cell imaging in the millisecond to second range.

Key words: Fluorescence, imaging, plasma membrane, caveolae, clathrin-coated pits, family of image correlation spectroscopy, confocal microscopy, protein dynamics.

1. Introduction

Cell membranes are structurally and dynamically quite heterogeneous (1–5). They are organized in domains in which certain proteins tend to exist as complexes or at higher concentrations. These domains include clathrin-coated pits, to which membrane proteins are thought to associate prior to internalization; caveolae, in which certain proteins are believed to exist prior to signaling;

D.B. Papkovsky (ed.), *Live Cell Imaging*, Methods in Molecular Biology 591,
DOI 10.1007/978-1-60761-404-3_21, © Humana Press, a part of Springer Science+Business Media, LLC 2010

and lipid rafts, where the lipid induced phase separation appears to selectively segregate specific signaling complexes. Current protein models predict that some proteins are trapped in domains where they can move rapidly in that area, but are released slower from the region. This data points toward the existence of membrane domains that are important for triggering specific event signaling. Signal transduction events depend on interactions of proteins in the membrane (6–13). It is crucial to understand what intermolecular interactions exist between proteins. Additionally, it is necessary to understand the aggregation and clustering of signaling molecules and receptors in these membrane domains. The family of image correlation spectroscopy (FICS) is a valuable powerful tool used to study these dynamics (14–19). The FICS consists of several imaging techniques, and in this chapter we will review three of them: image correlation spectroscopy (ICS), image cross-correlation spectroscopy (ICCS) and dynamic image correlation spectroscopy (DICS). ICS is a tool used to calculate the cluster density (CD), which is the average number of clusters per unit area. Additionally, ICS is used to determine the degree of aggregation (DA), which is the differential between cohesion within aggregates and adhesion between proteins of the plasma membrane proteins. ICCS allows one to measure colocalization of proteins of interest. DICS is used to analyze protein aggregate dynamics on the cell surface during live-cell imaging in the millisecond to second range.

FICS relies on three basic principles described here. First is collection of a large number of high-quality, high-magnified fluorescent images from different cells or objects of interest in which cellular conditions are experimentally controlled (usually 40). To obtain statistically significant data, each experiment should be performed at least three times for each experimental condition. Second, a computational analysis of the images is performed to extract the quantitative information. In this step, the two-dimensional (2D) correlation function is calculated. This function is subsequently fit to a known function to extract the desired parameters. During this process, background noise, white noise, aperture, and image size must be accounted. In the third step, the results are statistically analyzed based on fitted parameters, such as the correlation function amplitude and beam width, and derived parameters including the number density or fractions of associations. We use z scores, t-tests, and the F test to sort, evaluate, and compare each set of parameters. Then the data is evaluated and interpreted in the context of the biological system of interest (**Fig. 21.1**). This protocol describes the general method and process of FICS analysis. We also describe the protocol for analyzing the protein distribution of BRII receptor on the plasma membrane of C2C12 cells.

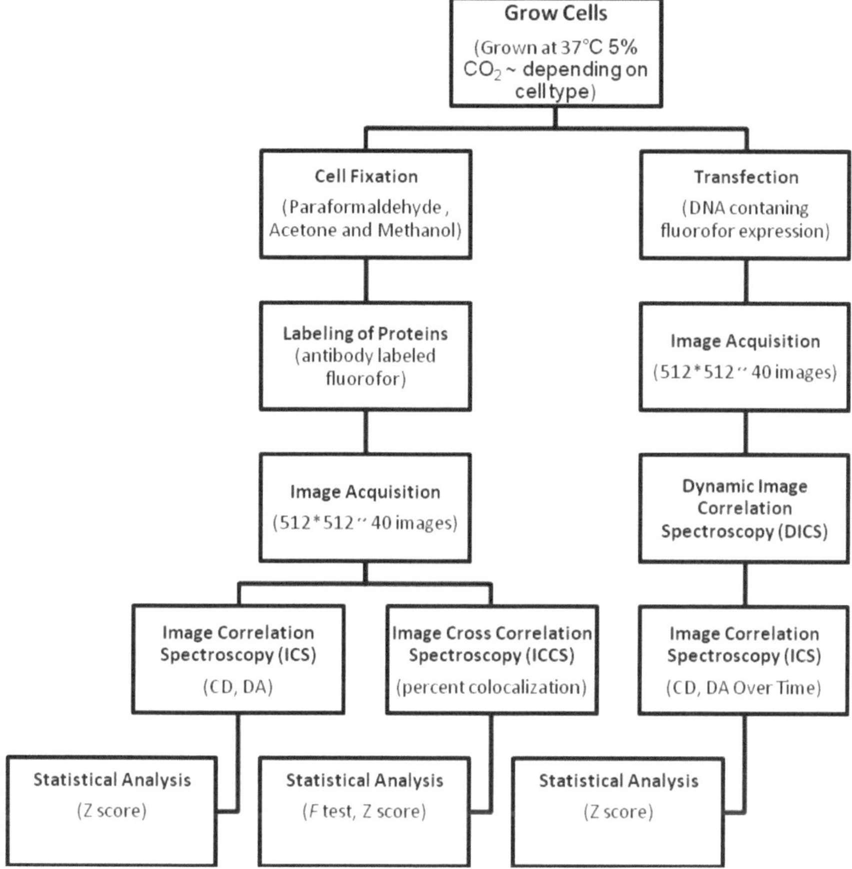

Fig. 21.1. The basic outline of steps required to get the data that is needed

2. Materials

2.1. Plasma Membrane Protein Labeling

1. Blocking peptides against BRII (Santa Cruz).
2. Polyclonal murine primary antibodies directed against BRII (Santa Cruz).
3. Secondary antibody, donkey antibody to goat immunoglobulin G (anti-goat IgG) conjugated with Rhodamine Red-x (RRX) labeled or Alexa Fluor 488 (Molecular Probes).
4. Plasmid DNA containing the HA-BRII (P.K.'s laboratory) (Nohe et al., 2002; (20)).

2.2. Cell Culture

1. C2C12 cells (American Tissue Culture Collection).
2. Dulbecco's modified Eagle's medium with 4 mM L-glutamine, 1.5 g/L sodium bicarbonate, and 4.5 g/L glucose (DMEM) (Invitrogen).

3. Fetal bovine serum (FBS) (Invitrogen).
4. Penicillin–streptomycin 100× (penicillin, 10,000 units/mL; streptomycin, 10,000 mg/mL) (Fisher Scientific).
5. Cell Culture Medium. Add 50 mL FBS and 5 mL of penicillin–streptomycin to 445 mL DMEM under sterile conditions, then sterile filter; store at 4°C. Warm media to 37°C before use.

2.3. Fixation and Mounting

1. Blocking Solution. Dissolve 0.3 g of bovine serum albumin (BSA) (Fisher) to 10 mL of phosphate-buffered saline (PBS, 8 g NaCl, 0.2 g KCl, 1.44 g Na_2HPO_4, 0.24 g KH_2PO_4; fill to 1 L and adjust pH to 7.2) for a final concentration of 3% BSA in PBS.
2. Lipofectamine 2000 (Invitrogen).
3. Ice-cold PBS (PBS held on ice).
4. Methanol (Fisher) stored in –20°C freezer.
5. Acetone (Fisher) stored in –20°C freezer.
6. 4.4% paraformaldehyde. Dissolve 4.4 g paraformaldehyde (Fisher) in 100 mL ddH_2O and add concentrated NaOH until it dissolves. Aliquot 9 mL into 15-mL conical tubes then freeze. To use thaw tube, add 1 mL of 10× PBS and adjust pH to 7.2 with HCl (Fisher).
7. Primary Antibody Labeling Solution. Add the Polyclonal murine primary antibodies directed against BRII (Santa Cruz) to 100 μL of blocking solution to the desired concentration.
8. Secondary Antibody Labeling Solution. Add the secondary antibody donkey antibody to goat immunoglobulin G to blocking solution for the final concentration desired.
9. Mounting Medium. Mix 20 g Gelvatol or Celvol 205 (Celanese Chemicals) in 80 mL of 140 mM NaCl (Fisher) and 20 mL of 10 mM sodium phosphate buffer (pH 7.2) and stir solution for 16 h at room temperature. Add 40 mL of glycerol (Fisher) and stir for an additional 16 h. Place solution in two 50-mL centrifuge tubes and centrifuge at 4000*g*. Remove the supernatant (pH between 6 and 7) and store in 10-mL aliquots in the freezer (–20°C).

2.4. Equipment

1. Confocal microscope (*see* **Note 1**).
2. ImageJ for image analysis (NIH, free download at http://rsb.info.nih.gov/ij/).
3. Incubator (set at 37°C, 5% CO_2).
4. Laminar flow culture hood.

5. PC with software for Fourier transforms, fitting, and plotting.
6. SigmaPlot (Systat Software Inc.) or Origin (OriginLab) (for plotting data).

3. Methods

3.1. Cell Culture and Preparation of Fixed Cells

If data for short-term analysis of receptor dynamics are acquired in the time range from 1 to 2 h, live cells are used. For the purpose of short-term analysis, transfection allows for fluorescently labeled proteins to be expressed in the membrane. For more long-term studies, it is far simpler to fix cells at various time points of interest. Doing so allows for visualization by means of immunofluorescence of the proteins of interest at a certain time point. For example, in monitoring of BRII on C2C12 cells, cells were first fixed with paraformaldehyde or acetone–methanol with a goat antibody against BRII and then fluorescently labeled Alexa 488 donkey anti-goat IgG (*see* **Note 2**) (21, 22). The following directions are for culturing C2C12 cells and for fixing these cells by acetone and methanol or with paraformaldehyde.

3.1.1. Cell Culture

1. Place sterile glass coverslips in 60-mm dishes.
2. Suspend C2C12 cells in cell culture medium ($\sim 5 \times 10^5$ cells per 2 mL of medium).
3. Place 2 mL of cell suspension into each dish (40% confluence).
4. Grow the cells at 37°C in a humidified 5% CO_2 to ~60% confluence (1–2 days).

3.1.2. Fixation of Cells with Acetone and Methanol

1. Aspirate medium and rinse plate with ice-cold PBS.
2. Add 2 mL of –20°C methanol.
3. Incubate for 5 min at –20°C.
4. Remove methanol.
5. Add 2 mL of –20°C acetone and incubate for 2 min on ice.

3.1.3. Paraformaldehyde Fixation

1. Remove media.
2. Wash three times with PBS.
3. Add 1 mL 4.4% paraformaldehyde fixation solution and incubate for 30 min.
4. Wash three times with PBS.

3.2. Labeling of Proteins on the Plasma Membrane

1. Remove fixative.
2. Add 1 mL of blocking solution.
3. Incubate for 30 min at room temperature.
4. Add 100 μL of primary antibody labeling solution to the coverslip.
5. Incubate cells with the solution for 30 min at room temperature.
6. Wash cells three times with PBS.
7. Add 100 μL of secondary antibody labeling solution to the coverslip.
8. Incubate cells with the solution for 30 min at room temperature.
9. Wash cells three times with PBS.
10. Mount coverslips on slides with mounting medium.

3.3. Transfection

1. Add DNA plasmid HA-BRII (~2 μL depending on purity and concentration) to DMEM (250 μL). Flip to mix.
2. Mix lipofectamine (10 μL) with DMEM (250 μL) and set for 5 min (no more than 25 min).
3. Place 2 mL of media in each dish.
4. Mix the solutions, DNA from step 1 and the lipofectamine from step 2 set for 20 min (stable for up to 6 h).
5. Add mixture (500 μL) from step 4 to dish.
6. Incubate for ~ 4 h (depending on cell survival rate).

3.4. Labeling

One must ensure that all proteins of interest on the plasma membrane are labeled (*see* **Note 3**). To minimize problems arising from insufficiently labeling the antibody, concentration must be optimized to ensure that all protein-binding sites are saturated with the antibodies directed against this site. The most reliable method of optimal antibody concentrations is determined by comparing the fluorescence intensity as a function of concentration from the measured cluster density. The observed cluster density will level off to a constant value at the point of saturation; at higher concentrations the intensity may increase because of nonspecific binding (**Fig. 21.2**).

3.5. Image Acquisition

3.5.1. Determine Cells to Image

We usually select cells that are spread out on the glass surface to allow for the collection of images of flat regions of the plasma membrane that are far away from the cell nucleus and organelles. Additionally, we select cells that express various concentrations of our protein of interest and compare the data obtained by FICS analysis with the expression level of our protein. For bone

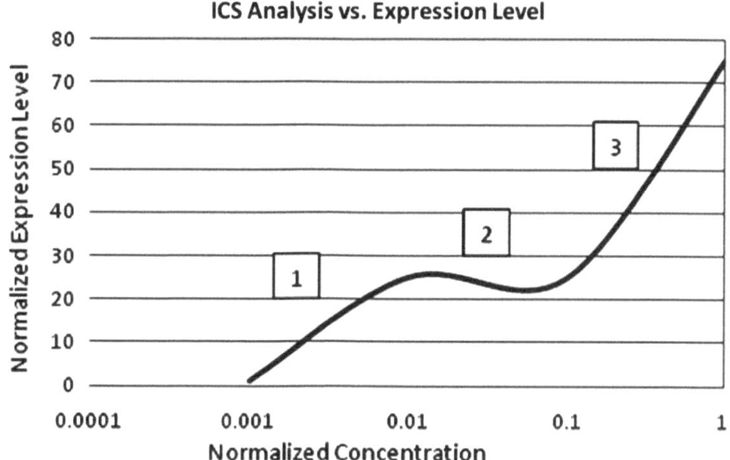

Fig. 21.2. Sample graph of normalized expression levels verses the concentration of protein of interest. (1) Antibody concentration is too low to saturate all binding sites. (2) Optimal antibody concentration with all sites saturated showing the best expression. (3) Nonspecific binding of the antibody.

morphogenetic proteins (BMP) receptors, our data showed no significant differences between the expression level of the receptors and the cluster density. However, for new proteins histogram analysis should be performed. By plotting the results of the ICS analysis versus the expression level of receptors, the influence of overexpression or protein concentration can be determined.

3.5.2. Image Collection

1. Select cells with representative labeling.
2. Collect high magnification images from flat regions of the cell membrane, far away from the nucleus or organelles (**Fig. 21.3**) and ensure that the membrane does not connect to neighboring cells. For the images to show contrast the size of each pixel should be matched to the size of the laser beam or the point spread function of the microscope. Typically images are collected at a magnification (or zoom factor) at which each pixel nominally has a resolution of $\sim 0.033\ \mu m$ (*see* **Note 4**).
3. Acquire images at about 1 s per image to minimize photobleaching. DIC images can be collected as fast as needed.
4. In most cases a minimum of 40 images of different cells for each experimental condition is needed to obtain statistically significant information.
5. Images that lack parts of the cell membrane or contain very high intensity clusters in only one part of the image are excluded from further analysis (*see* **Note 5**).

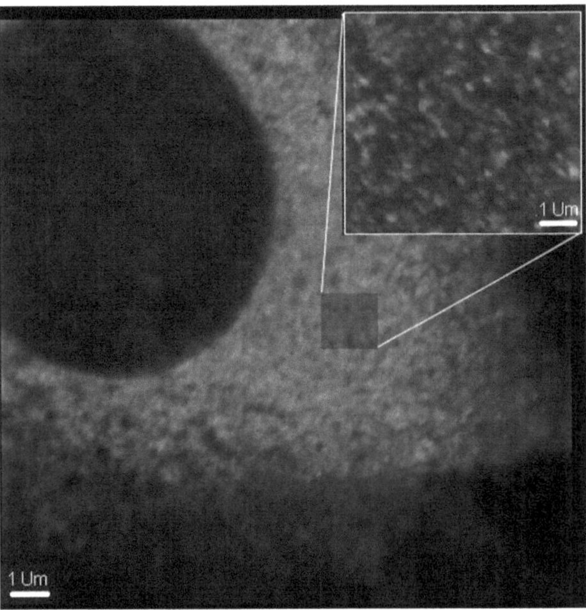

Fig. 21.3. Image of C2C12 cell expressing BRII; membrane is flat and the portion selected is far from the nucleus of the cell.

6. Determine the background intensity by collecting an image with the shutter to the laser closed. This is a measure of the dark counts or noise of the system in the absence of the signal and is used for comparison to nonlabeled cells and for calculations of corrections.
7. Repeat each experiment at least three times.

3.5.3. Image Analysis

All the imaging techniques described have four basic processes. The first is the calculation of the correlation function, the second involves using a 2D Gaussian function to fit the correlation function, the third is finding the amplitude of the correlation function, and the fourth involves calculating the cluster density (ICS, DICS), percentage colocalization (ICCS), or diffusion and flow coefficients (DICS). Each of these techniques can be used to extract data after the images are analyzed by autocorrelation or cross-correlation functions.

Analysis of ICS Images by Autocorrelation Using ImageJ and SigmaPlot

1. Determine the autocorrelation function (in ImageJ click on FFT/FDMath).
2. Crop the correlation function around its highest amplitude to gain a 64 × 64 square.
3. Save this cropped image as “txt image” in ImageJ.
4. Import the “txt image” file into SigmaPlot (file import ∗.∗).

5. Edit the 3D Gaussian fitting routine under statistics/regressionwizzard/3D/ Gaussian to match.
6. Perform Gaussian fitting using Eq. [1] in SigmaPlot with the standard nonlinear regression routines that are available (*see* **Notes 6** and **7**).

$$g(x, y) = g(0, 0) \exp\left(-\frac{x^2 + y^2}{w^2}\right) + g^0 \qquad [1]$$

7. Determine the amplitude of the autocorrelation function at the origin, $g(0,0)$ (in SigmaPlot, scroll the plot until you find the maximum in the center of the function).
8. Eliminate the contributions from white noise by subtracting the data from the three channels closest to zero, coordinates (0,0), (0,1), and (1,0) (in SigmaPlot, choose subtract) (*see* **Note 8**).
9. Calculate the cluster density per μm^2 for individual images, according to the following equation (Eq. [2]):

$$CD = \frac{1}{g(0, 0)\pi w^2} \approx \frac{\overline{N_c}}{\mu m^2} \qquad [2]$$

where $\overline{N_c}$ is an average measure of the number of clusters in the beam area (which is normalized to μm^2).

Analysis of ICCS and DICS Images by Cross-Correlation

1. Open the image pairs in ImageJ.
2. Calculate the cross-correlation function of each pair separately.
3. Create the joint power spectrum of the images (in ImageJ click on the process heading and then choose FFT/FDMath), followed by a reverse transform to obtain the cross-correlation function.
4. Crop the correlation function around its highest amplitude to gain a 64 × 64 square.
5. Save this cropped image as "txt image" in ImageJ.
6. Import the "txt image" file into SigmaPlot (file import *.*).
7. Search for the maximum of the function and fit from there if the maximum is within about 10 pixels from the origin. If it is farther than that, reject the data. Perform Gaussian fitting using Eq. [1] in SigmaPlot with the standard nonlinear regression routines that are available.
8. Determine the amplitude of the cross-correlation function at its maximum [called $g(0,0)$] (in SigmaPlot determine the maximum height).

9. For ICCS and DICS experiments, calculate the density of clusters (per μm^2) containing two different labeled proteins i and j, CD_{ij}, from a pair of images by dividing the amplitude of the cross-correlation function [$g_{ijt}(0,0)$] by the product of the amplitudes of the individual autocorrelation functions [$g_{it}(0,0)$ and $g_{jt}(0,0)$] according to the following equation:

$$CD_{ij} = \frac{g_{ijt}(0,0;0)}{g_{it}(0,0;0)g_{jt}(0,0;0)\pi w_i^2} = \frac{\bar{N}_{ijt}}{w_i^2} \qquad [3]$$

10. For ICCS experiments, calculate the fraction of colocalization, $F(i|j)$, from two pairs of images of the two proteins i and j according to the following equation:

$$[F(i|j) = \frac{CD_{ijt}}{CD_{it}} = \frac{g_{ijt}(0,0;0)}{g_{jt}(0,0;0)\pi w_i^2}] \qquad [4]$$

in which i and j are the two different proteins labeled and CD_{ijt} is the density of clusters of both molecules.

3.6. Statistical Treatment of Obtained Data

To find the data that will hold significance, we analyze as many images and correlation parameters as possible but at least 3 * 40 images can reasonably be judged to be good quality. From one cell to another there can and will be significant differences in membrane proteins, typically from about 20–30%. There are many cases in which one of the measured or fitted parameters will significantly skew data by artifacts, some of which include, for example, intracellular vesicles or antibody aggregates which may give rise to excessively bright or dark spots. To help eliminate such cases, we use a z score for each of the measured parameters [average intensity; correlation function amplitude, width, and offset]. ICS experiments allow us to apply the standard score [z score = (score – mean)/SD] to the calculated parameters (CD) and degree of aggregation. ICCS experiments data can be tested by means of F test. ICCS experiments test for the position of the maximum of the cross-correlation function and if it is within 10 pixels of the origin it is then acceptable. If the data from images have a maximum outside this region, the data from the image is rejected. If the maximum which arises from correlation of fluctuations is separated by a distance greater than the average beam width the data from that image is also removed.

4. Notes

1. An inverted microscope configuration allows measurements to be made on live cells grown on coverslips using oil-immersion microscope objectives from below. The numerical aperture will determine the optical resolution and the beam size, whereas the magnification of the objective will determine the field of view. However, the image size or scan range of the laser can be controlled independently by the "zoom factor" of the confocal microscope. In our work, we used a zoom factor of 10 in the FluoView 300, which led to the parameters of laser beams of 0.35 μm radius, with pixel dimensions of 0.03 μm in images of 512×512 pixels.
2. The fixation may interfere with membrane protein clustering. Therefore at least two different fixatives should be used and compared to each other.
3. Labeling only a subtype of proteins will cause a difference in intensity from one cell to another that reflects the different concentrations of the proteins on the respective cell surfaces in different microenvironments.
4. This corresponds to oversampling of about tenfold and ensures that the intensities in neighboring pixels are correlated because they share contributions from some of the same molecules or clusters. This also ensures that the shape of the correlation function is reasonably accurate.
5. Images with varying topography will result in part of the cell to be out of focus or regions of the cell where there are contributions from cytoplasmic material rather than from cell surface receptors.
6. Information on fitting can be obtained from the SigmaPlot Handbook.
7. Three fitting variables are found. It is then possible to enter the function of choice and extract: $g(0,0)$, w, and g_0. $g(0,0)$ is the amplitude of the correlation function at the origin. The w is known to be the beam width, but serves as an indicator of the quality of the data. If w is within about 30% of the expected value in for our system it is 0.35 μm this data is valid. g_0 is variable that measures the offset of the function.
8. Channel (0,0) should contain all the contributions from the white noise; we find that data in all channels are also often affected. Data channel (1,1) is the least affected.

References

1. Abrami L, Fivaz M, van der Goot FG. (2000) Surface dynamics of aerolysin on the plasma membrane of living cells. Int J Med Microbiol. 290(4–5):363–7.
2. Abumrad NA, Sfeir Z, Connelly MA, Coburn C. (2000) Lipid transporters: membrane transport systems for cholesterol and fatty acids. Curr Opin Clin Nutr Metab Care. 3(4):255–62.
3. Barr FA, Shorter J. (2000) Membrane traffic: do cones mark sites of fission? Curr Biol. 10(4):R141–4.
4. Basanez G. (2002) Membrane fusion: the process and its energy suppliers. Cell Mol Life Sci. 59(9):1478–90.
5. Bechinger B. (2000) Understanding peptide interactions with the lipid bilayer: a guide to membrane protein engineering. Curr Opin Chem Biol. 4(6):639–44.
6. Anderson RG, Jacobson K. (2002) A role for lipid shells in targeting proteins to caveolae, rafts, and other lipid domains. Science. 296(5574):1821–5.
7. Barnett-Norris J, Lynch D, Reggio PH. (2005) Lipids, lipid rafts and caveolae: their importance for GPCR signaling and their centrality to the endocannabinoid system. Life Sci. 77(14):1625–39.
8. Bathori G, Cervenak L, Karadi I. (2004) Caveolae -an alternative endocytotic pathway for targeted drug delivery. Crit Rev Ther Drug Carrier Syst. 21(2):67–95.
9. Brown DA, London E. (1998) Functions of lipid rafts in biological membranes. Annu Rev Cell Dev Biol. 14:111–36.
10. Cohen AW, Hnasko R, Schubert W, Lisanti MP. (2004) Role of caveolae and caveolins in health and disease. Physiol Rev. 84(4): 1341–79.
11. Dobrowsky RT. (2000) Sphingolipid signalling domains floating on rafts or buried in caves? Cell Signal. 12(2):81–90.
12. Fielding CJ, Fielding PE. (2004) Membrane cholesterol and the regulation of signal transduction. Biochem Soc Trans. 32(1):65–9.
13. Galbiati F, Razani B, Lisanti MP. (2001) Emerging themes in lipid rafts and caveolae. Cell. 106(4):403–11.
14. Brown CM, Roth MG, Henis YI, Petersen NO. (1999) An internalization-competent influenza hemagglutinin mutant causes the redistribution of AP-2 to existing coated pits and is colocalized with AP-2 in clathrin free clusters. Biochemistry.38(46): 15166–73.
15. Fire E, Brown CM, Roth MG, Henis YI, Petersen NO. (1997) Partitioning of proteins into plasma membrane microdomains. Clustering of mutant influenza virus hemagglutinins into coated pits depends on the strength of the internalization signal. J Biol Chem. 272(47):29538–45.
16. Nohe A, Petersen NO. (2004) Analyzing protein–protein interactions in cell membranes. Bioessays. 26(2):196–203.
17. Nohe A, Petersen NO. (2007) Image correlation spectroscopy. Sci STKE. 2007. DOI:10.1126/stke.4172007P17 417:l7.
18. Wiseman PW, Hoddelius P, Petersen NO, Magnusson KE. (1997) Aggregation of PDGF-beta receptors in human skin fibroblasts: characterization by image correlation spectroscopy (ICS). FEBS Lett. 401(1): 43–8.
19. Wiseman PW, Petersen NO. (1999) Image correlation spectroscopy. II. Optimization for ultrasensitive detection of preexisting platelet-derived growth factor-beta receptor oligomers on intact cells. Biophys J. 76(2):963–77.
20. Nohe A, Hassel S, Ehrlich M, Neubauer F, Sebald W. and Knaus P. (2002). The mode of BMP receptor oligomerization determines different BMP-2 signaling pathways. J. Biol. Chem. **15**:5330–8.
21. Nohe A, Keating E, Underhill TM, Knaus P, Petersen NO. (2003) Effect of the distribution and clustering of the type I A BMP receptor (ALK3) with the type II BMP receptor on the activation of signalling pathways. J Cell Sci. 116(16):3277–84.
22. Nohe A, Keating E, Underhill TM, Knaus P, Petersen NO. (2005) Dynamics and interaction of caveolin-1 isoforms with BMP-receptors. J Cell Sci. 118(3):643–50.

SUBJECT INDEX

D.B. Papkovsky (ed.), *Live Cell Imaging*, Methods in Molecular Biology 591,
DOI 10.1007/978-1-60761-404-3, © Humana Press, a part of Springer Science+Business Media, LLC 2010

G

H

I

L

M

O

P

R

S

T

V

Y

If you have any concerns about our products,
you can contact us on
ProductSafety@springernature.com

In case Publisher is established outside the EU,
the EU authorized representative is:
Springer Nature Customer Service Center GmbH
Europaplatz 3, 69115 Heidelberg, Germany

Printed by Libri Plureos GmbH
in Hamburg, Germany